Reproductive Physiology II

Publisher's Note

The *International Review of Physiology* remains a major force in the education of established scientists and advanced students of physiology throughout the world. It continues to present accurate, timely, and thorough reviews of key topics by distinguished authors charged with the responsibility of selecting and critically analyzing new facts and concepts important to the progress of physiology from the mass of information in their respective fields.

Following the successful format established by the earlier volumes in this series, new volumes of the *International Review of Physiology* will concentrate on current developments in neurophysiology and cardiovascular, respiratory, gastrointestinal, endocrine, kidney and urinary tract, environmental, and reproductive physiology. New volumes on a given subject generally appear at two-year intervals, or according to the demand created by new developments in the field. The scope of the series is flexible, however, so that future volumes may cover areas not included earlier.

University Park Press is honored to continue publication of the *International Review of Physiology* under its sole sponsorship beginning with Volume 9. The following is a list of volumes published and currently in preparation for the series:

Volume 1: **CARDIOVASCULAR PHYSIOLOGY** (A. C. Guyton and C. E. Jones)

Volume 2: **RESPIRATORY PHYSIOLOGY** (J. G. Widdicombe)

Volume 3: **NEUROPHYSIOLOGY** (C. C. Hunt)

Volume 4: **GASTROINTESTINAL PHYSIOLOGY** (E. D. Jacobson and L. L. Shanbour)

Volume 5: **ENDOCRINE PHYSIOLOGY** (S. M. McCann)

Volume 6: **KIDNEY AND URINARY TRACT PHYSIOLOGY** (K. Thurau)

Volume 7: **ENVIRONMENTAL PHYSIOLOGY** (D. Robertshaw)

Volume 8: **REPRODUCTIVE PHYSIOLOGY** (R. O. Greep)

Volume 9: **CARDIOVASCULAR PHYSIOLOGY II** (A. C. Guyton and A. W. Cowley, Jr.)

Volume 10: **NEUROPHYSIOLOGY II** (R. Porter)

Volume 11: **KIDNEY AND URINARY TRACT PHYSIOLOGY II** (K. Thurau)

Volume 12: **GASTROINTESTINAL PHYSIOLOGY II** (R. K. Crane)

Volume 13: **REPRODUCTIVE PHYSIOLOGY II** (R. O. Greep)

(Series numbers for the following volumes will be assigned in order of publication)

RESPIRATORY PHYSIOLOGY II (J. G. Widdicombe)
ENDOCRINE PHYSIOLOGY II (S. M. McCann)
ENVIRONMENTAL PHYSIOLOGY II (D. Robertshaw)

Consultant Editor: Arthur C. Guyton, M.D., Department of Physiology and Biophysics, University of Mississippi Medical Center

INTERNATIONAL REVIEW OF PHYSIOLOGY

Volume 13

Reproductive Physiology II

Edited by

Roy O. Greep, Ph.D.
Laboratory of Human Reproduction
and Reproductive Biology
Harvard Medical School

UNIVERSITY PARK PRESS
Baltimore • London • Tokyo

599
R425

UNIVERSITY PARK PRESS
International Publishers in Science and Medicine
Chamber of Commerce Building
Baltimore, Maryland 21202

Copyright © 1977 by University Park Press

Typeset by The Composing Room of Michigan, Inc.

Manufactured in the United States of America by Universal Lithographers, Inc.,
and The Optic Bindery Incorporated.

Library of Congress Cataloging in Publication Data

Main entry under title:

Reproductive physiology II.

  (International review of physiology; v. 13)
  Includes index.
  1. Reproduction. I. Greep, Roy Orval,
1905- II. Series. [DNLM: 1. Reproduction.
W1 IN834f v. 13 / WQ205 R429]
QP1.P62 vol. 13 [QP251]   599'.01'08s [599'.01'66]
ISBN 0-8391-1062-6   76-51740

Consultant Editor's Note

In 1974 the first series of the *International Review of Physiology* appeared. This new review was launched in response to unfulfilled needs in the field of physiological science, most importantly the need for an in-depth review written especially for teachers and students of physiology throughout the world. It was not without trepidation that this publishing venture was begun, but its early success seems to assure its future. Therefore, we need to repeat here the philosophy, the goals, and the concept of the *International Review of Physiology*.

The *International Review of Physiology* has the same goals as all other reviews for accuracy, timeliness, and completeness, but it also has policies that we hope and believe engender still other important qualities often missing in reviews, the qualities of critical evaluation, integration, and instructiveness. To achieve these goals, the format allows publication of approximately 2,500 pages per series, divided into eight subspecialty volumes, each organized by experts in their respective fields. This extensiveness of coverage allows consideration of each subject in great depth. And, to make the review as timely as possible, a new series of eight volumes is published approximately every two years, giving a cycle time that will keep the articles current.

Yet, perhaps the greatest hope that this new review will achieve its goals lies in its editorial policies. A simple but firm request is made to each author that he utilize his expertise and his judgment to sift from the mass of biennial publications those new facts and concepts that are important to the progress of physiology; that he make a conscious effort not to write a review consisting of an annotated list of references; and that the important material that he does choose be presented in thoughtful and logical exposition, complete enough to convey full understanding, as well as woven into context with previously established physiological principles. Hopefully, these processes will continue to bring to the reader each two years a treatise that he will use not merely as a reference in his own personal field but also as an exercise in refreshing and modernizing his whole store of physiological knowledge.

A. C. Guyton

Contents

Preface

It is a matter of record that scientific investigation of reproductive physiology got a very late start in comparison with that of all other major bodily systems. Prudery and taboos were the principal restraining influences. This neglect has come at the price of the many social, economic, and environmental problems arising out of the spiraling population expansion of the past quarter century, a phenomenon generated, ironically, by the life-saving benefits of research in other areas of medical research. It is also to be noted that significant financial support of research on reproduction by the United States Federal Government did not become available until the mid-1960s and that came about only as a result of the pressures of population problems. Despite this late beginning, knowledge of reproductive phenomena has been accruing at a very rapidly accelerating pace over the past decade. It will be apparent from the present volume that deep inroads have been made in the elucidation of many of the complex mechanisms that maintain or regulate the reproductive system. This rapid progress has been much abetted by skillful application of many of the highly sophisticated tools and techniques of modern biomedical science. This volume presents an account of the fascinating advances made in several areas of reproductive science that have only recently come into full bloom.

Each contributor is an outstanding authority in his particular area of research and has prepared a valuable account not only of the progress made but also of the current concepts and thinking that will serve to stimulate and guide further exploration of these critically important areas.

Excellent cooperation on the part of the contributors in holding to schedule for completion of manuscripts has made early and timely publication a reality. It is a pleasure to acknowledge my gratitude to the contributors for their fine work and to the MTP International Review of Science and University Park Press for expediting the production of *Reproductive Physiology II*.

R. O. Greep

Reproductive Physiology II

International Review of Physiology
Reproductive Physiology II, Volume 13
Edited by Roy O. Greep
Copyright 1977   University Park Press   Baltimore

1
Hypothalamic Gonadotropin-releasing Hormone and Reproduction

A. ARIMURA

Department of Medicine, Tulane University School of Medicine
and the Endocrine and Polypeptide Laboratories, Veterans Administration Hospital
New Orleans, Louisiana

The important role of hypophysial portal vessels as passageways for hormonal substances to the adenohypophysis was predicted by Green and Harris (1). Evidence for the view that neural impulses initiate secretion of gonadotropins via

This research in the author's laboratories was supported by Research Grants AM 09094, HD 06555, and AM 07467 from the United States Public Health Service and by the Veterans Administration.

1

neurohormonal substances is now overwhelming. Elaborated in neurons of the central nervous system, these regulating substances are transported to the pituitary gland by the hypophysial portal vessels (2).

Hypothalamic neurosubstances capable of stimulating the release of luteinizing hormone (LH) (3–5) and follicle stimulating hormone (FSH) (6–8) were reported by several laboratories. Secretions of LH and FSH were first considered to be regulated by discrete releasing factors or hormones, LH-releasing hormone (LHRH) and FSH-releasing hormone (FSHRH), respectively. This concept is still accepted by some investigators (9). However, following intensive purification, LHRH activity in the porcine (10) and ovine (11) hypothalamic extracts could not be separated from FSHRH activity. Porcine LHRH was isolated by Schally et al. (12), and its chemical structure was proposed by Matsuo et al. (13, 14) to be   pGlu–His–Trp–Ser–Tyr–Gly–Leu–Arg–Pro–Gly–NH$_2$.   Subsequently synthesized (15), this peptide proved to possess the same activity as natural LHRH (16). Ovine LHRH was characterized as possessing the same structure as porcine LHRH (17).

Synthetic LHRH, as well as natural LHRH, stimulated the release of both LH and FSH from the pituitary in vivo and in vitro. These findings led to the postulation that the secretions of LH and FSH are regulated by one hypothalamic releasing hormone, LHRH decapeptide, and that the differential release of these two gonadotropins is regulated by other modulators such as gonadal steroids (18).

The availability of synthetic LHRH prompted its use in reproductive physiology research. In addition, successful production of specific antiserum to LHRH made possible radioimmunoassays (RIA) for LHRH and immunohistochemical approaches. These techniques are helpful in providing a deeper insight into the localization and site of production of LHRH in the central nervous system. Neutralization of endogenous LHRH by passive immunization with the antiserum revealed the importance of this hormone in various physiological phenomena of the reproductive processes.

Several reviews on LHRH and its role in reproduction have already appeared (19–21); therefore, in this chapter, an overall survey of LHRH literature has been avoided. Rather, evidence for the fundamental importance of LHRH in reproductive physiology is presented, and the current problems of disagreement in this field are discussed.

LOCALIZATION AND ONTOGENY OF LHRH

Localization of LHRH in the Brain

Studies on localization of LHRH in the brain of various species were recently reviewed by Ramirez and Kordon (22) and Sawyer (23). A topographic study on LHRH in the brain was first reported by McCann (24), who used in vivo bioassay for LHRH activity. Subsequently, Watanabe and McCann (25) and Crighton et al. (26) described the localization of FSHRH and LHRH activities, respectively, using in vitro assay methods. LHRH activity was found in the median eminence, the arcuate nucleus, the suprachiasmatic nucleus, and the preoptic nucleus. FSHRH was also localized mainly in the median eminence region. These results obtained by bioassay were in strong agreement with those obtained by radio-immunoassay for LHRH in brain slices of three dimensions or small cylinders removed from frozen sections of the brain tissue. King et al. (27) determined LHRH in frozen rat brain sections cut serially in coronal, parasagittal, and horizontal planes and observed two prominent components—a caudal one in the arcuate-median eminence region and a rostral one in the prechiasmatic and preoptic areas. Similar results were also reported by Wheaton et al. (28). Figures 1 and 2 illustrate the results obtained by King et al. (27) for the assays of pooled coronal and parasagittal sections from diestrous female rats. Regions most rich in LHRH were confined to the arcuate-median eminence complex. The median eminence contained the larger amount of LHRH, but the arcuate nucleus also contained a significant amount. The rostral component of LHRH activity was much less significant than the caudal one. In the rostral region, the highest concentration of LHRH was observed in the prechiasmatic and preoptic areas in female rats. The rostral component became greater in the late afternoon, compared to the early afternoon, of diestrus. In the male, this subsidiary rostral zone of activity was small to marginal, as compared to females, and was located

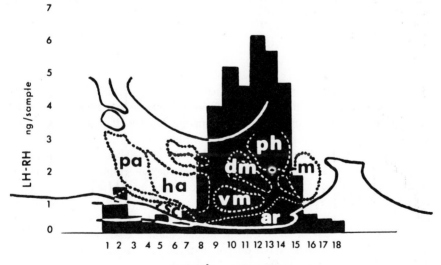

sample   number

Figure 1. LHRH content (ng/sample) in pooled coronal sections of brains from three diestrous female rats. Values for each sample are superimposed over a sagittal outline of hypothalamic nuclei. Sample 1 was taken just rostral to the anterior commisure. *ar,* arcuate nucleus; *dm,* dorsomedial nucleus; *ha,* hypothalamic area; *m,* mammillary nuclei; *pa,* preoptic area; *ph,* posterior hypothalamic area; *sc,* suprachiasmatic nucleus; *vm,* ventro-medial nucleus. Reprinted from King et al. (27), with permission of Cambridge University Press.

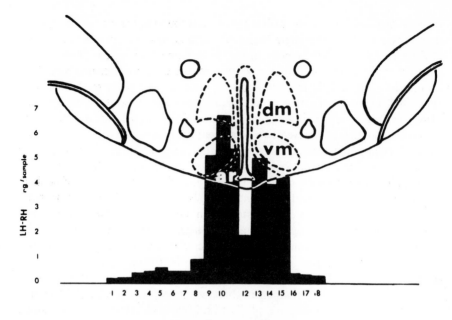

sample   number

Figure 2. LHRH content (ng/sample) in pooled parasagittal sections of brains from two diestrous female rats. Values for each sample are superimposed over a coronal outline at the level of the arcuate-median eminence region. *ar,* arcuate nucleus; *dm,* dorsomedial nucleus; *vm,* ventromedial nucleus. Reprinted from King et al. (27), with permission of Cambridge University Press.

slightly more caudally in the suprachiasmatic region. Palkovitz et al. (29) developed a refined technique which allows removal of isolated nuclei or even part of them by punching. This technique provided further insight into the precise site of origin of this hypothalamic hormone.

Table 1 presents the data on LHRH obtained by Palkovitz et al. (29). High concentrations of LHRH were found in the median eminence and arcuate nucleus. Other nuclei of the hypothalamus contained little LHRH. In this study, a very small fragment of the brain tissue was assayed for LHRH, and only in the regions with a substantially high concentration of LHRH was its presence revealed. The amount of LHRH in the median eminence was about twice that found in the arcuate nucleus, but the concentration in the median eminence was 7.7 times higher than that in the arcuate nucleus.

A considerable number of reports on the macrotopographic studies on LHRH using immunohistochemical methods with antiserum to synthetic LHRH preparations have appeared. In spite of different antisera and techniques by different investigators, all of these reports agree that immunoreactive LHRH was demonstrated in the external zone of the median eminence, especially in its lateral margin. However, there is considerable discordance in the findings of

Table 1. LHRH content of the hypothalamic nuclei of the
rat as measured by radioimmunoassay

	Protein (pg/μg)
Nucleus preopticus medialis	<0.05
Nucleus periventricularis	<0.05
Nucleus suprachiasmatis	Trace (<0.1)
Nucleus supraopticus	Trace (<0.1)
Nucleus hypothalamicus anterior	<0.05
Area hypothalamica lateralis anterior	<0.05
Nucleus paraventricularis	<0.05
Median eminence	22.4 ± 2.2[a]
Nucleus arcuatus	2.9 ± 0.8
Nucleus ventromedialis, pars medialis	Trace (<0.1)
Nucleus ventromedialis, pars lateralis	0.6 ± 0.5
Nucleus dorsomedialis	<0.05
Nucleus perifornicalis	<0.05
Area hypothalamic lateralis posterior	<0.05
Nucleus premamillaris dorsalis	<0.05
Nucleus premamillaris ventralis	<0.05

From Palkovitz et al. (29).

immunopositive perikarya which probably represent LHRH-producing cell bodies. Barry et al. (30), using an immunofluorescence method, described the presence of perikarya of neurons containing immunoreactive LHRH in the preoptic and septal areas of guinea pigs and rats. Most of the axons coming from these perikarya were incorporated in the hypothalamo-infundibular tract and terminated around the capillaries of the primary portal plexus of the median eminence. However, perikarya with LHRH were usually not demonstrated unless the possible synthesis was stimulated by castration and axoplasmic transport of LHRH was blocked by colchicine, suggesting that LHRH in the perikarya is present in very low concentrations.

Although Setalo et al. (31) could not demonstrate immunoreactive perikarya in the suprachiasmatic area in their earlier study, they recently (32) confirmed Barry's finding with the use of a different method with peroxidase as a marker for LHRH. In Nembutal-injected proestrous rats and frontally deafferented rats, they demonstrated LHRH-positive nerve cell bodies which scattered in the suprachiasmatic area between the anterior commissure and the optic chiasm and in the medial prechiasmatic area near the organum vasculosum of the lamina terminalis. Because the pathway of the LHRH-containing nerve fibers in the median eminence of the rats coincides with the course of the nerve fibers of the tuberoinfundibular tract, and because this tract receives numerous fibers from the arcuate nuclei, these nuclei were assumed to contain LHRH-producing cells. Naik (33) and Zimmerman et al. (34) demonstrated immunopositive perikarya in the arcuate nuclei, but other investigators (30–32, 35) failed to confirm this. According to Ramirez and Kordon (22), frontal deafferentation at the rostral end of the hypothalamus did not appear to reduce the amount of immunopositive LHRH in the rat median eminence. However, Setalo et al. (32) claimed that the completely isolated medial basal hypothalamus contained only a few LHRH-positive nerve fibers which entered the island in the most superficial layer of the median eminence. On the other hand, when only a frontal cut was made behind the optic chiasm, the median eminence-arcuate region retained a fair number of LHRH-positive fibers and terminals, but not LHRH-positive cell bodies. In any case, further evidence and data are needed to conclude whether LHRH-synthesizing cell bodies are located in the suprachiasmatic area or in the arcuate region or both. The possibility that synthesis of LHRH is completed in the axons, but not in the cell body, cannot be absolutely ruled out.

Furthermore, Zimmerman et al. (34) demonstrated immunoreactive LHRH in tanycytes of the median eminence in the mouse, suggesting the important role of tanycytes in transporting LHRH from the arcuate neurons to the portal plexus of the capillaries. However, other investigators (30–31, 35) failed to find immunoreactive LHRH in other species. Although the functional significance of the ependyma in regard to cyclic neuroendocrine events has recently received considerable attention, its physiological role in transport, secretion, and/or production of LHRH remains to be clarified.

Ontogeny of LHRH

Little is understood of the developmental interaction of the CNS-pituitary-gonadal axis, including the origins of the sexual dimorphism for gonadotropin release.

LHRH activity determined by radioimmunoassay in the hypothalamus was detected as early as day 18 of gestation in the rat, increasing 3-fold by the time of birth (36). Recently, Araki et al. (37) reported that in both male and female rats, immunoreactive LHRH in the hypothalamus increased suddenly on day 1 and rapidly thereafter until day 21. There was no difference in hypothalamic LHRH content between the sexes until at least 28 days of age.

King et al. (38) examined the development of nerve fibers with immunoreactive LHRH by using the technique of immunohistochemistry in the male and female rats during their critical postnatal period of sexual differentiation. Positive fibers were present in both sexes on the day of birth in both the arcuate-median eminence region and in the prechiasmatic area. Intensity of staining decreased, reaching a minimum at day 5. After this day, the number of LHRH-positive fibers increased in both areas until day 9, when it was equivalent in the female to the density observed in adult animals. However, in the male, the number of immunopositive fibers increased only slightly during this period. Although the pattern of change in total LHRH content in the hypothalamus during development is the same, regional differences in the development of LHRH-positive neurons between the sexes are apparent.

On the other hand, LHRH content as measured by radioimmunoassay in adult male rats was markedly higher than that in adult females (37). The pattern of increasing LHRH content during development in the whole hypothalamus was paralleled in the mid-hypothalamic region. A remarkable increase in LHRH was found in 35–37-day-old rats with closed vaginas, followed by a precipitous drop on the day of vaginal opening. Contrarily, no progressive increase in LHRH was demonstrated in the anterior hypothalamic region during development, although small increases were observed on days 14 and 35 (closed vagina). The interrelation between these two areas rich in LHRH is still not unequivocally delineated in either the adult or in the developing individual.

RELEASE MECHANISM

The concept that the factor which stimulates the release of ovulating hormone from the pituitary is an adrenergic agent was first proposed by Sawyer et al. (1949) (39). Using an exquisite fluorescence technique developed by Swedish investigators, high concentrations of catecholamines were found in the median eminence (40). Much of the fluorescence was found in dopaminergic fibers which originate in the arcuate nucleus and form the tuberoinfundibular tract. Fibers in this region have also been shown to be immunoreactive with LHRH. The dopaminergic nerve endings in the external zone of the median eminence came not only from the arcuate nucleus itself, but also from cell bodies in the periventricular zone above and anterior to the arcuate nucleus. Noradrenergic nerve endings in the internal zone of the median eminence arose outside the basal hypothalamus, presumably in the mid-brain (41). Palkovitz et al. (42) made a quantitative map of catecholamine distribution in the hypothalamus by determining the amount of catecholamine in punched cylinders of brain tissue using an enzyme-isotopic method. Dopamine (DA) was most concentrated in the median eminence and the arcuate nucleus, but was also present in the retrochiasmatic area, paraventricular nucleus, suprachiasmatic nuclei, and dorsomedial nuclei, although in lesser concentrations. Norepinephrine (NE) was concentrated in the paraventricular nuclei, the perifornical area, the retrochiasmatic area, and the ventral part of the dorsomedial nucleus. Its concentration in the median eminence was about half that of DA.

Although there is general agreement on the distribution of catecholamines in the hypothalamus, there has been some controversy regarding the role of DA and NE in LHRH release. Early reports from McCann's laboratories (43) indicated that DA stimulated the release of LHRH and prolactin release-inhibiting hormone (PIF). Intraventricular injection of DA elicited LH and FSH release and inhibited secretion of prolactin in rats. However, recent studies by Kalra and McCann (44) from the same laboratory used inhibitors of catecholamine synthetic pathways and suggested that NE is the main stimulator, whereas Ojeda et al. (45) showed that DA did not alter LH or FSH secretion, but did inhibit prolactin secretion. Sawyer (23) reported that intraventricular infusion of NE induced ovulation in the rabbit, but the infusion of DA failed to stimulate LH release and actually inhibited the stimulating effect of NE. Fuxe and Hokfelt (40) have also maintained that tuberoinfundibular DA inhibits the release of LHRH.

As mentioned before, LHRH-containing neurons are found in the median eminence and the arcuate nucleus. The course of positive LHRH fibers to the median eminence parallels that of tuberoinfundibular dopaminergic processes. However, a nearly complete destruction of the catecholaminergic terminals and all bodies in the arcuate nucleus and median eminence by intraventricular injection of 6-hydroxydopamine was not associated with a change in LHRH content (46). Therefore, LHRH-containing neurons can be considered to be distinct from catecholaminergic neurons.

Prostaglandin E (PGE) also appears to be involved in the release mechanism of LH and FSH by stimulating the release of LHRH. Intraventricular infusion of PGE increased the concentration of immunoreactive LHRH in the hypophysial portal blood and was associated with increased levels of LH and FSH in peripheral blood (47). In ovariectomized rats, intraventricular infusion of an inhibitor of prostaglandin synthetase (indomethacin) lowered plasma LH in ovariectomized rats (48). These data provide support for the view that prostaglandins are factors which trigger LHRH release.

Although these results indicate that both the monoaminergic system and PGE are involved in LHRH release, the way in which these two systems interact remains to be clarified. Ojeda et al. (49) reported that α,β-adrenergic receptor blockers or dopaminergic receptor blockers interfered with LH release induced by intraventricular infusion of PGE_2 in rats, implying that PGE_2 acts at a postsynaptic site to release LH by a direct effect on the LHRH-secreting neurons. PGE_2 appeared to have little direct effect on the pituitary gonadotrophs.

PITUITARY RESPONSIVENESS TO LHRH

Effect of Gonadal Steroids on Pituitary Response

Negative and positive feedback effects of gonadal steroids on gonadotropin secretion have been thought to take place primarily at the hypothalamic level. Döcke and Dörner (50) first proposed that increased sensitivity of the pituitary gland induced by the direct effect of gonadal steriods is at least partly responsible for the preovulatory surge of gonadotropins.

Arimura and Schally (51) first presented experimental evidence that the pituitary response to LHRH is higher at the time of the preovulatory surge of gonadotropin in sheep and hamsters. Aiyer et al. (52) observed that pituitary responsiveness to exogenous and endogenous LHRH increased markedly on the afternoon of proestrus in the rat. Subsequently, these findings have been confirmed by other investigators (53,54).

It is generally thought that the gradual increase in estrogen during the follicular stage triggers events which lead to ovulation. In experiments with rats, it was found that administration of 17β-estradiol on diestrus day 1 augmented the LH response to LHRH on diestrus day 2 (51). The effect of estrogen on pituitary responsiveness was, however, biphasic, with suppression occurring during the early period and augmentation during the later period (55, 56). In the rat, at least 6 hr had to elapse before the exogenously administered estrogen increased the pituitary sensitivity. During the estrous cycle in the rat, the rise in plasma 17β-estradiol reached a peak before the onset of LH surge. A potent antiestrogen, ICI 46474 (a triphenylethylene derivative), administered at 1700 hr of diestrus day 2, markedly reduced the LH and FSH response to exogenous LHRH on the afternoon of proestrus (57). The administration of antiestrogen to women also reduced the pituitary response to LHRH (58). These findings indicate that sensitization of the pituitary by a gradual increase in estrogen secretion during the follicular stage is, at least to some degree, responsible for the preovulatory surge of gonadotropin.

Other gonadal steroids appear to affect pituitary responsiveness by direct action or through the interplay with estrogen. Although regulation of gonadotropin secretion by negative feedback of gonadal steroids has been thought to operate at the hypothalamic level, many data have appeared which indicate direct pituitary suppression by progestins and androgens. Hillard et al. (59) reported that ovulation induced by intrapituitary infusion of LHRH in the rabbit was blocked by a single injection of norethindrone 15–24 hr before the infusion. Spies et al. (60) reported that a single subcutaneous injection of progesterone or chlormadinone blocked ovulation induced by intrapituitary infusion of median eminence extracts in rabbits. In addition, Arimura and Schally demonstrated that in cycling rats a single subcutaneous injection of 25 mg of progesterone blunted the LH response to LHRH injected into the carotid artery or the pituitary (51).

The suppressive effect of progesterone on LHRH-induced LH release appears to take place in a competitive manner. The ability of LHRH to overcome the blockade seems to be directly related to the dose of LHRH administered.

Hillard et al. (59) reported that in estrous rabbits, ovulation induced by intrapituitary infusion of LHRH was prevented by pretreatment with 2 mg of progesterone, but that ovulation following a large dose of LHRH was not affected by progesterone. Arimura and Schally (51) demonstrated that pretreatment of the diestrous rat with 5 mg of progesterone did not alter the LH response to a large dose of LHRH. However, 5 mg of progesterone given in combination with 20 μg of estradiol benzoate (EB) reduced the pituitary response to the same dose of LHRH. Direct suppressive effects of progesterone, testosterone, and their metabolites on pituitary responsiveness were also demonstrated in vitro (61).

Fink et al. (62) extensively studied the effect of steroid hormones on pituitary responsiveness. Ovariectomy on the morning or evening of diestrus day 2 resulted in a marked reduction of LH response to LHRH on the afternoon of proestrus in Nembutal-anesthetized rats. Yet neither ovariectomy nor a sham operation at 1230–1330 hr on proestrus affected the LH response to LHRH 5 hr later. A dosage of either 25 or 10 μg of estradiol benzoate injected subcutaneously immediately after ovariectomy on the morning of diestrus day 2 augmented the LH response to LHRH at 1330 hr proestrus. However, it failed to completely restore the response at 1830 hr on proestrus to that level observed in the sham-operated rats. However, the administration of estradiol benzoate immediately after the operation and of 2.5 mg of progesterone subcutaneously at 1300 hr of proestrus completely restored the gonadotropin response. Progesterone given alone immediately after the ovariectomy did not affect the response.

In contrast to LH, ovariectomy on diestrus day 2 increased the basal level of plasma FSH, but did not affect FSH response to LHRH on the afternoon of proestrus. Neither estradiol benzoate nor progesterone alone facilitated the FSH response. However, the FSH response was higher in animals treated in sequence with estradiol benzoate and progesterone than that observed in any group of ovariectomized or nonovariectomized animals.

Fink et al. (62) proposed that different mechanisms operate with respect to the first (before the onset of LH surge) and second (about the time of LH surge) phases of increased pituitary responsiveness. The development of the first phase probably depends solely on increased secretion of estrogen, whereas the development of the second phase probably requires ovarian and possibly adrenal progesterone, as well as the priming effect of LHRH. The sensitizing effect of progesterone appears to depend on a prior exposure of the hypothalamo-pituitary axis to estrogen (63,64). It should be remembered that simultaneous administrations of estradiol and progesterone suppress the pituitary sensitivity (51).

Priming Effect of LHRH

Fink et al. (62) proposed that LHRH itself sensitizes the pituitary. They injected two successive doses of LHRH at various intervals into the proestrous rats anesthetized with Nembutal. The response to the second injection was significantly greater than that to the first, and was greatest with the two injections of LHRH were separated by 60 min. The priming was not mediated by gonadal steroids, because the greater LH response to the second injection of LHRH was found in rats ovariectomized immediately before the first injection.

The priming effect of endogenous LHRH was also suggested by Gordon and Reichlin (65) and Zeballos and McCann (66).

The following findings (62) suggest that the magnitude of the priming effect of LHRH is dependent upon the degree of pituitary exposure to circulating steroids, especially 17β-estradiol. The priming effect was considerably less on days of the cycle other than proestrus in the rat. Ovariectomy on the morning of diestrus reduced the priming effect of LHRH on the afternoon of proestrus, and administration of estradiol benzoate at 1000 hr of diestrus day 1 increased the priming effect on diestrus day 2. Recently, Wang et al. (67) reported that the self-priming effect in women increased progressively with an increase in circulating estradiol. The poor initial LH response to LHRH in some patients with hypogonadotropic hypogonadism may be explained by lack of pituitary sensitization by endogenous LHRH.

Differential Release of LH and FSH

There are many biochemical and physiological findings which support the view that one hypothalamic hormone regulates the secretion of both LH and FSH under physiological conditions (68). However, in different experimental and physiological states, the secretion ratio of FSH/LH varies considerably (68). In animals and humans, a quick injection of LHRH stimulates significant release of LH alone, i.e., not accompanied by significant release of FSH. At the time of menstruation, plasma FSH increases without a concomitant rise in LH. If secretions of both LH and FSH are regualted by one hypothalamic hormone, LHRH decapeptide, other modulators must be present which act on pituitary LH and FSH cells in a differential way, thereby altering the FSH/LH ratio.

As mentioned in experiments by Fink et al. (62) (see under "Effect of Gonadal Steroids on Pituitary Response"), ovariectomy on the morning of diestrus day 2 reduced LH response to LHRH on the afternoon of proestrus in the rat, but did not modify FSH response. Estradiol benzoate administered immediately after ovariectomy restored the LH response, but did not change the FSH response. Nevertheless, sequential administration of estradiol and progesterone augmented both LH and FSH responses to LHRH.

Testosterone has also been implicated in FSH release in female rats (69). In male rats, the FSH response to LHRH was abolished 1 week after castration (70).

The duration of exposure to LHRH appears to be another factor modulating the FSH/LH ratio. A quick intravenous injection of LHRH in the rat stimulates significant release of LH without a concomitant release of FSH; however, a prolonged infusion of the same dose of LHRH induces considerable secretions of both LH and FSH (71). Peak blood LH levels are reached earlier in infusion than FSH levels, and LH levels start decreasing earlier than FSH levels (71). These observations suggest that LH cells respond to LHRH more rapidly, but exhaust more quickly than do FSH cells. Furthermore, the biological half-life of FSH is considerably longer than that of LH. LHRH intermittently infused into the rat increases circulating FSH levels without causing a concomitant rise in LH levels (72).

It is noteworthy that in prepubertal children, administration of LHRH leads to a small LH and a larger FSH response. When the child enters puberty (stage II) and as pubertal development continues (stages III, IV, and V), the LH response increases progressively, whereas the FSH response remains unchanged (73). Although the cause of these changes in the response of gonadotropins to LHRH during puberty are not delineated, it is conceivable that the elaboration of gonadal steriods and the priming effect of LHRH are responsible for these events.

MAJOR PHYSIOLOGICAL EVENTS MEDIATED BY LHRH

Puberty

The dominant change which triggers puberty is thought to occur in the hypothalamus; however, this does not mean that the hypothalamus functions independently, but, rather, that extrahypothalamic influences modulate hypothalamic activity. Presumably, central nervous pathways integrate certain influences which originate in the external environment at puberty. Internal environmental factors, particularly gonadal steroids, also influence the hypothalamus. Among various explanations of the onset of puberty, the hypothesis of the differential threshold of the negative feedback receptors to gonadal steroids may be the most attractive (74–76). According to this concept, the major event that precipitates puberty is an increase in the threshold of the inhibitory feedback center, the "gonadostat." In immature individuals, circulating levels of gonadotropins remain low, despite the low levels of circulating gonadal steroids. With the approach of puberty, the decreased sensitivity of the gonadostat should lead, initially, to increased release of LHRH, followed by increased gonadotropin secretion by the pituitary, and, finally, to augmented output of sex steroids by the gonads. At puberty, the gonadostat is reset at a higher level, and increased steroid levels in circulation no longer suppress the release of LHRH, and gonadotropin secretion continues. In addition, a change in pituitary sensitivity to LHRH and maturation and activation of the positive feedback mechanism ensue during the development of puberty.

Plasma LH levels of rats, both male and female, show two phases before puberty (76): the infantile phase, from 5 to 12–15 days of age, which is characterized by high levels of LH in plasma and low levels of LH in the pituitary, and the prepubertal phase from 15–35 days of age in the female rats and from 15–45 days of age in the male, when plasma LH levels are lower than those at the infantile phase and the pituitary content is higher. Measured by bioassay, hypothalamic LHRH activity increases shortly before the natural opening of the vagina and drops to low levels on the 1st day of natural opening in the rat. The sharp drop of hypothalamic LHRH activity is accompanied by an increase in plasma LH. In the male rat, hypothalamic LHRH is decreased, accompanied by an increase in plasma LH at the onset of puberty, when free spermatozoa appear in the seminiferous tubules (76).

In the female rat, plasma FSH levels are very high at about 15 days of age and then decrease as the animal approaches puberty. Pituitary FSH reaches the peak level at about 20 days of age and then decreases. By contrast, in the male rat, plasma FSH remains at the same low levels until 30 days of age and then increases. Pituitary FSH content approximately parallels the changes in plasma levels of FSH (76).

It should be remembered that the sensitivity of the pituitary gland to LHRH is modified at the onset of puberty, as well as LHRH secretion itself.

Debeljek et al. (77) demonstrated the variation in LH response to exogenous LHRH with age in rats. The magnitudes of LH responses in 25–60-day-old male rats were significantly greater than those in 15-day-old or 240-day-old rats. Maximal response was seen between 35 and 45 days of age, at which time some maturation phase spermatids could be seen. FSH response to LHRH was significantly enhanced in 25- and 45-day-old rats.

In contrast to male rats, augmented gonadotropin levels (pituitary and serum) and pituitary responsiveness to LHRH appear at an earlier age in the female. The greatest LH and FSH responses were observed in female rats 15 and 25 days of age, the lowest in rats 35 days of age (when 50% of the rats showed vaginal opening) (78). Lowered sensitivity may be partly due to increased plasma progesterone levels at this period (79) which exert a direct suppressive influence on the pituitary.

In both boys and girls, the increased pituitary responsiveness to LHRH may result in elevated levels of plasma FSH and LH at the onset of puberty (73,75). Increased secretion of LH associated with sleep is another characteristic feature in pubertal children (75). Although secretion of LHRH may increase, there has been no direct evidence for this. The self-priming effect of endogenous LHRH may also be responsible for the increased sensitivity of the pituitary at puberty.

Spermatogenesis and Follicular Maturation

It is generally accepted that FSH maintains the germinal epithelium, whereas LH facilitates the completion of spermatogenesis, stimulating the Leydig cells to produce testosterone (80). The administration of FSH to immature or mature hypophysectomized rats markedly increases the size of the testes, but does not accelerate the appearance of mature sperm or increase the secretory activity of the Leydig cells. For completion of spermatogenesis, an androgenic influence is needed which can be provided indirectly by the administration of LH or directly by the administration of testosterone (81).

Suppression of the release or blockade of the action of LHRH may eventually result in testicular atrophy and the arrest of spermatogenesis, because both LH and FSH secretions are probably regulated by hypothalamic LHRH. Following active immunization and the production of antibodies to LHRH, testicular weight in one male rabbit decreased to 6% of that of the untreated control rabbit. Interstitial cells and seminiferous tubules were markedly atrophied, and the latter were devoid of germ cells. Pituitary LH content was markedly reduced in this rabbit (82).

Spermatogenesis was greatly suppressed in the hypophysectomized male rat with a pituitary graft transplanted under the kidney capsule. Along the germinal line, only spermatogonia and some primary spermatocytes could be seen. The tubular diameter was decreased, and the interstitial cells had the appearance of fibroblast-like cells (83). Gonadotrophs of the pituitary graft showed atrophic change. When treated with 5 μg of LHRH twice daily for 2 months, considerable spermatogenesis was observed. Although the extent of stimulation varied among animals, the tubular diameter was enlarged, and spermatozoa could be observed in the tubular lumen in some animals. Leydig cells also showed evidence of stimulation. These results indicate the vital importance of hypothalamic LHRH in complete spermatogenesis. In patients with azoospermia and oligospermia, chronic treatment with LHRH has been reported to result in an increase in sperm counts (73).

Ovaries were atrophied and the follicular maturation was arrested in hypophysectomized female rats with a pituitary graft transplanted in an area remote from the hypothalamus. The ovaries contained primary follicles and a few growing follicles, but no andral follicles. Corpora lutea were absent in most ovaries, but some contained atrophic corpora lutea. Interstitial cells were also atrophic. Treatment with 5 μg LHRH twice daily for 51 days caused an increase in ovarian weight accompanied by a rise in serum FSH and estrogen levels. The ovaries of LHRH-stimulated animals showed several large Graafian follicles surrounded by well-developed thecal cells. Interstitial cells were normal in appearance and contained large nuclei and cytoplasm. The size and ultrastructural appearance of FSH gonadotrophs were most affected by the chronic treatment with LHRH in the pituitary graft. FSH cells were restored to normal appearance following LHRH treatment, whereas surviving FSH cells were very atrophic in the control hypophysectomized grafted rats. Although LHRH treatment stimulated LH cells slightly, the changes were not as consistent nor as striking as for the FSH cells. No stimulatory effects of LHRH on any of the other cells were observed (84).

Arrest of follicular maturation was also observed following passive immunization of cycling hamsters. In the ovaries of hamsters injected with the antiserum to LHRH on diestrus day 1, only preantral follicles developed on the prospective proestrus. Unlike the intact hamsters, these animals did not show a gradual rise in circulating estrogens as they approached proestrus and did not ovulate spontaneously or even after the injection of LHRH (85).

Ovulation

Neuroendocrine interplay in the hypothalamo-pituitary gonadal axis culminates at ovulation. It is generally accepted that LHRH secretion is increased at proestrus in animals and at mid-cycle in primates and humans, thereby inducing the preovulatory surge of LH and FSH secretaion. At that time, the sensitivity of the pituitary is increased by estrogen and the self-priming effect of LHRH. In monkeys anesthetized with phencyclidine hydrochloride, blood collected from the sectioned pituitary stalk contained immunoreactive LHRH in levels ranging from a low of 20–75 pg/ml to a high of 800 pg/ml at mid-cycle. There was also a pulsatile increase of LHRH release (86). On the other hand, an injection of the antiserum to LHRH in cycling rats on the morning of proestrus completely suppressed the preovultory surge of both LH and FSH, and prevented ovulation (87,88). In hamsters, an injection of the antiserum to LHRH at noon of proestrus also blocked spontaneous ovulation, but administration of LHRH to these rats induced ovulation (85). These results indicate that the preovulatory surge of LH and FSH is indeed caused by an increased secretion of endogenous LHRH, accompanied by an increased sensitivity of the pituitary. The inhibition of the preovulatory surge of gonadotropins by passive immunization with the antibody to LHRH may have resulted from the nullification of the action of endogenous LHRH. An increase of immunoreactive LHRH in the peripheral blood at proestrus in animals and at mid-cycle in women was reported, but specificity of radioimmunoassay for LHRH in peripheral plasma has been questioned (89).

Gestation

Although the importance of pituitary LH in the implantation of fertilized ova and the maintenance of fetuses during early pregnancy in animals has been well documented, the extent of dependency of LH secretion on hypothalamic LHRH during pregnancy has remained undefeated. Recently, this author's group demonstrated the importance of endogenous LHRH in both implantation and maintenance of fetuses during early pregnancy (90–92).

In the first experiment, a proestrous rat was caged with two male rats which had been proven fertile. If vaginal lavage on the following morning contained spermatozoa, that day was designated as pregnancy day 1. Anit-LHRH sheep gamma globulin (anti-LHRHG) was administered intravenously to the rats once daily from day 1 through day 7 or from day 3 to day 5 of the pregnancy. Viable sites were hardly distinguishable on day 8, when an exploratory laparotomy was performed; however, they became distinguishable on day 14 in some of these animals. On the other hand, administration of normal sheep gamma globulin (NSG) did not interfere with normal implantation. Inhibition of implantation was observed even when a single injection of anti-LHRHG was administered to the rat on day 4, but not on days 3 or 5. The suppressive effect of anti-LHRHG was completely prevented by a simultaneous injection of two doses of 1 μg of LHRH in 16% gelatine/0.9% saline and was nearly completely reversed by a single dose of 1 μg of estradiol (91).

The effect of anti-LHRHG on gestation after implantation was also examined. After pregnancy was confirmed by a laparotomy on day 7, the rats were given anti-LHRHG or NSG daily from days 7–11. In all the anti-LHRHG-treated rats, resorption of fetuses occurred, as indicated by a second exploratory laparotomy on day 14. Yet, no vaginal bleeding accompanied this resorption. Gestation was not disturbed in the NSG-treated animals. It was found that a single injection of anti-LHRHG on day 9 or 10 caused complete resorption of fetuses in all the treated rats. Injection on day 8, 11, or 12 was only partially effective (92). These results indicate that the critical days are days 9 and 10, when LHRH is essential for the maintenance of viable fetuses.

Administration of LHRH twice daily from days 9–12 completely overcame the effect of anti-LHRH. Treatment with 4 mg of progesterone daily from days 7–12 of gestation also prevented the resorption of fetuses by anti-LHRHG. Serum progesterone levels were elevated on day 7 of gestation and reached a peak on days 13–14 in the control animals. Injection of anti-LHRHG on day 9 or 10 drastically reduced plasma progesterone levels on day 11 and thereafter. The suppression of progesterone was slight or absent when the anti-LHRHG was injected on day 12.

Therefore, there is a good correlation between the termination of gestation and the decrease in serum progesterone. Because, in the control animals, circulating LH levels during days 9 and 10 of pregnancy declined to the point of being barely detectable by our radioimmunoassay, the possible reduction of LH levels by anti-LHRHG could not be confirmed. However, the results of these studies indicate that the hypothalamic LHRH plays an important role in the early pregnancy of rats. On day 4, LHRH, probably through LH, stimulates estrogen secretion, which is essential for implantation of fertilized ova, and on days 9 and 10, LHRH stimulates progesterone secretion, which is essential for the maintenance of viable fetuses (92). However, whether this is the case for primates and humans remains to be determined.

CONCLUSION

Hypothalamic LHRH regulates the secretions of pituitary LH and FSH, thereby controlling almost all reproductive processes. Thus, processes such as the onset of puberty, spermatogenesis, follicular maturation, the periodicity of the estrous or menstrual cycle, ovulation, and gestation are controlled through the alteration of secretion rates of LH, FSH, and gonadal steroids. Gonadal steroids exert an inhibitory or stimulatory effect through negative or positive feedback mechanisms, respectively, on the release of LHRH, as well as on the pituitary LH and FSH responses to LHRH. LHRH itself appears to have a priming effect on pituitary sensitivity and is needed not only for stimulating gonadotropin release, but also for maintaining relatively low levels of LH and FSH secretions. The latter function is particularly important during the early to middle period of gestation, before pituitary hormones are replaced by placental gonadotropins and mammotropic hormones. Any events which block the synthesis, release,

and/or action of LHRH will eventually interfere with these physiological reproductive processes.

ACKNOWLEDGMENTS

The author is indebted to Dr. J. King and Miss G. Farley for their help in
preparing the manuscript.

REFERENCES

1. Green, J. D., and Harris, G. W. (1947). The neurovascular link between the neurohypo-
    physis and adenohypophysis. J. Endocrinol. 5:36.
2. Harris, G. W. (1955). Neural Control of the Pituitary Gland. Edward Arnold, London.
3. McCann, S. M. (1962). A hypothalamic luteinizing hormone-releasing factor. Am. J.
    Physiol. 202:395.
4. Campbell, H. J., Fever, G., Garcia, J., and Harris, G. W. (1961). The infusion of brain
    extracts into the anterior pituitary gland and the secretion of gonadotrophic hormone.
    J. Physiol. (London) 157:30.
5. Kobayashi, T., Kobayashi, T., Kigawa, T., Mizuno, M., and Amenomori, Y. (1963).
    Influence of rat hypothalamic extract on gonadotropic activity of cultivated anterior
    pituitary cells. Endocrinol. Jap. 10:16.
6. Mittler, J. C., and Meites, J. (1964). *In vitro* stimulation of pituitary follicle stimulating
    hormone release by hypothalamic extract. Proc. Soc. Exp. Biol. Med. 117:309.
7. Igarashi, M., and McCann, S. M. (1964). A new sensitive bioassay for follicle-
    stimulating hormone (FSH). Endocrinology 74:440.
8. Kuroshima, A., Ishida, Y., Bowers, C. Y. and Schally, A. V. (1965). Stimulation of
    release of follicle-stimulating hormone by hypothalamic extracts *in vitro* and *in vivo*.
    Endocrinology 76:614.
9. Bowers, C. Y., Currie, B. L., Johansson, K. N. G., and Folkers, K. (1973). Biological
    evidence that separate hypothalamic hormones release the follicle stimulating and
    luteinizing hormones. Biochem. Biophys. Res. Commun. 50:20.
10. Schally, A. V., Arimura, A., Redding, T. W., Debeljuk, L., Carter, W., Dupont, A., and
    Vilchez-Martinez, J. A. (1975). Re-examination of porcine and bovine hypothalamic
    factors for additional LH and FSH-releasing activities. Endocrinology 98:380.
11. Amoss, M., Burgus, R., Blackwell, R., Vale, W., and Guillemin, R. (1971). Purification,
    amino acid composition and N-terminus of the luteinizing hormone-releasing factor
    (LRF) of ovine origin with recent studies on its biological activity. *In* B. Saxena, C.
    Billing, and H. Gandy (eds.), Gonadotropin, pp. 26–31. Wiley Interscience, New York.
12. Schally, A. V., Arimura, A., Baba, Y., Nair, R. M. G., Matsuo, H., Redding, T. W.,
    Debeljuk, L., and White, W. F. (1971). Isolation and properties of the FSH and
    LH-releasing hormone. Biochem. Biophys. Res. Commun. 43:383.
13. Matsuo, H., Baba, Y., Nair, R. M. G., Arimura, A., and Schally, A. V. (1971). Structure
    of the porcine LH- and FSH-releasing hormone. I. The proposed amino acid sequence.
    Biochem. Biophys. Res. Commun. 43:1334.
14. Baba, Y., Matsuo, H., and Schally, A. V. (1971). Structure of porcine LH- and
    FSH-releasing hormone. II. Confirmation of the proposed structure by conventional
    sequential analyses. Biochem. Biophys. Res. Commun. 44:459.
15. Matsuo, H., Arimura, A., Nair, R. M. G., and Schally, A. V. (1971). Synthesis of the
    porcine LH- and FSH-releasing hormone by the solid-phase method. Biochem. Biophys.
    Res. Commun. 45:822.
16. Arimura, A., Matsuo, H., Baba, Y., Debeljuk, L., Sandow, J., and Schally, A. V.
    (1972). Stimulation of release of LH by synthetic LHRH *in vivo*. I. Comparative study
    of natural and synthetic hormone. Endocrinology 90:163.
17. Burgus, R., Butcher, M., Ling, N., Monahan, M., Rivier, J., Fellows, R., Amoss, M.,

Blackwell, R., Vale, W., and Guillamin, R. (1971). Structure moleculaire du facteur hypothalamique (LRF) d'origine ovine controlant la secretion de l'hormone gonadotrope hypophysaire de luteinisation (LH).

18. Schally, A. V., Arimura, A., Kastin, A. J., Matsuo, H., Baba, Y., Redding, T. W., Nair, R. M. G., Debeljuk, L., and White, W. F. (1971). The gonadotropin-releasing hormone: a single hypothalamic polypeptide regulates the secretion of both LH and FSH. Science 173:1036.

19. Schally, A. V., Kastin, A. J., and Arimura, A. (1971). Hypothalamic follicle-stimulating hormone (FSH) and luteinizing hormone (LH)-regulating hormone: structure, physiology and clinical studies. Fertil. Steril. 22:703.

20. Schally, A. V., Kastin, A. J., and Arimura, A. (1975). The hypothalamus and reproduction. Am. J. Obstet. Gynecol. 122:857.

21. Buillemin, R., and Burgus, R. (1972). The hormones of the hypothalamus. Sci. Am. 227:24.

22. Ramirez, V. D., and Kordon, C. (1975). Localization and intracellular distribution of hypothalamic hormones: studies on luteinizing hormone-releasing hormone (LHRH). In M. Motta, G. Grosginani, and L. Martini (eds.), Hypothalamic Hormones, pp. 57–74. Academic Press, London.

23. Sawyer, C. H. (1975). Some recent developments in brain pituitary ovarian physiology. Neuroendocrinology 17:97.

24. McCann, S. M. (1962). A hypothalamic luteinizing-hormone releasing factor. Am. J. Physiol. 202:395.

25. Watanabe, S., and McCann, S. M. (1968). Localization of FSH-releasing factor in the hypothalamus and neurohypophysis as determined by in vitro assay. Endocrinology 82:664.

26. Crighton, D. B., Schneider, H. P. G., and McCann, S. M. (1970). Localization of LH-releasing factor in the hypothalamus and neuropophysis as determined by an in vitro method. Endocrinology 87:323.

27. King, J. A., Arimura, A., and Williams, T. H. (1975). Localization of luteinizing hormone-releasing hormone in rat hypothalamus using radioimmunoassay. J. Anat. 120:275.

28. Wheaton, J. E., Krulich, L., and McCann, S. M. (1975). Localization of luteinizing hormone-releasing hormone in the preoptic area and hypothalamus of the rat using radioimmunoassay. Endocrinology 97:30.

29. Palkovitz, M., Arimura, A., Brownstein, M., Schally, A. V., and Saavedra, J. M. (1974). Luteinizing hormone-releasing hormone (LHRH) content of the hypothalamic nuclei in the rat. Endocrinology 96:554.

30. Barry, J., Dubois, M. P., and Poulain, (1973). LRF-producing cells of the mammalian hypothalamus. Z. Zellforsch. Mikrosk. Anat. 146:351.

31. Setalo, G., Vigh, S., Schally, A. V., Arimura, A., and Flerko, B. (1975). LHRH-containing neural elements in the rat hypothalamus. Endocrinology 96:135.

32. Setalo, G., Vigh, S., Schally, A. V., Arimura, A., and Flerko, B. Immunohistological study of the origin of LHRH-containing nerve fibers of the rat hypothalamus. Brain Res. In press.

33. Naik, D. V. (1975). Immunoreactive LHRH neurons in the hypothalamus identified by light and fluorescent microscopy. Cell. Tissue Res. 157:423.

34. Zimmerman, E. A., Hsu, K. C., Ferin, M., and Kozlowski, G. P. (1974). Localization of gonadotropin-releasing hormone (Gn-RH) in the hypothalamus of the mouse by immunoperoxidase. Endocrinology 95:1.

35. King, J. C., Parsons, J. A., Erlandsen, S. L., and Williams, T. H. (1974). Luteinizing hormone-releasing hormone (LHRH) pathway of the rat hypothalamus revealed by the unlabeled antibody peroxidase-anti-peroxidase method. Cell Tissue Res. 153:211.

36. Eskay, R. L. (1974). Immunoreactive LRH and TRH in the fetal, neonatal and adult rat brain. In Program of the 56th Annual Meeting of the Endocrine Society, Abstract 55, pp. A–83, The Endocrine Society. Oklahoma City, Oklahoma.

37. Araki, S., Toran-Allerand, D., Ferin, M., and Vande Wiele, R. L. (1975). Immunoreactive gonadotropin-releasing hormone (Gn-RH) during maturation in the rat: ontogeny of regional hypothalamic differences. Endocrinology 97:693.

38. King, J. C., Gerall, A. A., Elkind, K. E., and Fishback, J. B. (1975). The LHRH system in the postnatal rat. *In* Program of the Fifth Annual Meeting, Society for Neuroscience, Abstract 677, p. 436.

39. Sawyer, C. H., Markee, J. E., and Townsend, B. F. (1949). Cholinergic and adrenergic components in the neurohumoral control of the release of LH in the rabbit. Endocrinology 44:18.

40. Fuxe, K., and Hokfelt, T. (1969). Catecholamines in the hypothalamus and the pituitary gland. *In* W. F. Ganong and L. Martini (eds.), Frontiers in Neuroendocrinology, pp. 47–96. Oxford University Press, London.

41. Jonsson, G., Fuxe, K., and Hokfelt, T. (1972). On the catecholamine innervation of the hypothalamus, with special reference to the median eminence. Brain Res. 40:271.

42. Palkovitz, M., Brownstein, M., Saavedra, J. M., and Axelrod, J. (1974). Neuroendocrine and dopamine content of hypothalamic nuclei of the rat. Brain Res. 77:137.

43. McCann, S. M., Kalra, R. S., Kalra, S. P., Donoso, A. O., Bishop, W., Schneider, H. P. G., Fawcett, C. P., and Krulich, L. (1972). The role of monoamines in the control of gonadotropin and prolactin secretion. *In* B. B. Saxena, C. G. Beling, and H. M. Gandy (eds.), Gonadotropins, pp. 49–60. Wiley Interscience, New York.

44. Kalra, S. P., and McCann, S. M. (1973). Effect of drugs modifying catecholamine synthesis on LH release induced by preoptic stimulation in the rat. Endocrinology 93:356.

45. Ojeda, S. R., Harms, P. G., and McCann, S. M. (1974). Possible role of cyclic AMP and prostaglandin E_1 in the dopaminergic control of prolactin release. Endocrinology 95:1694.

46. Kizer, J. S., Arimura, A., Schally, A. V., and Brownstein, M. J. (1975). Absence of luteinizing hormone-releasing hormone (LHRH) from catecholaminergic neurons. Endocrinology 96:523.

47. Eskay, R. L., Waberg, J., Mical, R. S., and Porter, J. C. (1975). Prostaglandin E_2-induced release of LHRH into hypophysial portal blood. Endocrinology 97:816.

48. Ojeda, S. R., Harms, P. G., and McCann, S. M. (1975). Effect of inhibitors of prostaglandin synthesis on gonadotropin release in the rat. Endocrinology 97:843.

49. Ojeda, S. R., Harms, P. G., and McCann, S. M. (1974). Mechanism of action and physiological role of prostaglandin E_2 (PGE_2) in hypothalamic control of LH release. *In* Program of the 56th Annual Meeting of the Endocrine Society, Abstract 149, pp. A–130, The Endocrine Society. Oklahoma City, Oklahoma.

50. Döcke, F., and Dörner, G. (1965). The mechanism of the induction of ovulation by oestrogens. J. Endocrinol. 33:491.

51. Arimura, A., and Schally, A. V. (1972). Physiological and clinical studies with natural and synthetic luteinizing hormone-releasing hormone (LHRH). Med. J. Osaka. Univ. 23:77.

52. Aiyer, M. S., Greig, F., and Fink, G. (1973). Changes in sensitivity of the anterior pituitary gland to synthetic luteinizing hormone releasing factor during the oestrous cycle of the rat.

53. Krey, L. C., Butler, W. R., Weiss, G., Weick, R. F., Dierschke, D. J., and Knobil, E. (1973). Influence of endogenous and exogenous gonadal steroids on the actions of synthetic LRF in the Rhesus monkey. *In* C. Gual and E. Rosenberg (eds.), Hypothalamic Hypophysiotropic Hormones, pp. 39–47. Excerpta Medica, Amsterdam.

54. Yen, S. S. C., VandenBerg, G., Rebar, R., and Ehara, Y. (1972). Variation of pituitary responsiveness to synthetic LRF during different phases of the menstrual cycle. J. Clin. Endocrinol. Metab. 35:931.

55. Vilchez-Martinez, J. A., Arimura, A., Debeljuk, L., and Schally, A. V. (1974). Biphasic effect of estradiol Benzoate on the pituitary responsiveness to LHRH. Endocrinology 94:1300.

56. Libertun, C., Orias, R., and McCann, S. M. (1974). Biphasic effect of estrogen on the sensitivity of the pituitary to luteinizing hormone-releasing factor (LRF). Endocrinology 94:1094.

57. Aiyer, M. S., Fink, G., and Greig, F. (1974). Changes in the sensitivity of the pituitary gland to luteinizing hormone releasing factor during the oestrous cycle of the rat. J. Endocrinol. 60:47.

58. Wang, C. F., and S. S. C. Yen. (1975). Direct evidence of estrogen modulation of pituitary sensitivity to luteinizing hormone-releasing factor during the menstrual cycle. J. Clin. Invest. 55:201.

59. Hillard, J., Schalla, A. V., and Sawyer, C. H. (1971). Progesterone blockade of the ovulatory response to intrapituitary infusion of LHRH in rabbits. Endocrinology 88:730.

60. Spies, H. G., Stevens, K. R., Hillard, J., and Sawyer, C. H. (1969). The pituitary as a site of progesterone and chlormadinone blockade of ovulation in the rabbit. Endocrinology 84:277.

61. Schally, A. V., Redding, T. W., and Arimura, A. (1973). Effect of sex steroids on pituitary responses to LH- and FSH-releasing hormone in vitro. Endocrinology 93:893.

62. Fink, G., Aiyer, M. S., Jamieson, M. G., and Chiappa, S. A. (1975). Factors modulating the responsiveness of the anterior pituitary gland in the rat with special referenve to gonadotropin releasing hormone (GnRH). In M. Motta, P. G. Crosignani, and L. Martini (eds.), Hypothalamic Hormones, pp. 139–160. Academic Press, London.

63. Döcke, F., and Dörner, G. (1966). Facilitative action of progesterone in the induction of ovulation by oestrogen, J. Endocrinol. 36:209.

64. Brown-Grant, K., and Naftolin, F. (1972). Facilitation of luteinizing hormone secretion in the female rat by progesterone. J. Endocrinol. 53:37.

65. Gordon, J. H., and Reichlin, S. (1974). Changes in pituitary responsiveness to luteinizing hormone releasing factor during the rat estrous cycle. Endocrinology 94:974.

66. Zeballos, G., and McCann, S. M. (1975). Alterations during the estrous cycle in the responsiveness of the pituitary to subcutaneous administration of synthetic LH-releasing hormone (LHRH). Endocrinology 96:1377.

67. Wang, C. F., Lasley, B. L., and Yen, S. S. C. (1975). The mechanism for the development of the preovulatory gonadotropin surge. In Program of the 57th Annual Meeting of the Endocrine Society, p. 99. The Endocrine Society, Bethesda, Maryland.

68. Arimura, A., and Schally, A. V. (1974). Hypothalamic LH- and FSH-releasing hormone: re-evaluation of the concept that one hypothalamic hormone controls the release of LH and FSH. In M. Kawakami (ed.), Biological Rhythms in Neuroendocrine Activity, pp. 73–90. Igaku Shoin, Tokyo.

69. Gay, V. L., and Tomacari, R. L. (1974). Follicle-stimulating hormone secretion in the female rat: cyclic release is dependent on circulating androgen. Science 184:75.

70. Debeljuk, L., Arimura, A., and Schally, A. V. (1973). Effect of estradiol on the response to LHRH in male rats at different times after castration. Proc. Soc. Exp. Biol. Med. 143:1164.

71. Arimura, A., Debeljuk, L., and Schally, A. V. (1972). Stimulation of FSH release in vivo by prolonged infusion of synthetic LHRH. Endocrinology 91:529.

72. Vilchez-Martinez, J. A., Arimura, A., and Schally, A. V. (1974). Effect of intermittent infusion of LH-releasing hormone on serum LH and FSH levels in immature male rats. Proc. Soc. Exp. Biol. Med. 148:913.

73. Mortimer, C. H., Besser, G. M., and McNeilly, A. S. (1975). Gonadotrophin releasing hormone therapy in the induction of puberty, potency, spermatogenesis and ovulation in patients with hypothalamic pituitary-gonadal dysfunction. In M. Motta, P. G. Crosignani, and L. Martini (eds.), Hypothalamic Hormones, pp. 325–336. Academic Press, London.

74. Davidson, J. M. (1974). Hypothalamic-pituitary regulation of puberty, evidence from animal experimentation. In M. M. Grumbach, G. D. Grave, and F. E. Mayer (eds.), Control of the Onset of Puberty, pp. 79–103. John Wiley, New York.

75. Grumbach, M. M., Roth, J. C., Kaplan, S. L., and Kelch, R. P. (1974). Hypothalamic pituitary regulation of puberty: evidence and concepts derived from clinical research. In M. M. Grumbach, G. D. Grave, and F. E. Mayer (eds.), Control of the Onset of Puberty, pp. 79–103. John Wiley, New York.

76. Ramirez, V. D. (1973). Endocrinology of puberty. In R. O. Greep (ed.), Handbook of Physiology, Vol. II, Section 7, pp. 1–28, Williams & Wilkins, Baltimore.

77. Debeljuk, L., Arimura, A., and Schally, A. V. (1972). Studies on the pituitary responsiveness to luteinizing hormone releasing hormone (LHRH) in intact male rats of different ages. Endocrinology 90:585.

78. Debeljuk, L., Arimura, A., and Schally, A. V. (1972). Pituitary responsiveness to LH-releasing hormone in intact female rats of different ages. Endocrinology 90:1499.

79. Dohler, K. D., and Wuttke, W. (1974). Serum LH, FSH, prolactin and progesterone from birth to puberty in female and male rats. Endocrinology 94:1003.

80. Gemzell, C., and Roos, P. (1966). The physiology and chemistry of follicle-stimulating hormone. *In* G. W. Harris and B. T. Donovan (eds.), The Pituitary Gland, pp. 492–517. Butterworth, London.

81. Means, A. R. (1975). Biochemical effects of follicle-stimulating hormone on the testis. *In* Handbook of Physiology, Vol. V, Section 7, pp. 203–218. Williams & Wilkins, Baltimore.

82. Arimura, A., Sato, H., Kumasaka, T., Worobec, R. B., Debeljuk, L., Dunn, J., and Schally, A. V. (1973). Production of antiserum to LH-releasing hormone (LHRH) associated with gonadal atrophy in rabbits: development of radioimmunoassay for LHRH. Endocrinology 93:1092.

83. Debeljuk, L., Arimura, A., Shiino, M., Rennels, E. G., and Schally, A. V. (1973). Effects of chronic treatment with LHRH in hypophysectomized pituitary grafted rats. Endocrinology 92:921.

84. Arimura, A., Debeljuk, L., Shiino, M., Rennels, G., and Schally, A. V. (1973). Follicular stimulation by chronic treatment with synthetic LH releasing hormones (LHRH) in hypophysectomized female rats bearing pituitary graft. Endocrinology 92:1507.

85. de la Cruz, A., Arimura, A., de la Cruz, K., and Schally, A. V. Effect of administration of anti-LHRH serum on gonadal function during the estrous cycle in the hamster. Endocrinology. In press.

86. Carmel, P. C., Araki, S., and Ferin, M. (1975). Prolonged stalk portal blood colection in rhesus monkeys. *In* Program of the 57th Annual Meeting of the Endocrine Society, p. 104. The Endocrine Society, Bethesda, Maryland.

87. Arimura, A., Debeljuk, L., and Shally, A. V. (1974). Blockade of preovulatory surge of gonadotropins, LH and FSH, and of ovulation by anti-LHRH in rats. Endocrinology 95:323.

88. Koch, Y., Chobsieng, P., Zor, R., Fridkin, M., and Lindner, H. R. (1973). Suppression of gonadotropin secretion and prevention of ovulation in the rat by antiserum to synthetic gonadotropin-releasing hormone. Biochem. Biophys. Res. Commun. 55:623.

89. de la Cruz, K., and Arimura, A. (1975). Evidence for the presence of immunoreactive plasma LHRH which is unrelated to LHRH decapeptide. *In* Program of 57th Annual Meeting of the Endocrine Society, p. 103. The Endocrine Society, Bethesda, Maryland.

90. Arimura, A. *In vivo* methods for studying the actions of hypothalamic hormones with special reference to their antisera as tools for investigation. *In* F. Labrie, J. Meites, and G. Pelletier (eds.), Hypothalamus and Endocrine Functions, pp. 387–396. Plenum Press, New York.

91. Arimura, A., Nishi, N., and Schally, A. V. (1976). Delayed implantation caused by administration of sheep immunogamma globulin against LHRH in the rat. Proc. Soc. Exp. Biol. Med. 152:71–75.

92. Nishi, N., Arimura, A., de la Cruz, K. G., and Schally, A. V. (1976). Termination of pregnancy by sheep anti-LHRH gamma globulin in rats. Endocrinology 98:1024.

International Review of Physiology
Reproductive Physiology II, Volume 13
Edited by Roy O. Greep
Copyright 1977   University Park Press   Baltimore

2
Brain Monoamines and Reproductive Function

J. D. FERNSTROM AND R. J. WURTMAN

Laboratory of Neuroendocrine Regulation
Department of Nutrition and Food Science
Massachusetts Institute of Technology
Cambridge, Massachusetts

Some of the studies described in this chapter were supported in part by Grant
AM-11237 from the National Institute of Health to R.J.W.

The secretion of the gonadotropins and prolactin by the pituitary is modulated by a variety of control systems in the body that include as components the pituitary itself, the gonads, the mammary glands, and portions of the central and peripheral nervous systems. One such control system, for example, mediates the secretion of prolactin induced by suckling; another may be responsible for the elevation of serum prolactin levels that accompanies sleep; still another is involved in the stimulation of luteinizing hormone (LH) secretion that follows gonadectomy and its suppression by high circulating levels of estradiol. Probably the greatest single task of the neuroendocrinologist is to describe these systems, first by identifying all of their components, and second, by explaining their operation from infancy through old age.

The components of these neuroendocrine systems include at least three cell types: 1) neurons (peripheral and central); 2) neuroendocrine transducer cells, within the hypothalamus, that respond to signals from brain neurons by secreting hormones into the pituitary portal vascular system; and 3) gland cells, that secrete hormones or other cell-specific products (e.g., milk) in response to another hormonal signal.

Each group of neurons that participates in a neuroendocrine system can be characterized *anatomically,* in terms of the neurons that project to it, and those with which its axons make synapses; *physiologically,* by its responses to neurotransmitter molecules, or to hormones or other circulating compounds, that impinge on it; and *biochemically,* by the particular neurotransmitter molecule that its neurons release. This chapter discusses the neurons involved in reproductive function from the biochemical perspective: it describes the neurochemical and neuropharmacological methods used to identify their neurotransmitters and how they participate in systems controlling gonadotropin secretion. Special emphasis is placed on the monoamine neurotransmitters (dopamine [DA], norepinephrine [NE], and serotonin [5-HT]), even though these compounds are used by only a modest fraction of all brain neurons, simply because at present most of the information available concerning the neuropharmacology of reproductive function relates to these transmitters.

It should be emphasized that neurotransmitters per se do not control any process in the body; rather, the neurons that release them do (or, better, systems that include neurons that release neurotransmitters do). There is no basis for suggesting, for example, that dopamine controls prolactin secretion. However, considerable evidence supports the view that cells which release dopamine are components of the systems that normally control prolactin secretion. An awareness of the difference between these two statements may help the reader to cope with the numerous ambiguities in the literature, such as the evidence that the same neurotransmitter, released at two different loci within the brain, can both facilitate and suppress gonadotropin secretion.

This chapter begins with a brief description of the monoamine neurotransmitters in the central nervous system and the drugs most often used to examine their participation in neuroendocrine systems. It then attempts to summarize our present knowledge of neurotransmitter involvement in the secretion of prolactin, LH, and follicle-stimulating hormone (FSH). Finally, it reviews the effects of gonadotropins and gonadal steroids on brain monoamine metabolism.

IDENTIFICATION AND
LOCALIZATION OF BRAIN NEUROTRANSMITTERS

The demonstration that a compound functions as the neurotransmitter released at a particular synapse requires several types of evidence: 1) the compound must be shown to be localized within the presynaptic terminals; 2) it must be released when the nerves that store it are stimulated; 3) the physiological effect of applying the compound to postsynaptic effector cells must be identical with that observed following stimulation of the presynaptic neuron; and 4) drugs that enhance or diminish its availability to receptors must cause the expected functional changes at the synapses where it functions as a neurotransmitter. All of these criteria have been satisfied in the identification of the neurotransmitters used by the autonomic nervous system. However, the difficulty of examining single central nervous system synapses has precluded the definitive demonstration that any compound functions as a central neurotransmitter. At present, the strongest evidence in favor of this role is available for dopamine, norepinephrine, serotonin, and, to a lesser extent, acetylcholine.

The regional distribution of catecholaminergic neurotransmitters in the brain and spinal cord was determined imprecisely, using biochemical assays, many years ago; however, the more recent application of histochemical fluorescence methods has made it possible for these compounds to be visualized within specific neuronal tracts and cell bodies (1–3). With the use of this approach, *dopamine* appears to be concentrated in three major groups of brain neurons: *nigro-neostriatal neurons* terminating in the basal ganglia; *meso-limbic neurons,* with cell bodies localized in the upper brainstem and axon terminals in the nucleus accumbens, olfactory tubercles, and cerebral cortex; and *tubero-infundibular neurons,* located entirely within the hypothalamus and having their most prominent set of terminals in the external layer of the median eminence (4, 5). Nerve tracts that appear, by histochemical fluorescence assay, to use *norepinephrine* as a neurotransmitter are also grouped into three main fiber bundles: a *descending* system, with cell bodies localized in the medulla oblongata, whose axons course through the spinal cord to form synapses with ventral and dorsal horn neurons; and *two ascending tracts,* one with cell bodies located in the locus coeruleus and terminals in the cerebral cortices and the hippocampus, and a second tract with cell bodies in the medulla and pons and axons terminating largely within the hypothalamus. Within the hypothalamus, norepinephrine can be detected by microenzymatic assay in each nucleus, with the highest concentrations in the median eminence and arcuate nucleus (6).

Other investigators (7) have attempted to identify central noradrenergic neurons by use of an immunofluorescent antibody to dopamine β-oxidase, the enzyme that catalyzes the conversion of DA to NE (and presumably is present only in noradrenergic neurons). Their findings suggest a different distribution pattern than that mapped out with the use of histochemical fluorescence; i.e., a single ascending fiber system, originating in three groups of cell bodies within the pons and medulla and distributing terminals to most of the forebrain.

Epinephrine-containing cell bodies have also been identified (based on their content of an enzyme unique to the production of epinephrine) near the locus coeruleus; the exact distribution of their terminals remains to be determined. Because both norepinephrine- and epinephrine-containing nerves contain dopamine β-hydroxylase, it is not possible to distinguish nerve terminals using one or the other of these neurotransmitters by currently available immunochemical methods. Moreover, the histochemical fluorescence approach, as generally used, simply fails to distinguish between dopamine, norepinephrine, or epinephrine.

Serotonin-containing neurons are only poorly visualized with the histochemical fluorescence method; however, available data seem adequate to justify the conclusion that virtually all of the serotonin cell bodies in brain are confined to the mid-line raphe nuclei of the medulla and pons. Fibers from these cell bodies descend to terminate in the spinal cord gray matter and ascend in the brain (mainly via the medial forebrain bundle, along with dopaminergic and noradrenergic axons) to innervate most of the telencephalon and diencephalon. Most hypothalamic nuclei contain serotonin (and thus, probably, 5-HT terminals), as assayed by microenzymatic techniques (8). The highest concentrations are found in the arcuate and suprachiasmatic nuclei, with smaller, but nonetheless substantial, amounts being presented in the median eminence.

No satisfactory histochemical techniques are presently available for localizing brain neurons that contain acetylcholine, a situation that may change if and when satisfactory immunofluorescent methods are developed (probably based on antibodies directed against choline acetyltransferase). Histochemical techniques are available for staining acetylcholinesterase, the enzyme that hydrolyzes acetylcholine; however, there is little basis for believing that this enzyme is confined only to cholinergic neurons or to neurons receiving cholinergic synapses. Biochemical measurements of acetylcholine or of choline acetyltransferase activity and determinations of choline uptake by brain homogenates in vitro provide evidence that some neurons in the caudate nucleus, thalamus, and visual somato-sensory, cerebellar, and limbic cortices contain this neurotransmitter. Some specific central cholinergic pathways have also been tentatively identified (e.g., motor neuron collaterals to spinal cord Renshaw cells, a septo-hippocampal tract, and a habenulo-interpenducular tract).

FATE OF A TYPICAL BRAIN NEUROTRANSMITTER

Even though particular neurotransmitters may be involved in very different functional roles, all appear to share similar metabolic fates in brain neurons and synapses. Each is 1) synthesized from a circulating precursor or precursors (often an amino acid) through the action of a rate-limiting enzyme; 2) stored within one or more intracellular functional pools (which may be associated with subcellular organelles known as synaptic vesicles); 3) released into synapses as a result of nerve stimulation; 4) complexed with pre- or postsynaptic receptors (which causes modifications in the flux of one or more ions across plasma membranes); and, 5) inactivated by reuptake into its neuron of origin, or by enzymatic degradation.

Brain neurons have been shown to contain all of the precursors, enzymes, and cofactors needed for the biosynthesis of dopamine, norepinephrine, serotonin, and acetylcholine. The *essential* precursor for each transmitter cannot be made by the brain, and thus must be derived from the diet (e.g., phenylalanine, tryptophan, choline) or from peripheral synthesis (e.g., tyrosine or choline). The enzymes that catalyze neurotransmitter synthesis are made in the cell body of the neuron, and then are transported in the axon to the nerve terminals, where most of the neurotransmitter molecules are synthesized. Many factors can influence the rate of neurotransmitter synthesis, e.g., the abilities of neurons to take up sufficient amounts of precursor compounds, the availability of cofactors, the amount or activity of the rate-limiting biosynthetic enzyme, and hormonal effects.

When the neuron is depolarized physiologically (by presynaptic inputs) or stimulated electrically, or when calcium enters its terminals, neurotransmitter is released. The quantity released into synapses per unit of time depends theoretically on both the number of times that the neuron is depolarized (the number of action potentials) and the number of molecules released per action potential. This latter number—the molecules in a single "quantum"—may or may not be fairly constant among particular groups of neurons. Following their release into synapses, some of the neurotransmitter molecules survive long enough to interact with receptors on the postsynaptic effector cells. If sufficient numbers of neurotransmitter molecules are released from one or more presynaptic excitatory neurons per unit of time (i.e., if sufficient excitatory postsynaptic potentials are generated), the postsynaptic effector cell will become depolarized and, if it is also a neuron, will generate an action potential. The precise ways in which neurotransmitters interact with the postsynaptic effector cells to produce depolarization (or, in the case of "inhibitory" neurons, hyperpolarization and inhibitory postsynaptic potentials) are poorly understood. However, it seems likely that these interactions involve specific receptors, which, when stimulated, modify the inward or outward fluxes of particular ions and which also mediate longer term functional changes (e.g., in the synthesis of enzyme proteins by mechanisms that may involve such "second messengers" as the cyclic nucleotides). The sensitivity of postsynaptic receptors to intrasynaptic neurotransmitter molecules may vary with such factors as 1) the extent to which they are already occupied; 2) actions of hormones (e.g., thyroxine); and 3) the effects of other compounds released into the synapses along with the neurotransmitter (e.g., adenine derivatives, storage proteins, and peptides).

Once neurotransmitter molecules are released into synapses, they are rapidly inactivated by enzymes (e.g., cholinesterase) or by being taken up again within their neuron of origin (e.g., catecholamines, serotonin). Neurotransmitters released into peripheral synapses may also be inactivated by being "washed out" into the circulation.

CATECHOLAMINE NEUROTRANSMITTERS

The catecholamines are synthesized from the amino acid tyrosine. Sources of this amino acid to the brain include: 1) the tyrosine in dietary protein, 2) tyrosine produced from the hepatic hydroxylation of phenylalanine, or 3) tyrosine entering the blood from reservoirs (i.e., free or peptide-bound amino acid) in peripheral tissues. The initial step in catecholamine biosynthesis involves the hydroxylation of L-tyrosine to form L-dihydroxyphenylalanine (L-dopa), a catechol amino acid. This reaction step is catalyzed by tyrosine hydroxylase, an enzyme specifically localized within catecholaminergic neurons. Under most circumstances, this hydroxylation step appears to rate limit catecholamine biosynthesis. The L-dopa is rapidly decarboxylated to dopamine through the catalytic action of aromatic L-amino acid decarboxylase; hence, little (if any) L-dopa is normally released as such from the neuron, and virtually none is found in the circulation. Unlike tyrosine hydroxylase, aromatic L-amino acid decarboxylase is found in many brain cells that do not normally synthesize catecholamines; hence, the administration of L-dopa causes most neurons to become transiently "dopaminergic." Brain neurons capable of synthesizing norepinephrine or epinephrine also contain an enzyme, dopamine β-hydroxylase, that catalyzes the addition of a hydroxyl group to the β-carbon of dopamine. Because this enzyme is found only in cells that produce norepinephrine or epinephrine, it has been used (as described above) as a histochemical or immunohistochemical "marker" for such neurons in brain. Norepinephrine is further converted to epinephrine within a small number of brain neurons through the action of phenylethanolamine N-methyltransferase (PNMT).

Because tissue catecholamine concentrations remain relatively constant, even when there are major increases in the rates at which they are being used, it has long been suspected that feedback mechanisms exist which couple the rate of catecholamine synthesis to the rate of uses. This coupling is currently thought to be mediated by variations in the rate at which tyrosine is hydroxylated by the enzyme tyrosine hydroxylase. At least three factors may control this rate of hydroxylation: 1) the amount of tyrosine hydroxylase enzyme protein; 2) the activity of the enzyme (which is increased allosterically when the neurons containing it are depolarized or which is directly inhibited by its major end products, dopamine or norepinephrine); and 3) the extent to which the enzyme is saturated with respect to its substrate, tyrosine, and its cofactors, oxygen and tetrahydropterin.

The end-product inhibition of tyrosine hydroxylase is effected by a hypothetical "physiologically active" catecholamine pool, presumably located within the cytoplasm, which impairs the binding of the pteridine cofactor to the enzyme molecule. When catecholamine release is accelerated through nerve stimulation, this pool presumably becomes depleted, thereby lessening inhibition of the tyrosine hydroxylase enzyme. Brain catecholamine synthesis has recently been shown also to be subject to "open loop" control by changes in brain tyrosine concentration. Injections of amino acids that do or do not modify brain tyrosine levels cause parallel changes in the rates at which rat brains can accumulate dopa after decarboxylase inhibition. In some dopaminergic neurons in the brain (e.g., those that project to the basal ganglia), changes in dopamine synthesis and release, caused by varying brain tyrosine levels, are rapidly compensated by feedback mechanisms that modify tyrosine hydroxylase activity; hence, after a short while, there is no longer any net change in neurotransmitter formation. This self-correcting mechanism may or may not characterize all catecholaminergic neurons in the brain, so it cannot yet be stated whether alterations in hypothalamic tyrosine would, by altering dopamine synthesis, modify the release of prolactin.

Catecholamine molecules, released into brain synapses following neuronal depolarization, either interact with receptors or are rapidly inactivated by being taken up into the presynaptic neuron. Once such reuptake occurs, the neurotransmitter can be stored for possible release or be deaminated through the action of monoamine oxidase. The aldehydes formed by this reaction may then be oxidized to acids (e.g., homovanillic acid or dihydroxyphenylacetic acid from dopamine, vanillylmandelic acid from norepinephrine), or reduced to alcohols (e.g., 3-methoxy-4-hydroxyphenylglycol from norepinephrine). Relatively small proportions of the catecholamines released into the synapse may also undergo a different metabolic transformation: the m-hydroxyl group on the catechol nucleus can be methylated through the action of catechol O-methyltransferase to form 3-methoxytyramine (dopamine) or normetanephrine (norepinephrine).

Most of the catecholamine molecules synthesized within brain neurons are apparently never liberated into synapses. Instead, they are deaminated, and these physiologically inert products then leave the cell and can be O-methylated. They subsequently enter the cerebrospinal fluid, from which a probenecid-sensitive mechanism transports them into the bloodstream for excretion into the urine, or they pass directly into the brain capillaries.

SEROTONIN

The initial step in the synthesis of serotonin in the brain involves the 5-hydroxylation of the amino acid tryptophan, a process catalyzed by the enzyme tryptophan hydroxylase. Tryptophan is an essential amino acid and cannot be made in the mammalian body; hence, brain tryptophan must ultimately derive from dietary protein. Tryptophan hydroxylase appears to be confined to those brain neurons that synthesize and contain serotonin. Like tyrosine hydroxylase, this enzyme uses a reduced pteridine cofactor; however, unlike the catecholamines, serotonin does not compete with the cofactor for a binding site on the hydroxylase enzyme, and thus does not suppress its own biosynthesis by end-product inhibition. 5-Hydroxytryptophan, like L-dopa, is rapidly decarboxylated by aromatic L-amino acid decarboxylase to form serotonin (5-HT).

Several lines of evidence indicate that the amount of tryptophan available to the brain can control serotonin synthesis, i.e., by controlling the rate at which tryptophan is hydroxylated: 1) injections of small doses of tryptophan, which increase serum and brain concentrations of tryptophan to an extent similar to those normally occurring diurnally, increase brain serotonin synthesis; 2) many drugs (amphetamine, lithium, dibutyryl cyclic AMP, reserpine) and other manipulations (electroconvulsive shock, and exposure to a high ambient temperature) that increase brain tryptophan concentrations also increase brain serotonin synthesis; and 3) various large neutral amino acids (e.g., tyrosine, phenylalanine, valine, leucine, and isoleucine), that compete with tryptophan for transport into brain from the serum, decrease brain tryptophan concentrations and suppress serotonin synthesis. Because the sole source of tryptophan for mammals is the diet, it might be expected that food consumption would, depending on its composition, modify serum and brain tryptophan levels and, ultimately, brain serotonin synthesis. This is indeed the case: a) animals chronically fed artificial or natural diets (e.g., corn protein) that contain inadequate amounts of tryptophan have reduced serum and brain concentrations of the amino aid and decreased brain concentrations of serotonin; b) the consumption of a high-carbohydrate, protein-free meal, by eliciting the secretion of insulin, increases serum and brain tryptophan and brain serotonin; and c) the consumption of various artificial diets causes changes in brain serotonin which parallel their effects on the ratio of serum tryptophan concentration to the sum of the serum concentrations of the other large neutral amino acids. The control by tryptophan of brain serotonin synthesis is unaffected by other treatments that increase or decrease serotonin levels (e.g., monoamine oxidase inhibitors) and is independent of the rate at which serotoninergic neurons depolarize.

The mechanisms by which serotonin is stored intracellularly in brain neurons are, at present, unclear; however, it seems likely that a major fraction of the transmitter is not confined within synaptic vesicles. The turnover of brain serotonin is well characterized by a single exponential decay curve. This suggests that the monoamine is stored in a single metabolic compartment. Serotonin released from nerve terminals (physiologically or after the administration of reserpine) is inactivated by reuptake, and then deaminated and oxidized to form 5-hydroxyindoleacetic acid (5-HIAA).

ACETYLCHOLINE

Although acetylcholine (ACh) is not, strictly speaking, a monoamine and its specific roles in gonadotropin secretion are even less well understood than those of the catecholamines and serotonin, recent advances should make it much easier to manipulate central cholinergic synapses and, thus, to study their involvement in neuroendocrine function. Hence, this transmitter is discussed briefly here.

Brain neurons synthesize acetylcholine from two precursors, choline and acetyl coenzyme A, in a reaction catalyzed by choline acetyltransferase. In rat brain, the concentrations of acetyl coenzyme A (acetyl-CoA) and choline are well below those needed to saturate choline acetyltransferase. Hence, increases in brain choline (e.g., following its injection or consumption in the diet for 3 or more days) cause parallel changes in acetylcholine synthesis, and apparently, release. The brain appears incapable of synthesizing choline de novo; moreover, an intravenous pulse of radioactively labeled choline is very rapidly incorporated into brain ACh, confirming that serum choline is the physiological precursor of brain ACh. The uptake of choline from serum into brain bears a linear relation to serum choline concentration; this process is apparently mediated by a low affinity uptake system, which differs from the high affinity system specifically localized within the terminals of cholinergic brain neurons. (The high affinity system probably functions to take up intrasynaptic choline formed from the rapid hydrolysis of ACh; this allows for its reutilization.)

Within the neuron, ACh appears to be localized within at least two compartments, i.e., the cytoplasm and synaptic vesicles. Acetylcholine synthesis apparently occurs in the cytoplasm. If ACh is released from brain slices (e.g., by increasing local K^+ concentrations), its synthesis is stimulated. The concentration of ACh at the site of synthesis may competitively inhibit choline acetyltransferase with respect to choline, and noncompetitively with respect to acetyl-CoA. Newly synthesized acetylcholine molecules may be released preferentially following nerve stimulation.

Once released into the synaptic cleft, acetylcholine is rapidly hydrolyzed by acetylcholinesterase to form choline and acetate. There do not appear to be reuptake mechanisms for the recapture of synaptic acetylcholine molecules; however, such a mechanism probably does exist for the choline liberated by the action of cholinesterase on acetylcholine. Most probably, blood choline levels (and thus, the diet) determine *total,* steady-state brain choline levels while the choline reuptake mechanism in presynaptic terminals allows this scarce precursor to be conserved and reused by the brain.

DRUGS FREQUENTLY USED TO EXAMINE INVOLVEMENT OF BRAIN MONOAMINES IN GONADOTROPIN SECRETION

Seven families of drugs have most often been used to examine the participation of monoaminergic neurons in the control of neuroendocrine mechanisms involved in reproduction. They are discussed below.

Neutral Amino Acids

Tryptophan and Tyrosine   These amino acids are the true circulating precursors of the monoamine neurotransmitters. The metabolic consequences of administering them alone can be quite different from those that follow their consumption as constituents of natural proteins; hence, when injected, they must be thought of as drugs. As discussed above, giving tryptophan or tyrosine accelerates the synthesis of serotonin and the catecholamines within just those brain neurons that normally produce them (because only these cells contain tryptophan hydroxylase or tyrosine hydroxylase); therefore, these amino acids are probably among the most specific drugs currently available for modifying transmitter release at monoaminergic synapses. However, their specificity is *not* total; for example, each amino acid inhibits the uptake by brain of the other and thus the administration of tyrosine suppresses brain serotonin synthesis.

L-Dopa and L-5-Hydroxytryptophan These two amino acids are the immediate precursors of the monoamines. They are widely used by neuroendocrinologists and are among the least specific that affect monoaminergic synapses. Each can be decarboxylated to its respective monoamine (dopamine and serotonin) wherever the enzyme aromatic L-amino acid decarboxylase (AAAD) exists (i.e., in most cells in the body). Thus, dopa administration accelerates the release from brain neurons of not just dopamine, but also of serotonin and norepinephrine, probably because the large amounts of dopamine that form within 5-HT and NE nerve terminals compete with the "true transmitters" for storage sites. L-Dopa and L-5-HTP (5-hydroxytryptophan) also transiently disaggregate brain polyribosomes and suppress brain protein synthesis. Moreover, the *O*-methylation of exogenous dopa also depletes the brain of the methyl donor *S*-adenosylmethionine (SAM), thus slowing the metabolism of endogenous dopamine, NE, histamine, and other methyl acceptors.

A synthetic amino acid, *L-threodihydroxyphenylserine* (DOPS), can, at theoretically, be decarboxylated in brain to form norepinephrine; hence, although this compound is not truly an NE precursor (since it probably is not formed in vivo), its utility as a neuropharmacological tool is discussed here. DOPS works much better in theory than in practice; most investigators who have actually measured brain NE after DOPS administration have failed to detect the expected increases, probably because this amino acid is a poor substrate for AAAD or because it penetrates the blood-brain barrier poorly. Small increases in brain NE have been observed in animals pretreated with drugs that block the decarboxylation of DOPS in peripheral tissues. At the present time, it seems prudent to refrain from attributing any neuroendocrine effects of DOPS specifically to the enhancement of noradrenergic transmission. Indeed, the physiological actions of the drug could result from peripheral mechanisms (e.g., activation of sympathetic synapses) or from inhibition of the uptake of other neutral amino acids into the brain.

Leucine, Isoleucine, and Valine The branched chain amino acids compete with circulating tryptophan and tyrosine for transport into the brain; hence, their administration lowers brain tryptophan and tyrosine levels, thereby suppressing the synthesis of serotonin and the catecholamines. As far as is now known, leucine, isoleucine, and valine are not themselves substrates or inhibitors for any of the enzymes involved in monoamine synthesis and metabolism.

α-Methyl-m-tyrosine and α-Methyldopa The monoamines formed in vivo from the decarboxylation of these α-methylated amino acids are not substrates for monoamine oxidase so they tend to persist in the body for relatively long periods of time. The initial in vivo products are α-methyl-*m*-tyramine and α-methyldopamine; subsequently, both can be β-hydroxylated by the dopamine β-oxidase present in synaptic vesicles of noradrenergic neurons. The biological activity of the products of α-methyl-*m*-tyrosine presumably derives from their storage and release in place of norepinephrine and their failure to activate noradrenergic receptors to the same degree. In contrast, the biological effects of the α-methylnorepinephrine formed from α-methyldopa are currently thought to derive from its ability to stimulate noradrenergic receptors (e.g., in brainstem centers, causing a fall in blood pressure). As might be expected, these agents have a number of other effects on monoamine metabolism in addition to those outlined above. For example, they compete with tryptophan and tyrosine for uptake into brain and inhibit the activity of aromatic L-amino acid decarboxylase to a variable extent; as a substrate for catechol *O*-methyltransferase, α-methyldopa depletes the central nervous system of SAM.

α-Methyl-p-tyrosine and p-Chlorophenylalanine These agents are potent inhibitors of tyrosine and tryptophan hydroxylase, respectively. Thus, their administration to rats causes major decreases in brain catecholamines and serotonin. Unfortunately, each agent also inhibits the synthesis of the other monoamines, at least transiently, because, as neutral amino acids, both lower the brain levels of tryptophan and tyrosine. This effect is less of a problem with *p*-chlorophenylalanine than with α-methyl-*p*-tyrosine, inasmuch as the inhibition of tryptophan hydroxylase lasts much longer than the effect on brain tryptophan.

A number of investigators have used these or other inhibitors of monoamine synthesis to examine the effects of neuroendocrine manipulations on brain neurons by using a particular monoamine neurotransmitter. For example, the effects of castration on brain catecholamine synthesis and turnover have been examined by measuring the rate at which brain norepinephrine levels decreased after α-methyl-*p*-tyrosine was injected into castrated or control animals. The paradigm is based on the assumption that an accelerated decline in brain norepinephrine (or dopamine) implies that, prior to drug administration, catecholamine synthesis had been occurring at a more rapid rate. This approach is fraught with more than the usual number of pitfalls: 1) the neuroendocrine manipulation being studied may, quite independently, have modified the fate of the α-methyl-*p*-tyrosine and, consequently, the extent of the enzyme inhibition attained in vivo; 2) the profound reductions in brain catecholamine levels produced by the inhibitors probably cause such large changes in the rates at which catecholaminergic neurons fire and in other aspects of synaptic dynamics as to obscure those produced by the neuroendocrine manipulation being studied.

Enzyme Inhibitors

Inhibitors of Aromatic L-Amino Acid Decarboxylase MK-486 (Carbidopa) was designed, and is currently in use, to block the decarboxylation of exogenous L-dopa outside the brain without affecting its conversion to dopamine in the brain itself (although it is not yet known whether decarboxylation in any brain region, such as the hypothalamus, is affected by giving MK-486). Large doses of the drug (100 mg/kg in rats) also suppress the decarboxylation of endogenous dopa in peripheral sympathetic neurons and might thus modify sympathetic function.

High doses of RO4-4602 (800 mg/kg) inhibit the decarboxylation of L-dopa and L-5-HTP both in peripheral tissues and in the brain; they can thus be used, for example, to determine whether the neuroendocrine effect of an amino acid drug (e.g., the suppression of prolactin secretion by L-dopa) results from an action on the brain of the monoamine (dopamine) formed from the drug's decarboxylation. Such doses are also useful in estimating the rates at which brain neurons synthesize catecholamines and serotonin after various treatments. This is done by measuring the rates at which brain tissue accumulates L-dopa and L-5-HTP for 30–60 min after RO4-4602 is administered, i.e., when the accumulation is occurring at linear rates and when the resulting reductions in brain catecholamines and serotonin are probably small enough not to perturb synaptic function. Lower doses of RO4-4602 (50–100 mg/kg) also act largely or exclusively in the periphery, like MK-486. Both agents are hydrazines and presumably bind pyridoxal phosphate; both might, therefore, be expected to interfere with a large number of enzymatic reactions affecting amino acids. Both drugs elevate plasma concentrations of tyrosine and tryptophan, presumably by slowing the metabolism of these amino acids by pyridoxine-dependent enzymes in the liver.

Monoamine Oxidase Inhibitors These compounds elevate brain monoamine levels; it is thus possible, but not yet proven, that they might cause corresponding increases in the amounts of the neurotransmitter molecules that are released into synapses when monoaminergic neurons depolarize. Oxidative deamination is apparently not involved in the inactivation of catecholamines in vivo (i.e., their removal from the synaptic cleft); this process depends largely on reuptake into presynaptic terminals, and to a lesser extent, on *O*-methylation. Monoamine oxidase (MAO) may, however, be involved in the inactivation of serotonin; there appears to be a close relationship between the amount of 5-HT released into synapses and the rate at which its chief metabolite, 5-HIAA, is formed in the brain.

MAO inhibitors can be classified according to whether or not they are hydrazines (*iproniazid, pheniprazine, nialamide,* and *isocarboxazid* are; *tranylcypromine* and *pargyline* are not) and by their spectrum of actions on the various MAO isozymes (now generally classified as types A and B). Unfortunately, no MAO inhibitor is available which selectively blocks the deamination of one of the brain monoamines without also affecting the fates of the others.

Inhibitors of Dopamine β-Oxidase Disulfiram (tetraethylthiuram), *FLA-63* [bis(4-methyl-1-homopiperazinyl-thiocarbonyl disulfide)], and other similarly acting compounds or their in vivo metabolites inhibit the conversion of dopamine to norepinephrine by chelating the copper needed by dopamine β-oxidase (DBH). Of course, as chelating agents, they also affect the activities of many other enzymes (e.g., aldehyde dehydrogenase). The interpretation of the neuroendocrine consequences of their administration is complicated by this nonspecificity, and also by the largely untested possibility that central noradrenergic receptors might be able to respond to the dopamine that would be released in place of norepinephrine after the inhibition of DBH. (Noradrenergic receptors in the rat pineal are effectively stimulated by dopamine in vitro. Moreover, the dopamine-sensitive adenylate cyclase in brain slices prepared from the basal ganglia does respond to norepinephrine.)

Uptake Blockers

These agents prolong the survival of monoamine neurotransmitters within the synaptic cleft and thus potentiate their actions. They act by suppressing the reuptake of transmitter molecules into the presynaptic terminals of origin, and thus have the effect of increasing the likelihood that a given transmitter molecule will interact with its postsynaptic receptors. When these drugs are given in sufficient doses, they so prolong the activation of receptors by intrasynaptic neurotransmitter molecules that feedback mechanisms are activated which completely shut off further release of the transmitter. Lilly-110140 (*fluoxetine hydrochloride*) appears to be a specific inhibitor of 5-HT reuptake, and Lilly 94939 (*nisoxetine*) almost as specifically suppresses the uptake of brain norepinephrine. *Chlorimipramine* and *desmethylimipramine* (DMI) act with less specificity on the uptake of 5-HT and NE, respectively, whereas *imipramine* blocks the uptake of all three amines. *Benztropine,* an antihistamine, and *nomitensin* inhibit dopamine reuptake in vivo and in vitro.

Receptor Agonists

Drugs can activate postsynaptic monoaminergic receptors, either directly or indirectly, by causing the presynaptic terminal to release neurotransmitter molecules into the synaptic cleft. Examples of direct acting central agonists are *apomorphine* and *piribidil* (ET-495) for dopaminergic receptors, *clonidine* for noradrenergic receptors, and *quipazine* for serotoninergic receptors. The drugs acting on dopamine receptors are relatively specific. Terminals of some dopaminergic neurons contain presynaptic receptors which, when activated, inhibit the release of dopamine into synapses. Their existence can generate paradoxical responses to drugs like apomorphine. Low drug doses, acting predominantly on the presynaptic receptors, can, by suppressing dopamine release, actually *decrease* the activation of postsynaptic receptors.

Indirect acting receptor agonists include d-*amphetamine,* which liberates catecholamines, and *fenfluramine,* which causes the release of serotonin. Amphetamine also potentiates central catecholaminergic transmission by interfering with catecholamine reuptake (and, possibly, by weakly inhibiting MAO); it may also have some activity as a direct receptor agonist. Fenfluramine may be toxic to serotoninergic neurons.

Receptor Blockers

Very good drugs exist for blocking dopamine receptors, such as the butyro-phenones *pimozide* and *haloperidol*. In contrast, virtually no agents have really been shown by rigorous pharmacological techniques to block central norepine-phrine receptors, and the neuroendocrinologist who uses such peripherally active agents as *dibenzyline, phentolamine,* and *propranolol* is well advised to interpret the results conservatively. A number of drugs can be shown to block peripheral serotoninergic receptors and to produce physiological changes compatible with the blockade of central serotonin receptors; these include *cyproheptadine, methysergide,* and *methiothepin.* However, electrophysiological studies raise some doubts about where these drugs act in vivo. They appear to block the responses of only certain neurons to locally applied serotonin; these neurons tend, unfortunately, to be the ones 1) that do not normally receive serotonin-ergic synapses, and, 2) the ones in which the amine causes excitation, not inhibition (which is thought to be the normal in vivo response to the amine).

Neurotoxins

The presynaptic uptake mechanism which normally functions to inactivate intrasynaptic dopamine or norepinephrine can also be used to kill catechola-minergic neurons. When the synthetic dopamine analogue *6-hydroxydop-amine (6-OHDA)* is placed in the cerebrospinal fluid (e.g., by intraventricular or intracisternal injection), terminals of dopaminergic and noradrenergic neurons concentrate this material so effectively that it or one of its metabolites kills them, but not other neurons lacking the uptake mechanism. Noradrenergic neurons are highly susceptible to this agent; dopaminergic neurons are less so, some groups in the hypothalamus apparently responding not at all. The reduction in brain dopamine produced by 6-OHDA administration can be potentiated by repeated treatments and by giving MAO inhibitors (which slow the metabolism of the drug); its specificity can be enhanced by the coadministration of DMI (which inhibits its uptake into noradrenergic neurons). The amino acid analogue of 6-OHDA, *6-hydroxydopa,* reportedly causes selective destruction of noradren-ergic neurons.

Serotonin-containing brain neurons can similarly be destroyed by administering *5,7-dihydroxytryptamine* or *5,6-dihydroxytryptamine* by the intraventricular or intracisternal routes, or *p-chlorophenylalanine* or *fenfluramine* systemically.

Inhibitors of Monoamine Storage

Reserpine, the prototype of this category of drug, apparently interferes with the ability of the synaptic vesicles in catecholaminergic neurons to maintain very high concentrations of dopamine or norepinephrine relative to those present in the cytoplasm. In peripheral sympathetic neurons, the administration of reserpine causes a depletion of norepinephrine, but apparently does not enhance the release of the neurotransmitter into the synapse (e.g., it fails to elevate blood pressure). Because the norepinephrine depletion is associated with increased production of its deaminated metabolites, it is generally held that, after reserpine, catecholamines are released into the *cytoplasm* of the presynaptic bouton and immediately deaminated (and inactivated) by monoamine oxidase. Reserpine administration also elevates the brain levels of the deaminated metabolites of dopamine and norepinephrine; hence, even though there is no direct evidence that reserpine fails to liberate catecholamines into brain synapses, it is generally held that in brain, as in sympathetic nerves, reserpine's net effect is to *decrease* catecholaminergic transmission. The effect of reserpine on serotoninergic transmission is harder to evaluate. The drug does deplete the brain of serotonin and does cause the accumulation of 5-HIAA; however, this latter effect seems to require that serotonin be released from the neuron, taken up again, and then deaminated in the cytoplasm (i.e., the coadministration of a drug that blocks serotonin *reuptake* suppresses the reserpine-induced rise in brain 5-HIAA). *Tetrabenazine,* like reserpine, depletes the brain of monoamines.

BRAIN MONOAMINES AND CONTROL OF PROLACTIN SECRETION

Studies of the effects of central monoamine transmitters on prolactin secretion have frequently started from the assumption that these compounds normally influence prolactin release indirectly, by modifying the secretion from the hypothalamus of a pituitary tropic factor (in this instance, an inhibitory substance, prolactin release-inhibitory factor (PIF)). This notion may not be entirely valid, inasmuch as considerable evidence, discussed below, suggests that at least one monoamine neurotransmitter, dopamine, may itself be released into pituitary portal vessels to inhibit prolactin (PL) secretion directly.

Among the animal models that have been employed to study the role of monoamine neurons in the control of prolactin secretion are 1) normal and castrated males and females; 2) normal females on the morning and afternoon of proestrus; 3) normal males injected with drugs into the third ventricle or the pituitary portal circulation; and 4) lactating females. Each of these paradigms is discussed below for each of the monoamine neurotransmitters, followed by separate sections dealing with in vitro studies and studies on humans. It should become apparent that many of these experiments have attributed considerably more biochemical specificity to the brain-active drugs that affect prolactin secretion than now seems warranted.

Catecholamine Neurons

Data from a large number of pharmacological studies have been interpreted as showing that the release of dopamine from brain neurons inhibits, whereas norepinephrine either stimulates or fails to affect, prolactin release.

Normal and Castrated Animals

Females on Morning of Proestrus Lu and his associates (9, 10) injected normal females on the morning of proestrus with reserpine, chlorpromazine, α-methyl-*p*-tyrosine (AMPT), α-methyl-*m*-tyrosine (AMMT), α-methyldopa (AMD), *d*-amphetamine, pargyline, or dopa, and measured plasma prolactin levels for several hours thereafter. Each of the first six drugs raised plasma prolactin concentrations, whereas dopa or pargyline depressed circulating hormone levels. Dopamine, norepinephrine, or epinephrine, when administered intraperitoneally or into the carotid artery, failed to affect plasma prolactin. These results suggested to the authors that the above drugs acted by modifying central dopamine or noradrenergic transmission and did not themselves *directly* affect pituitary prolactin release or cause the release of catecholamines into the pituitary portal circulation.

As described above, all of the drugs used in this study could be expected to modify the metabolism of brain serotonin, as well as that of the catecholamine neurotransmitters. If it is assumed that the drugs affected prolactin via a catecholaminergic mechanism, then the rise in prolactin after *d*-amphetamine is paradoxical, inasmuch as this agent might be expected to increase dopamine release (and thus to act like dopa). The interpretation of all such studies ultimately requires two kinds of corroborative evidence: direct neurochemical measurements (e.g., of levels of monoamine neurotransmitters in the hypothalamus and portal vascular system or of their turnover rates in the hypothalamus) and neuropharmacological confirmation (e.g., by showing that the effects of all agents thought to act by releasing dopamine, norepinephrine, or serotonin can be blocked by administering the appropriate receptor antagonists).

In later experiments (11), haloperidol, a dopamine receptor blocker, was shown to elevate plasma prolactin levels when administered on the morning of proestrus. This result was consistent with the view that a catecholamine-releasing neuron inhibits prolactin release and suggested that dopamine might be the particular catecholamine mediating this inhibitory signal. The hypothalamic content of PIF (assayed indirectly by measuring the release of prolactin by pituitaries incubated in the presence of hypothalamic extracts) fell in these animals, leading the authors to suggest that haloperidol increases plasma prolactin indirectly by reducing the amount of PIF in the hypothalamus available for release into the pituitary portal system. This interpretation was but one of several consistent with the data. For example, the injected drug might also have been present in the hypothalamic extracts from treated animals and might itself have stimulated pituitary prolactin release in vitro. Alternatively, haloperidol might have lowered hypothalamic dopamine levels (e.g., by feedback activation of dopaminergic neurons).

Castrated Males and Females Data similarly implicating a catecholaminergic neuron as an inhibitor of prolactin secretion have been obtained in experiments using castrated animals. AMPT has been observed to elevate plasma prolactin rapidly in castrated male and female rats (12). Concurrent with this rise, the rate of dopamine and norepinephrine synthesis falls (13). The administration of dopa 1 hr after AMPT not only blocks the rise in plasma prolactin, but depresses circulating hormone levels to very low values within 1 hr. Diethyldithiocarbamate, an inhibitor of dopamine β-hydroxylase, administered in lieu of AMPT, did not modify the lowering effect of dopa on plasma prolactin, again suggesting that dopaminergic and not noradrenergic neurons mediate an inhibitory effect on prolactin secretion (no evidence was provided, however, that the conversion of dopa to norepinephrine had actually been blocked). In this study, the administration of DOPS slightly potentiated the effect of AMPT and by itself raised plasma prolactin. This was interpreted as showing that noradrenergic neurons might stimulate prolactin secretion; however, as described above, there is little basis for believing that DOPS administration actually changes the levels or release of brain NE significantly. (Data on its effects on hypothalamic NE would presumably be very useful.) Moreover, AMPT and dopa might both have been acting via serotoninergic neurons. AMPT would be expected to decrease serotonin synthesis and release, whereas dopa would, by a false transmitter mechanism, enhance the release of the indoleamine from serotoninergic neurons.

It was subsequently found that all of the above effects also occurred in rats that had undergone median eminence lesioning 1 week earlier (14). This finding suggested that the drugs, or their decarboxylated products, perhaps formed intravascularly or in the liver or kidneys, act directly on the pituitary. One such decarboxylated product would, of course, be dopamine itself.

In subsequent studies in male rats, it was shown that dopa had to undergo peripheral decarboxylation in order to suppress prolactin secretion. Intravenous dopa reduced the initially high plasma prolactin levels in male rats whose anterior pituitaries had been implanted into a kidney; this effect of dopa was blocked in animals pretreated with the decarboxylase inhibitor MK-486 (15). In intact humans, MK-486 potentiates the depressing effects of dopa on prolactin secretion (16), suggesting that DA acts within the brain to modify prolactin release by the pituitary. The explanation for this apparent difference in drug action is not yet known.

Castrated animals have also been injected with pimozide, either subcutaneously or into the hypothalamus or anterior pituitary (17). The dopamine receptor blocker elevated plasma prolactin regardless of its site of injection. The greatest and most prolonged rise followed subcutaneous injection (10-fold over a 3-day period); hypothalamic and pituitary implants raised plasma hormone levels 4 and 24 hr after injection, respectively. The response to pituitary implants suggests that this tissue contains a dopamine receptor inhibitory to prolactin secretion; the response to hypothalamic implants could be mediated by dopamine receptors in this brain region, or by blood-borne delivery of the drug to pituitary receptors.

Studies in normal and ovariectomized female monkeys (18) and in normal male rats (19) further support the role of dopamine as an inhibitory transmitter in prolactin secretion. Apomorphine, a dopamine receptor agonist, rapidly blocked the rise in plasma prolactin induced either by perphenazine or thyrotropin-releasing hormone (TRH). Moreover, the dopamine receptor blockers haloperidol and clozapine (clozapine also affects cholinergic receptors) stimulated prolactin secretion.

Ovariectomized female rats, treated with estrogens, have also been used in pharmacological studies on prolactin secretion (20, 21). Pimozide was found to elevate, and apomorphine or dopa to depress, the already elevated plasma levels of prolactin in these animals. Clonidine, an α-adrenergic agonist, raised plasma prolactin, indicating that norepinephrine might stimulate the secretion of this hormone. However, this view was thought not to be sustained by the finding that phenoxybenzamine, phentolamine, and propranolol all increased plasma prolactin as well. (As described above, none of the latter agents has clearly been shown to affect norepinephrine receptors in the brain.)

Normal Male Rats with Drugs Injected into Third Ventricle or Pituitary Portal Circulation   Abundant in vitro evidence, to be discussed below, indicates that dopamine directly inhibits prolactin secretion from the pituitary; hence, if dopamine is released into the pituitary portal circulation, it could itself be the PIF. Kamberi and his associates (22), however, failed to detect a fall in serum prolactin after infusing dopamine into the pituitary portal vessels. In contrast, male rats receiving dopamine (1.25 μg dissolved in saline solution, via cannula) into the third ventricle did exhibit the expected fall in serum prolactin levels; moreover, portal blood from these animals also depressed serum prolactin when infused into the portal vessels of other rats (23). Infusions of norepinephrine and epinephrine into the third ventricle or into the pituitary portal circulation were without effect, except in extremely high doses (100 μg). The failure of dopamine, infused directly into the portal vessels, to suppress prolactin secretion was later attributed by other investigators (24) to oxidation of the amine: if dopamine was dissolved in 5% glucose instead of saline solution, it markedly reduced prolactin secretion when infused into the pituitary portal vessels. (Norepinephrine also depressed hormone secretion when so administered.)

Suckling-induced Rise in Plasma Prolactin in Lactating Female Rats   Suckling induces a rapid increase in plasma prolactin concentrations in lactating female rats separated from their pups for several hours. Presumably, the rise in hormone levels is effected by stimulation of a neural circuit which ultimately inhibits PIF-secreting cells in the hypothalamus; this circuit includes sensory neurons in the nipple, secondary sensory pathways in the spinal cord, as well as hypothalamic components. If a catecholamine-containing neuron is directly or tangentially involved anywhere along this pathway, then it should be possible to modify the suckling-induced rise in maternal plasma prolactin levels by giving drugs that act on catecholaminergic neurons and synapses. Thus, the injection of apomorphine, which partially blocks the increase in prolactin elicited by chlorpromazine injection, also completely blocks the prolactin response to suckling in lactating rats (25). In other studies (26), an injection of dopa, just prior to suckling, suppresses the anticipated prolactin rise.

Thus, these results support the participation of dopaminergic neurons in the reflex activation of prolactin secretion by suckling. An interesting view as to the physiological nature of this involvement was recently offered by Voogt and Carr (27), who observed that the suckling-induced rise in plasma prolactin was followed by an increase in the hypothalamic synthesis of [^{3}H] dopamine (but not [^{3}H] norepinephrine) from [^{3}H] tyrosine. They suggested that dopamine-containing neurons may function to turn off prolactin secretion following a period of stimulated secretion.

In Vitro Studies of Pituitary Prolactin Secretion Dopamine, added to pituitary cultures, suppresses prolactin release into the medium (28,29). This observation provided the initial basis for the hypothesis that dopamine itself might act as the prolactin inhibitory factor in vivo. Several laboratories have subsequently studied the action of dopamine on the pituitary in vitro, focusing mainly on the pharmacological identification of a dopamine receptor. One group (30) has found that preincubation of pituitaries with haloperidol or perphenazine blocks the effect of dopamine on prolactin secretion. They also observed that apomorphine, like dopamine, reduced prolactin release into the medium and that this effect is blocked by preincubation with perphenazine. Other investigators (31) have also blocked the apomorphine effect with pimozide, another dopamine receptor blocker.

Shaar and Clemens (32) have recently used chemical methods to show that the active principle in hypothalamic extracts that suppresses pituitary prolactin release in vitro is a catecholamine. If they preincubated the hypothalamic extracts with monoamine oxidase, the extracts lost the ability to reduce pituitary prolactin release. That the monoamine oxidase was responsible for the effect, rather than an unknown accompanying enzyme metabolizing compounds other than monoamines, was demonstrated by adding iproniazid, an inhibitor of MAO, to the preincubation medium: this drug blocked the loss of PIF activity. PIF activity could also be removed by passing the extracts over an alumina column. Alumina absorbs catechols, and it was observed that when the column was washed with 0.2 N acetic acid, the solution normally used to elute catechols from alumina, the inhibitory activity could be recovered. These results provide support for the hypothesis that PIF is a catecholamine; that PIF is not a peptide or protein is further indicated by the finding that the activity of hypothalamic extracts was not lost by preincubation with pepsin, a proteolytic enzyme.

Tashjian and his associates (33) observed that TRH induces the release of prolactin from pituitaries in culture. More recently, it has been observed that dopamine antagonizes this action of TRH (34,35). It is not yet known whether the actions of TRH and dopamine are mediated by the same pituitary "receptor." One approach to this question might be to obtain dose-response data for the effect of TRH on prolactin release in vitro in the presence and absence of dopamine. A Michaelis-Menten plot of the data should indicate whether the compounds are competitive or noncompetitive antagonists, and thus suggest if the two act at the same locus on the pituitary cell membrane.

Studies on Human Prolactin Secretion   Many of the pharmacological agents used to determine whether adrenergic synapses participate in the control of prolactin secretion are also used clinically and have been found (sometimes as a side effect) to modify plasma prolactin levels in humans. Dopa administration, for example, reduces plasma prolactin concentrations chronically, when administered to 1) females with galactorrhea, who have elevated levels of the hormone (36), and 2) male and female subjects afflicted with Parkinson's disease (37). In this study, it was noted that at *each* administration of the drug, prolactin levels were transiently depressed, even below the already low baseline values. A single injection of the catechol amino acid, or of apomorphine, also reduced prolactin in males and females with chronically high plasma levels of the hormone (four of the subjects in this study had prolactin-secreting tumors) (38). These effects are also sometimes observed in humans with *normal* plasma prolactin concentrations (39,40) and appear to be enhanced by pretreatment with MK-486, a peripheral decarboxylase inhibitor (16). This latter finding supports the notion that dopamine acts within the brain to modify pituitary prolactin secretion. (The difficulty in demonstrating the effect of dopa in normal subjects may simply reflect the fact that their plasma prolactin concentrations are often quite low.) Humans given pimozide (2–4 mg/day for 1–4 days) exhibit elevated plasma prolactin concentrations (41). The administration of TRH or chlorpromazine raises human plasma prolactin levels (39,42,43). As in experimental animals, intravenous infusion of dopamine (44) (5 μg/kg/min) or the administration of dopa (39) blocks the TRH-induced rise in prolactin.

Serotonin Neurons

Drugs thought to affect the activity of serotoninergic neurons or synapses also modify prolactin secretion; in general, their effects suggest that neurons using this transmitter enhance prolactin secretion.

Normal and Castrated Animals

Females on Morning of Proestrus   The injection of 5 HTP rapidly elevates plasma prolactin levels in proestrus female rats (43). L-Tryptophan also appeared to stimulate secretion; however, the increase in plasma prolactin did not attain statistical significance. Although 5 HT injection had no effect on plasma hormone levels in proestrus females, 5 HTP administration did elevate prolactin concentrations when administered to females with intrarenal pituitary implants. Thus, 5 HTP might exert at least part of its stimulatory effect directly at the pituitary level. The injection of Lilly-51641, originally (but no longer) thought to be an inhibitor of the MAO specific to serotonin, reduced plasma prolactin (9). This contradictory result is best explained by the nonspecificity of the MAO inhibitor, the administration of which probably also elevates brain dopamine levels.

Melatonin, a pineal hormone derived from serotonin, also raised plasma prolactin levels when injected into proestrus female rats (45). Other experiments (vide infra) in intact and pinealectomized rats also suggest that the pineal stimulates prolactin release by means of melatonin secretion (45).

Normal and Castrated Males and Females 5-HTP injection also acutely elevates serum prolactin levels in male rats; the effect can be potentiated by Lilly-110140 (fluoxetine), which alone has no effect on prolactin (20,46). Methysergide blocks the increases in plasma prolactin that follow stress or the administration of 5-HTP in male rats (47). Methysergide alone had no effect on prolactin secretion in these animals, but in apparent contradiction, was found to stimulate release of the hormone when administered to lactating or ovariectomized female rats (48). Finally, *p*-chlorophenylalanine (PCPA) decreased plasma prolactin levels in male rats and abolished the normal daily rhythm in plasma hormone levels (49). A similar decrease in prolactin after PCPA was also observed in ovariectomized, estrogen-primed female rats (20).

As indicated above, melatonin administration stimulates prolactin secretion. This finding is compatible with the observation that melatonin raises brain serotonin levels (and probably accelerates its turnover) (50). Other data support the possibility of pineal involvement in prolactin secretion. The exposure of male and female rats to constant dark or light elevates or depresses plasma prolactin levels, respectively (45), but both effects are blocked by prior pinealectomy. In other studies (51, 52), the nocturnal rise in plasma prolactin seen in normal animals failed to occur in pinealectomized rats.

Injections into Pituitary Portal Vessels Serotonin or melatonin, administered into the third ventricle of otherwise normal male rats, rapidly increased plasma prolactin concentrations. If these compounds were infused instead into the pituitary portal circulation, no enhancement of prolactin secretion resulted (53). Although these results indicate that serotonin must act within the brain to effect prolactin secretion, it is possible that, as suggested for dopamine and norepinephrine (24), the amine is oxidized on administration into the portal vessels, explaining its failure to affect the pituitary.

In a similar study (54), *N*-acetylserotonin also reportedly stimulated pituitary prolactin secretion when administered into the third ventricle.

Suckling-induced Rise in Plasma Prolactin Serotoninergic neurons have also been implicated as components of the neuronal circuit mediating the suckling-induced rise in maternal plasma prolactin. If lactating females are injected with PCPA at least 48 hr before the suckling test, no increase in plasma prolactin accompanies the stimulus (55). Moreover, if 5-HTP is given to the PCPA-pretreated animals 2 hr before they are allowed access to their pups, the effect of PCPA is blocked and a normal prolactin surge ensues. Methysergide also reportedly blocks the suckling-induced rise in prolactin in lactating females; however, when the drug was given twice to the dam, once before suckling and again soon after suckling begins, a paradoxical *surge* in prolactin concentrations reportedly ensued (48). These data were offered as evidence that methysergide must have additional sites of action unrelated to blockade of serotonin receptors through which it produces the paradoxical increase in plasma prolactin. (Methysergide apparently blocks stress-induced increases in serum prolactin (47); hence the rise in prolactin following the second injection into suckling dams was probably not simply a response to the stress of repeated injections.)

In Vitro Studies of Pituitary Prolactin Secretion   Methysergide reportedly fails to affect prolactin release from cultured pituitaries (47, 48). Apparently, no other studies have been published on the in vitro effects of serotonin-active agents on prolactin secretion.

Studies on Human Prolactin Secretion   MacIndoe and Turkington (56) have found that tryptophan (5–10 g), administered intravenously over a 20-min period, elicits a 10-fold increase in plasma prolactin concentrations within 30 min after infusion begins. This effect was not found when other neutral amino acids were infused and could be inhibited at least partially by the prior adminis-tration of methysergide. Methysergide has also been reported to cause a pro-found depression in plasma prolactin levels during sleep, when administered repeatedly for a period of 48 hr (57). 5-Hydroxytryptophan was also found by one group of investigators (58), but not by another (59), to stimulate prolactin secretion; the effect of 5-HTP was blocked with cyproheptadine (58).

BRAIN MONOAMINES AND CONTROL OF FSH AND LH SECRETION

Only fragmentary information is available concerning the precise roles of mono-aminergic neurons in controlling the secretions of the pituitary gonadotropic hormones FSH and LH; most of this information relates to catecholamine neurons.

Catecholamine Neurons

General Pharmacologic Approach   The most popular experimental approach for examining the involvement of catecholaminergic neurons in FSH and LH secretion has used, drug-induced changes in ovulation and in serum gonadotropin levels. AMPT injections, for example, block spontaneous ovulation completely in otherwise normal female rats (60, 61). Consistent with this observation is the finding that AMPT blocks the preovulatory surge in serum LH and FSH in ovariectomized rats primed with estrogen and progesterone (62). The drug also partially blocks the increase in serum LH observed after electrical stimulation of the preoptic area on the morning of proestrus (63). AMPT by itself, however, does not lower LH and FSH levels in the sera of gonadectomized females (12, 13). These results have been interpreted as supporting the existence of a stimulatory catecholamine synapse in the reflex pathways controlling the LH and FSH release associated with ovulation. Similar conclusions have been drawn from the study of gonadal hypertrophy following hemiovariectomy: AMPT blocks the ovarian hypertrophy that occurs in hemiovariectomized rats (64).

Serum LH and FSH levels rise significantly within 16 hr after male rats are castrated (65). The injection of AMPT soon after castration (2–8 hr) blocks these increases. In chronically castrated male rats, however, AMPT injection does not modify serum FSH or LH.

The injection of diethyldithiocarbamate (DDC), a drug that reduces DBH activity, blocks the FSH and LH surges that are elicited in ovariectomized female rats by priming them with estrogen and progesterone (62); it also partially blocks the surge in LH that follows preoptic stimulation in normal female rats on the morning of proestrus (63). In castrated male rats, DDC injection reduces serum LH, but not FSH (12, 65).

Experiments involving the use of AMPT or DDC usually include groups of animals given dopa plus AMPT or DOPS plus DDC. If the amino acid drug reverses the endocrine effect of the AMPT or DDC, this supposedly affirms the involvement of dopamine or norepinephrine, respectively, in the endocrine mechanism. It should be noted that in almost all of the neuroendocrine papers involving these drugs, no measurements were actually made of brain dopamine and norepinephrine; this plus the multiplicity of effects of dopa (and the possible lack of conversion of DOPS to norepinephrine in the brain) make many studies uninterpretable.

Dopa administration to AMPT-pretreated females prevents the AMPT from blocking ovulation; DOPS injection, however, fails to suppress this action of AMPT (61). Thus, dopaminergic neurons might be important in controlling (stimulating) gonadotropin release. Dopa did *not,* however, reverse the blockade by AMPT of the LH and FSH surges that follow estrogen-progesterone "priming" of ovariectomized female rats (62), whereas DOPS reversed the effects of either AMPT or DDC (an effect which cannot now be interpreted). In female rats on the morning of proestrus, preoptic stimulation rapidly elevated serum LH (63). AMPT and DDC pretreatment partially blocked this effect. Dopa or DOPS administration to AMPT-treated rats partially reversed the blockade by AMPT of LH release. Only DOPS, however, partly modified the effects of DDC on LH secretion (63). This again was taken as evidence that noradrenergic neurons mediate the preoptic stimulation-induced rise in serum LH.

In hemiovariectomized, prepubertal female rats, AMPT blocked ovarian hypertrophy; dopa or DOPS administration prevented this blockade (64). These observations were interpreted as supporting the involvement of a noradrenergic synapse in this process. Finally, in acutely castrated male rats, AMPT blocked the rise in LH and FSH seen 16–18 hr after gonadectomy (65), an effect that was not clearly modified by dopa or DOPS. Moreover, dopa and DOPS did not reverse the fall in serum LH observed if these animals were given DDC 18 hr after castration.

Conservatively interpreted, these studies suggest that noradrenergic and/or dopaminergic neurons are involved in the control of gonadotropin secretion.

The FSH-LH surge in ovariectomized females primed with estrogen and progesterone was blocked by treatment with phenoxybenzamine or haloperidol, but not propranolol (62). Similarly, pimozide antagonized the increase in plasma luteinizing hormone-releasing factor (LRF) observed in hypophysectomized female rats (66), whereas haloperidol, another dopamine receptor antagonist, reduced plasma FSH and LH if injected on the morning of proestrus and also blocked ovulation (11). These studies were also interpreted as supporting a role for noradrenergic and dopaminergic nneurons in the control of gonadotropin secretion.

Microinjection of Drugs into the Brain Ventricles or Substance Iontophoresis of dopamine into the third ventricle of male rats elicited rapid (within 10 min) elevations in both serum LH (67) and FSH (68) concentrations. Norepinephrine or epinephrine, similarly administered, was without effect. None of the transmitters modified plasma FSH or LH levels if injected into the pituitary portal circulation; this was taken as evidence that dopamine acts within the brain. In other experiments, pituitary stalk blood, collected from donor rats previously injected with dopamine into the third ventricle, was found to induce FSH and LH secretion when infused into the pituitary portal vessels of other rats (23). Such blood also caused the release of FSH from pituitaries cultured in vitro (69).

In support of the view that dopamine acting within the hypothalamus causes gonadotropin release, Schneider and McCann (70) reported that implants of dopamine (or, to a much lesser extent, norepinephrine) into the third ventricle stimulated LRF secretion. However, other investigators (71) noted that a ventricular infusion of norepinephrine, but not dopamine, could trigger an LH surge and cause ovulation in estrogen-primed rabbits. Moreover, if dopamine was injected concurrently with norepinephrine, the ovulation-inducing effect of the latter transmitter could be blocked. Similar observations were also made on female rats (72). The injection of norepinephrine, but not of dopamine, into the third ventricle induced ovulation. However, this result stands in contrast to the observation that dopamine or norepinephrine, injected into the median eminence, blocked ovulation in proestrus rats (73).

The injection of pimozide into the hypothalamus (but not the pituitary) reduced plasma LH in ovariectomized females (17). However, the intracisternal administration of 6-hydroxydopamine (which has little effect on most dopaminergic neurons in the hypothalamus (74)) blocked the compensatory ovarian hypertrophy that follows hemiovariectomy (66). These findings implicate both dopaminergic and noradrenergic neurons in the control of gonadotropin secretion.

Even though the bulk of evidence points to a role for central catecholaminergic neurons in LH and FSH release, the nature of their participation cannot be deduced from available data. As described above, observations from different laboratories using different but related paradigms have not infrequently been at odds. To some extent, these discrepancies undoubtedly reflect the nonspecificity of the drugs used for testing catecholaminergic neurons, the failure of most investigators to test more than one or two drug doses, the relative lack of information concerning the loci of hypothalamic catecholamine (e.g., where dopaminergic and noradrenergic—and even adrenergic—neurons originate and terminate), and the likelihood that variations in endocrine state (e.g., estrogen priming) modify the functional activity of such neurons.

Gonadotropin Release In Vitro   The addition of dopamine to cultures of pituitaries and hypothalamic-median eminence fragments induced FSH release into the medium (75). The effect was absent if the brain fragments were omitted from the cultures (29, 75). In these studies, neither serotonin, norepinephrine, nor epinephrine elicited FSH secretion. These findings were interpreted as supporting a role for hypothalamic dopaminergic neurons in stimulating follicle-stimulating hormone-releasing factor (FRF) secretion. Other investigators (76), however, found that dopamine depressed both LH and FSH release in a super-fusion system containing pituitaries and median eminence fragments, suggesting that dopaminergic neurons *block* FRF (and LRF) release.

Human Studies   Many drugs that modify plasma prolactin levels in humans do not alter FSH or LH. Thus, apomorphine or chlorpromazine administration fails to change plasma LH and FSH (40); dopa (77) and clonidine (78) are also without effect on plasma gonadotropin levels. Pimozide, administered orally, however, does reduce plasma LH (but not FSH) levels (41).

Serotonin Neurons

General Pharmacological Approach   Very few data are available concerning the participation of serotoninergic neurons in the control of gonadotropin secretion. *p*-Chlorophenylalanine has been used in most of the studies performed to date. Its administration to hemiovariectomized female rats did not modify compensatory ovarian hypertrophy (64), and its injection into male rats 65 hr before castration failed to block the postcastration increases in plasma LH and FSH (65). It also failed to affect the LH and FSH surges that follow the estrogen-progesterone priming of ovariectomized female rats (62), and it produced no acute effects on plasma LH or FSH concentrations in normal or castrated male rats (12). (In general, no chemical evidence was provided in these studies that the PCPA modified brain serotonin and, at the same time, did not affect other brain neurotransmitters.)

In contrast to the above observations, Labhsetwar (60) reported that PCPA, given late in diestrus, blocked ovulation in every case. In these studies, 5-HTP injection during proestrus or diestrus had no effect on ovulation; however, this agent blocked ovulation in females pretreated with an inhibitor of monoamine oxidase. These findings are thus internally inconsistent, and may perhaps reflect the involvement of nonserotoninergic neurons.

Microinjection Studies   Serotonin or melatonin administration into the third ventricle, but not into the pituitary portal vessels, has been observed to reduce both plasma FSH and LH concentrations (53, 67). In other studies, however, 5-HT injection into the third ventricle elevated plasma LH (54). *N*-Acetylsero-tonin had no effect on FSH or LH in normal male rats when administered alone into the third ventricle, but it blocked the rise in plasma LH caused by infusing hypertonic saline solution into the third ventricle (54). (This compound is the immediate precursor of melatonin in the pineal, but probably cannot be converted to melatonin in the hypothalamus.) Melatonin or serotonin administration produced no clear effects on the saline-induced secretion of LH. Finally, animals pretreated 10 days earlier with intraventricular 5,6-dihydroxytryptamine (a drug that can destroy serotoninergic neurons) exhibited elevated plasma LH concentrations; this effect was noted in both normal and gonadectomized male and female rats (79).

In Vitro Studies   The incubation of pituitaries and hypothalamic-median eminence tissue, in the presence of serotonin, had no effect on FSH release (75), but it reduced LH secretion into the medium.

Human Studies   The only study known to us is that of MacIndoe and Turkington (56), who showed that intravenous tryptophan administration (5–10 g over a 20-min period) significantly reduced plasma LH and FSH concentrations.

EFFECTS OF GONADOTROPINS AND
GONADAL HORMONES ON BRAIN MONOAMINES

A number of investigators have examined the relationships between circulating gonadotropin and/or gonadal hormone levels and the metabolism of monoamine neurotransmitters (chiefly norepinephrine) in the brain. Hormone levels have either been allowed to vary physiologically (i.e., with the estrous cycle) or have been modified experimentally (i.e., by castration, with or without replacement therapy). Underlying such studies is the hope that the systems which control gonadotropin secretion include at least one set of monoaminergic neurons, or, failing this, that the behavioral responses produced by the steroid hormones involve monoaminergic neurons.

The specific goal of such studies—an estimation of whether more or less of a transmitter is being released at a particular synapse, causing greater or lesser activation of postsynaptic receptors—cannot really be attained using methods currently available. Instead, the investigator obtains indirect evidence—for example, that presynaptic neurons contain more of the transmitter or synthesize it faster after a particular endocrine manipulation—and hopes that the transmission of information across the synapses is proportional to what is actually being measured. The interpretation of such data is further complicated by some of the lacunae, cited above, in our knowledge of the anatomy of these brain systems (e.g., are releasing factor cells directly innervated by monoaminergic synapses? Must the monaminergic neurons that participate in the control of gonadotropin secretion necessarily even be hypothalamic?). Furthermore, the techniques that have been used to estimate brain neurotransmitter synthesis rates all have major flaws. (A full discussion of these techniques and their pitfalls is beyond the scope of this chapter; some are described by Anton-Tay and Wurtman (80)).

Gonadectomy was probably first shown to modify brain monoamine metabolism by Donoso and his associates (81); these investigators reported that norepinephrine levels in the anterior hypothalamus were significantly elevated (and those of dopamine reduced) 10 days after rats were castrated. In subsequent publications (82, 83), they showed that treatment of spayed rats with estradiol plus progesterone could restore the concentrations of both neurotransmitters to normal. Studies in our laboratory and elsewhere have provided evidence that removal of the gonads also increases the synthesis and/or turnover of norepinephrine in whole brain or in particular regions. Thus, a) the rate at which exogenous [^{3}H] norepinephrine disappeared from whole brain or from all regions examined was accelerated 6 days after castration (84); b) prior castration increased the rate at which [^{3}H] norepinephrine accumulated in whole brains of animals given [^{3}H] tyrosine intraperitoneally (96), in hypothalami of animals given [^{3}H] tyrosine intraventricularly (83), or in anterior hypothalami of rats given [^{3}H] tyrosine by intraventricular cannula (85); c) the administration of ovine follicle-stimulating hormone produced changes similar to castration among intact (86) or gonadectomized (96) rats, whereas treatment with estradiol alone, or with estradiol plus progesterone, reversed these changes (85, 96); d) the decline in hypothalamic norepinephrine following treatment with AMPT was accelerated in ovariectomized rats (87); and e) the intensity of the catecholamine fluorescence in tubero-infundibular neurons was increased following castration and could be lowered in castrated animals by administering estradiol, alone or with progesterone (88).

In related experiments, one group of investigators noted that the tyrosine hydroxylase activity of the whole hypothalamus was increased when measured 2–60 days after castration (89) whereas another group observed a significant elevation only in the median eminence (90). The former increase could be blocked by progesterone (which was also found to inhibit tyrosine hydroxylase activity in vitro (91)), while the increase in the median eminence was suppressed by in vivo treatment with testosterone. Other investigators have found depressed whole brain norepinephrine levels in castrated male rats (92, 93), or, using the fall in brain catecholamines after AMPT treatment, have concluded that castration slows the turnover of dopamine in the tubero-infundibular neurons and of norepinephrine elsewhere in the hypothalamus (4).

During the rat estrous cycle, norepinephrine levels in the anterior and middle hypothalamus are lowest in estrus and highest in proestrus (82). Moreover, the rate at which the whole brain synthesizes [^{3}H] catechols (i.e., its [^{3}H] catechol content, corrected for precursor specific activity, 10 min after rats receive [^{3}H] tyrosine intraperitoneally) exhibits similar variations, peaking during proestrus (94). (Paradoxically, whole brain norepinephrine levels are lowest during this phase.) The activity of monoamine oxidase in the hypothalamus is lowest in estrus and twice as high during proestrus and diestrus (95). Dopamine turnover in hypothalamic tubero-infundibular neurons was reportedly high in diestrus and slowed during proestrus (4).

Taken together, the above studies suggest that events occurring before or during proestrus accelerate the synthesis and turnover of a catecholamine (probably norepinephrine) within at least some brain neurons and may slow the synthesis and release of a catecholamine (possible dopamine) at other brain loci (e.g., the tubero-infundibular nuclei). These events could be entirely neural, or the *immediate* signals causing the changes could be endocrine (e.g., gonadotropins acting on the brain). The data also suggest that high levels of circulating steroids can suppress brain catecholamine synthesis, at least in animals unaccustomed to normal steriod concentrations (i.e., gonadectomized rats). Although it is possible to make models in which, for example, hypothalamic norepinephrine is excitatory to, and dopamine inhibitory to, gonadotropin release, available data provide no real basis for choosing among such models, nor do they provide compelling proof that hormone-dependent changes in brain catecholamine synthesis and turnover are involved in the central regulation of gonad function, or even in mediating the behavioral or other neurophysiological effects of the hormones.

REFERENCES

1. Dahlstrom, A., and Füxe, K. (1964). Evidence for the existence of monoamine-containing neurons in the central nervous system. Acta Physiol. Scand. (Suppl. 232) 62:1.
2. Füxe, K. (1965). Evidence for the existence of monoamine neurons in the central nervous system. IV. The distribution of monoamine nerve terminals in the central nervous system. Acta Physiol. Scand. (Suppl. 247) 64:39.
3. Andén, N. E., Dahlstrom, A., Füxe, K., Larsson, K., Olson, L., and Ungerstedt, U. (1966). Ascending monoamine neurons to the telencephalon and diencephalon. Acta Physiol. Scand. 67:313.
4. Füxe, K., and Hökfelt, T. (1969). Catecholamines in the hypothalamus and the pituitary gland. *In* W. F. Ganong and L. Martini (eds.), Frontiers in Neuroendocrinology, pp. 47–96. Oxford University Press, New York.
5. Jonsson, G., Füxe, K., and Hökfelt, T. (1972). On the catecholamine innervation of the hypothalamus, with special reference to the median eminence. Brain Res. 40:271.
6. Palkovits, M., Brownstein, M., Saavedra, J. M., and Axelrod, J. (1974). Norepinephrine and dopamine content of hypothalamic nuclei of the rat. Brain Res. 77:137.
7. Swanson, L. W., and Hartman, B. K. (1975). The central adrenergic system. An immuno-fluorescence study of the localization of cell bodies and their efferent connections in the rat utilizing dopamine-β-hydroxylase as a marker. J. Comp. Neurol. 163:467.
8. Saavedra, J. M., Palkovits, M., Brownstein, M. J., and Axelrod, J. (1974). Serotonin distribution in the nuclei of the rat hypothalamus and preoptic region. Brain Res. 77:157.
9. Lu, K. H., and Meites, J. (1971). Inhibition of L-dopa and monoamine oxidase inhibitors of pituitary prolactin release; stimulation by methyldopa and *d*-amphetamine. Proc. Soc. Exp. Biol. Med. 137:480.
10. Lu, K. H., Amenomori, Y., Chen, C. L., and Meites, J. (1970). Effects of central acting drugs on serum and pituitary prolactin levels in rats. Endocrinology 87:667.
11. Dickerman, S., Kledzik, G., Gelato, M., Chen, H. J., and Meites, J. (1974). Effects of haloperidol on serum and pituitary prolactin, LH and FSH, and hypothalamic PIF and LRF. Neuroendocrinolgy 15:10.
12. Donoso, A. O., Bishop, W., Fawcett, C. P., Krulich, L., and McCann, S. M. (1971).

Effects of drugs that modify brain monoamine concentrations on plasma gonadotropin and prolactin levels in the rat. Endocrinology 89:774.

13. Carr, L. A., Conway, P. M., and Voogt, J. L. (1975). Inhibition of brain catecholamine synthesis and release of prolactin and luteinizing hormone in the ovariectomized rat. J. Pharmacol. Exp. Ther. 192:15.

14. Donoso, A. O., Bishop, W., and McCann, S. M. (1973). The effects of drugs which modify catecholamine synthesis on serum prolactin in rats with median eminence lesions. Proc. Soc. Exp. Biol. Med. 143:360.

15. Donoso, A. O., Banzan, A. M., and Barcaglioni, J. C. (1974). Further evidence on the direct action of L-dopa on prolactin release. Neuroendocrinology 15:236.

16. Frantz, A. G., Habif, D. V., Hyman, G. A., Suh, H. K., Sassin, J. F., Zimmerman, E. A., Noel, G. L., and Kleinberg, D. L. (1973). Physiological and pharmacological factors affecting prolactin secretion, including its suppression by L-dopa in the treatment of breast cancer. In Proceedings of the International Symposium on Human Prolactin. Brussels, June 12–14, 1973; International Congress Series No. 308, pp. 273–290. Excerpta Medica, Amsterdam.

17. Ojeda, S. R., Harms, P. G., and McCann, S. M. (1974). Effect of blockade of dopaminergic receptors on prolactin and LH release: median eminence and pituitary sites of action. Endocrinology 94:1650.

18. Gala, R. R., and Jaques, S. (1975). The influence of 2 Br-α-ergo-cryptine (CB-154) and apomorphine on induced prolactin secretion in the crab-eating monkey (Macaca fascicularis). Endocrine Res. Commun. 2:95.

19. Meltzer, H. Y., Daniels, S., and Fang, V. S. (1975). Clozapine increases rat serum prolactin levels. Life Sci. 17:339.

20. Chen, H. J., and Meites, J. (1975). Effect of biogenic amines and TRH on release of prolactin and TSH in the rat. Endocrinology 96:10.

21. Lawson, D. M., and Gala, R. R. (1975). The influence of adrenergic, dopaminergic, cholinergic, and serotoninergic drugs on plasma prolactin levels in ovariectomized, estrogen-treated rats. Endocrinology 96:313.

22. Kamberi, I. A., Mical, R. S., and Porter, J. C. (1971). Effect of anterior pituitary perfusion and intraventricular injection of catecholamines on prolactin release. Endocrinology 88:1012.

23. Kamberi, I. A., Mical, R. S., and Porter, J. C. (1971). Hypophysial portal vessel infusion: in vivo demonstration of LRF, FRF, and PIF in pituitary stalk plasma. Endocrinology 89:1042.

24. Takahara, J., Arimura, A., and Shally, A. V. (1974). Suppression of prolactin release by a purified porcine PIF preparation and catecholamines infused into a rat hypophysial portal vessel. Endocrinology 95:462.

25. Smalstig, E. B., Sawyer, B. D., and Clemens, J. A. (1974). Inhibition of rat prolactin release by apomorphine in vivo and in vitro. Endocrinology 95:123.

26. Chen, H. J., Mueller, G. P., and Meites, J. (1974). Effects of L-dopa and somatostatin on suckling-induced release of prolactin and GH. Endocrine Res. Commun. 1:283.

27. Voogt, J. L., and Carr, L. A. (1974). Plasma prolactin levels and hypothalamic catecholamine synthesis during suckling. Neuroendocrinology 16:108.

28. Macleod, R. M., Fontham, E. H., and Lehmeyer, J. E. (1970). Prolactin and growth hormone production as influenced by catecholamines and agents that affect brain catecholamines. Neuroendocrinology 6:283.

29. Quijada, M., Illner, P., Krulich, L., and McCann, S. M. (1973/74). The effect of catecholamines on hormone release from anterior pituitaries and ventral hypothalami incubated in vitro. Neuroendocrinology 13:151.

30. Macleod, R. M., and Lehmeyer, J. E. (1974). Studies on the mechanism of the dopamine-mediated inhibition of prolactin secretion. Endocrinology 94:1077.

31. Smalstig, E. B., Sawyer, B. D., and Clemens, J. A. (1974). Inhibition of rat prolactin release by apomorphine in vivo and in vitro. Endocrinology 95:123.

32. Shaar, C. J., and Clemens, J. A. (1974). The role of catecholamines in the release of anterior pituitary prolactin in vitro. Endocrinology 95:1202.

33. Tashjian, A. H., Barowsky, N. J., and Jensen, D. K. (1971). Thyrotropin releasing

hormone: direct evidence for stimulation of prolactin production by pituitary cells in culture. Biochem. Biophys. Res. Commun. 43:516.

34. Takahari, N., Arimura, A., and Shally, A. V. (1974). Effect of catecholamines on the TRH-stimulated release of prolactin and growth hormone from sheep pituitaries *in vitro*. Endocrinology 95:1490.

35. Hill-Samli, M., and Macleod, R. M. (1974). Interaction of thyrotropin-releasing hormone and dopamine on the release of prolactin from the rat anterior pituitary *in vitro*. Endocrinology 95:1189.

36. Turkington, R. W. (1972). Inhibition of prolactin secretion and successful therapy of the Forbes–Albright syndrome with L-dopa. J. Clin. Endocrinol. Metab. 34:306.

37. Malarkey, W. B., Cyrus, J., and Paulson, G. W. (1974). Dissociation of growth hormone and prolactin secretion in Parkinson's disease following chronic L-dopa therapy. J. Clin. Endocrinol. Metab. 39:229.

38. Martin, J. B., Lal, S., Tolis, G., and Friesen, H. G. (1974). Inhibition by apomorphine of prolactin secretion in patients with elevated serum prolactin. J. Clin. Endocrinol. Metab. 39:180.

39. Friesen, H., Guyda, H., Hwang, P., Tyson, J. E., and Barbeau, A. (1972). Functional evaluation of prolactin secretion: a guide to therapy. J. Clin. Invest. 51:706.

40. Lal, S., de la Vega, C. E., Sourkes, T. L., and Friesen, H. G. (1973). Effect of apomorphine on growth hormone, prolactin, luteinizing hormone and follicle-stimulating hormone levels in human serum. J. Clin. Endocrinol. Metab. 37:719.

41. Collu, R., Jequier, J. C., Leboeuf, G., Letarte, J., and Ducharme, J. R. (1975). Endocrine effects of pimozide, a specific dopaminergic blocker. J. Clin. Endocrinol. Metab. 41:981.

42. Plummer, N. A., Thody, A. J., Burton, J. L., Goolamali, S. K., Huster, S. S., Cole, E. N., and Boyns, A. R. (1975). The effect of chlorpromazine on the secretion of immunoreactive β-MSH and prolactin in man. J. Clin. Endocrinol. Metab. 41:380.

43. Lu, K. H., and Meites, J. (1973). Effects of serotonin precursors and melatonin on serum prolactin release in rats. Endocrinology 93:152.

44. Besses, G. S., Burrow, G. N., Spaulding, S. W., and Donabedian, R. K. (1975). Dopamine infusion acutely inhibits the TSH and prolactin response to TRH. J. Clin. Endocrinol. Metab. 41:985.

45. Relkin, R. (1972). Effects of variations in environmental lighting on pituitary and plasma prolactin levels in the rat. Neuroendocrinology 9:278.

46. Krulich, L. (1975). The effect of a serotonin uptake inhibitor (Lilly-110140) on the secretion of prolactin in the rat. Life Sci. 17:1141.

47. Marchlewska-Koj, H., and Krulich, L. (1975). The role of central monoamines in the stress-induced prolactin release in the rat. Fed. Proc. 34:252.

48. Gallo, R. V., Rabii, J., and Moberg, G. P. (1975). Effect of methysergide, a blocker of serotonin receptors, on plasma prolactin levels in lactating and ovariectomized rats. Endocrinology 97:1096.

49. Mulloy, A. L., and Moberg, G. D. (1975). Effects of p-chlorophenylalanine and raphe lesions on diurnal prolactin release in the rat. Fed. Proc. 34:251.

50. Anton-Tay, F., Chou, C., Anton, S., and Wurtman, R. J. (1968). Brain serotonin concentration: elevation following intraperitoneal administration of melatonin. Science 162:277.

51. Rønnekleiv, O. K., and McCann, S. M. (1975). Effects of pinealectomy, anosmia and blinding on serum and pituitary prolactin in intact and castrated male rats. Neuroendocrinology 17:340.

52. Rønnekleiv, O. K., Krulich, L., and McCann, S. M. (1973). An early morning surge of prolactin in the male rat and its abolition by pinealectomy. Endocrinology 92:1339.

53. Kamberi, I. A., Mical, R. S., and Porter, J. C. (1971). Effects of melatonin and serotonin on the release of FSH and prolactin. Endocrinology 88:1288.

54. Porter, J. C., Mical, R. S., and Cramer, O. M. (1971/72). Effect of serotonin and other indoles on the release of LH, FSH, and prolactin: hormones and antagonists. Gynecol. Invest. 2:13.

55. Kordon, C., Blake, C. A., Terkel, J., and Sawyer, C. H. (1974). Participation of

serotonin-containing neurons in the suckling-induced rise in plasma prolactin levels in lactating rats. Neuroendocrinology 13:213.

56. MacIndoe, J. H., and Turkington, R. W. (1973). Stimulation of human prolactin secretion by intravenous infusion of L-tryptophan. J. Clin. Invest. 52:1972.

57. Mendelson, W. B., Jacobs, L. S., Reichman, J. D., Othmer, E., Cryer, P. E., Triveder, B., and Daughaday, W. H. (1975). Methysergide: suppression of sleep-related prolactin secretion and enhancement of sleep-related growth hormone secretion. J. Clin. Invest. 56:690.

58. Kato, Y., Nakai, Y., Imura, H., Chihara, K., and Ohgo, S. (1974). Effect of 5-hydroxy-tryptophan (5-HTP) on plasma prolactin levels in man. J. Clin. Endocrinol. Metab. 38:695.

59. Handwerger, S., Plonk, J. W., Lebovitz, H. E., Bivens, C. H., and Feldman, J. M. (1975). Failure of 5-hydroxytryptophan to stimulate prolactin and growth hormone secretion in man. Horm. Metab. Res. 7:214.

60. Labhsetwar, A. P. (1972). Role of monoamines in ovulation: evidence for a serotoninergic pathway for inhibition of spontaneous ovulation. J. Endocrinol. 54:269.

61. Kordon, C. (1971). Involvement of catecholamines and indoleamines in the control of pituitary gonadotropin release. J. Neurovasc. Relat. (Suppl. X) 32:41.

62. Kalra, P. S., Kalra, S. P., Krulich, L., Fawcett, C. P., and McCann, S. M. (1972). Involvement of norepinephrine in transmission of the stimulatory influence of progesterone on gonadotropin release. Endocrinology 90:1168.

63. Kalra, S. P., and McCann, S. M. (1973). Effects of drugs modifying catecholamine synthesis on LH release induced by preoptic stimulation in the rat. Endocrinology 93:356.

64. Müller, E. E., Cocchi, D., Villa, A., Zambotti, F., and Fraschini, F. (1972). Involvement of brain catecholamines in the gonadotropin-releasing mechanism(s) before puberty. Endocrinology 90:1267.

65. Ojeda, S. R., and McCann, S. M. (1973). Evidence for participation of a catecholaminergic mechanism in the post-castration rise in plasma gonadotropins. Neuroendocrinolgy 12:295.

66. Zolovick, A. J. (1972). Role of central sympathetic neurons in the release of gonadotropin after hemiovariectomy. J. Endocrinol. 52:201.

67. Kamberi, I. A., Mical, R. S., and Porter, J. C. (1970). Effect of anterior pituitary perfusion and intraventricular injection of catecholamines and indoleamines on LH release. Endocrinology 87:1.

68. Kamberi, I. A., Mical, R. S., and Porter, J. C. (1971). Effect of anterior pituitary perfusion and intraventricular injection of catecholamines on FSH release. Endocrinology 88:1003.

69. Kamberi, I. A., Mical, R. S., and Porter, J. C. (1970). Follicle stimulating hormone releasing activity in hypophysial portal blood and elevation by dopamine. Nature 227:714.

70. Schneider, H. P. G., and McCann, S. M. (1970). Release of LH-releasing factor (LRF) into the peripheral circulation of hypophysectomized rats by dopamine and its blockage by estradiol. Endocrinology 87:249.

71. Sawyer, C. H., Hilliard, J., Koniematsu, S., Scaramuzzi, R., and Blake, C. A. (1974). Effects of intraventricular infusions of norepinephrine and dopamine on LH release and ovulation in the rabbit. Neuroendocrinology 15:328.

72. Tima, L., and Flerko, B. (1974). Ovulation induced by norepinephrine in rats made anovulatory by various experimental procedures. Neuroendocrinology 15:346.

73. Craven, R. P., and McDonald, P. G. (1971). The effect of intrahypothalamic infusions of dopamine and noradrenaline on ovulation in the adult rat. Life Sci. 10:1409.

74. Hadreen, J. C., and Chalmers, J. P. (1972). Neuronal degeneration in rat brain induced by 6-hydroxydopamine: a histological and biochemical study. Brain Res. 47:1.

75. Kamberi, I. A., Schneider, H. P. G., and McCann, S. M. (1970). Action of dopamine to induce release of FSH-releasing factor (FRF) from hypothalamic tissue in vitro. Endocrinology 86:278.

76. Miyachi, Y., Mecklenburg, R. S., and Lipsett, M. B. (1973). In vitro studies of pituitary–median eminence unit. Endocrinology 93:492.

77. Mims, R. B., Stein, R. B., and Bethume, J. E. (1973). The effect of a single dose of L-dopa on pituitary hormones in acromegaly, obesity and in normal subjects. J. Clin. Endocrinol. Metab., 37:34.

78. Lal, S., Tolis, G., Martin, J. B., Brown, G. M., and Guyda, H. (1975). Effect of clonidine on growth hormone, prolactin, luteinizing hormone, follicle-stimulating hormone, and thyroid-stimulating hormone in the serum of normal men. J. Clin. Endocrinol. Metab. 41:827.

79. Ladosky, W., and Noronha, J. G. L. (1974). Further evidence for an inhibitory role of serotonin in the control of ovulation. J. Endocrinol. 62:677.

80. Anton-Tay, F., and Wurtman, R. J. (1971). Brain monoamines and the control of anterior pituitary function. In L. Martini and W. F. Ganong (eds.), Frontiers in Neuroendocrinology, pp. 45–66. Oxford University Press, New York.

81. Donoso, A. O., Stefano, F. J. E., Biscardi, A. M., and Cukier, J. (1967). Effects of castration on hypothalamic catecholamines. Am. J. Physiol. 212:737.

82. Stefano, F. J. E., and Donoso, A. O. (1967). Norepinephrine levels in the rat hypothalamus during the estrus cycle. Endocrinology 81:1405.

83. Donoso, A. O., de Gutierrez Moyano, M. B., and Santolaya, R. C. (1969). Metabolism of noradrenaline in the hypothalamus of castrated rats. Neuroendocrinology 4:12.

84. Anton-Tay, F., and Wurtman, R. J. (1968). Norepinephrine: turnover in rat brains after gonadectomy. Science 159:1245.

85. Bapna, J., Neff, N. H., and Costa, E. (1971). A method for studying norepinephrine and serotonin metabolism in small regions of rat brain: effect of ovariectomy on amine metabolism in anterior and posterior hypothalamus. Endocrinology 89:1345.

86. Anton-Tay, F., Pelham, R., and Wurtman, R. J. (1969). Increased turnover of [^{3}H]-norepinephrine in rat brain following castration or treatment with ovine follicle-stimulatory hormone. Endocrinology 84:1489.

87. Coppola, J. A. (1971). Brain catecholamines and gonadotropin secretion. In L. Martini and W. F. Ganong (eds.), Frontiers in Neuroendocrinology, pp. 129–143. Oxford University Press, New York.

88. Lichtensteiger, W., Korpela, K., Langemann, H., and Keller, P. J. (1969). The influence of ovariectomy, estrogen, and progesterone on the catecholamine content of hypothalamic nerve cells in the rat. Brain Res. 16:199.

89. Beattie, C. W., Rodgers, C. H., and Soyka, L. (1972). Influence of ovariectomy and ovarian steroids on hypothalamic tyrosine hydroxylase activity in the rat. Endocrinology 91:276.

90. Kizer, J. S., Palkovits, M., Zivin, J., Brownstein, M. J., Saavedra, J. M., and Kopin, I. J. (1974). The effect of endocrinological manipulations on tyrosine hydroxylase and dopamine-beta-hydroxylase activities in individual hypothalamic nuclei of the adult male rat. Endocrinology 95:799.

91. Beattie, C. W., and Soyka, L. F. (1973). Influence of progestational steroids on hypothalamic tyrosine hydroxylase activity in vitro. Endocrinology 93:1453.

92. Montgomery, R. L., and Christian, E. L. (1973). Influence of chorionic gonadotropin on brain amine levels in male rats. Pharmacol. Biochem. Behav. 1:735.

93. Bernard, B. K., and Paolino, R. M. (1973). Brain norepinephrine levels and turnover rates in castrated mice isolated for 13 months. Experientia 29:221.

94. Zschaeck, L. L., and Wurtman, R. J. (1973). Brain [^{3}H] catechol synthesis and the vaginal estrous cycle. Neuroendocrinology 11:144.

95. Holzbauer, M., and Youdim, M. B. H. (1973). The oestrous cycle and monoamine oxidase activity. Br. J. Pharmacol. 48:600.

96. Anton-Tay, F., Anton, S. M., and Wurtman, R. J. (1970). Mechanism of changes in brain norepinephrine metabolism after ovariectomy. Neuroendocrinology 6:265.

International Review of Physiology
Reproductive Physiology II, Volume 13
Edited by Roy O. Greep
Copyright 1977   University Park Press   Baltimore

3
Physiological Aspects of the Steroid Hormone- Gonadotropin Interrelationship

K. BROWN-GRANT

Faculty of Medicine
Memorial University of Newfoundland
St. John's, Newfoundland, Canada A1C 5S7

The general outline of the functional interplay between the brain, the anterior pituitary gland, and the gonads was established by 1955 (1). The progressive filling in of detail over the next decade or so is well illustrated in two important summarizing papers written in 1968 and published in 1969 (2, 3). The basis of progress to that date in the understanding of the steroid control of gonadotropin secretion had been mainly the ingenious design of experiments using the responses of the appropriate end organ of the animal itself as indirect indices of changes in the rate or pattern of pituitary and gonadal hormone secretion. Pure steroid hormones were available for injection, as well as relatively pure pituitary or placental gonadotropins from a few species. The postulated hypothalamic gonadotropin-releasing (and -inhibiting) factors were being sought intensively, but only relatively crude extracts of median eminence tissue were available for physiological studies. Feed forward was mainly studied by electrical or electrochemical stimulation of the brain and feedback by the implantation of target organs or target organ hormones in the brain or pituitary. Opinion at the time still generally favored the brain, rather than the pituitary, as the site of feedback action of the gonadal steroids, although dissenting views had been expressed (4). In contrast to the situation for the thyroid and adrenocortical hormones, accurate measurement of physiological levels of gonadal and gonadotropic hormones in small volumes of peripheral plasma was either impossible or impracticable.

Between 1968 and 1971, the range of possible investigations altered dramatically as a result of two major developments. One was the explosive growth of competitive protein binding methodology which rapidly provided assays of adequate specificity and sensitivity for the measurement of gonadal steroids and gonadotropins in plasma (5). The second was the isolation, and subsequent analysis of a tripeptide hypothalamic releasing factor for thyroid-stimulating hormone (TSH), later shown to stimulate prolactin secretion as well, and a decapeptide releasing factor for luteinizing hormone (LH) and follicle-stimulating hormone (FSH) (6).

The newer information on steroid and gonadotropin concentrations was reviewed in 1971 with particular reference to the control of ovulation in different species (7). Much of the work was observational rather than experimental, descriptions of the time course and sequence of changes throughout the cycle. The conclusion reached was that estrogen, secreted by the ripening follicle, was the primary agent involved in initiating the ovulatory surge of gonadotropin secretion. The first section of this chapter will deal with the more recent evidence supporting this view and with some aspects of estrogen-progesterone interaction. One major problem that will not be considered is the anomalous situation of an increasing production of estrogen by the follicle at a time in the cycle when plasma gonadotropin concentrations are either constant or even decreasing slightly. It seems likely that intraovarian changes are involved, and there are some interesting suggestions in the proceedings of a recent symposium (8). The second topic to be considered is how estrogen acts to trigger the ovulatory surge and to what extent similar mechanisms may be involved under other circumstances. One of the most interesting recent findings has been the demonstration of an episodic pattern of gonadotropin secretion; this is considered in relation to the nature of negative feedback control under physiological conditions. Finally, some of the problems raised by recent studies of changes in steroid-gonadotropin relationships during development will be discussed.

STEROID CONTROL OF OVULATORY SURGE OF GONADOTROPINS

Evidence for an increase in the estradiol concentration in peripheral plasma preceding the ovulatory surge of LH in rats, sheep, and women has now been extended to other species, and new experimental findings have been reported that reinforce the concept of a causal relationship between the two events.

Laboratory Animals

Work on the rat has recently been reviewed in detail (9). In the hamster, preovulatory estrogen levels have been found to be high (180 pg/ml) compared with about 30 pg/ml in the rat. In addition to a sustained rise between late diestrus and proestrus that precedes the ovulatory surge of LH and FSH, a second peak, following in time and dependent upon the release of LH at proestrus, was also observed (10). In contrast to the rat, peripheral plasma progesterone concentration did not increase until after the onset of LH release and appeared to be related to the onset of behavioral estrus, rather than ancillary to the action of estrogen on LH release (11). Exogenous progesterone did not advance the onset of the ovulatory surge of LH when given early on the day of proestrus, although it delayed LH release when given at diestrus. A second sustained rise in FSH was seen in the hamster cycle that began late in proestrus and persisted throughout the day of estrus but, unlike the ovulatory peak, was not associated with raised LH levels (12). Only a reduced second peak of FSH was seen in the next "cycle" in intact hamsters treated with a long acting estrogen that produced sustained high plasma estrogen levels although FSH, LH, and progesterone peaks of normal magnitude were observed on the day of expected proestrus (13). This suggests that a late decrease in estrogen levels may normally contribute to the second extended period of FSH secretion. A facilitatory action of estrogen on LH release was demonstrated in ovariectomized hamsters, but not in castrated males. The estrogen-induced LH surges in the female occurred at the same time of day as the ovulatory surge under similar lighting conditions (14). Whether FSH and prolactin also increased in the experiments on ovariectomized hamsters is not known; increases do occur in comparable experiments on female, though not male, rats (15).

Only incomplete information is available for the guinea pig, data on estrogen levels not being available, but surprisingly, there does not appear to be an FSH surge in association with the LH surge at mid-cycle (16). The only other example of an ovulatory surge of LH occurring without an increase in FSH is in the rabbit, a reflex ovulator (17).

Farm Animals

Stimulation of both LH and FSH release by exogenous estrogen in the sheep is now well established (18). The estrogen-induced surge of LH, although it can be blocked by sodium pentobarbitone or reserpine (19), does not occur at a particular time in the light-dark cycle but rather at a set time after estrogen administration (20). This is in striking contrast to the hamster, as discussed above, and to the rat, in which LH release occurs only during a limited period of the light-dark cycle (21). As in the rodents, however, there is a clear sex difference in that this facilitatory effect of estrogen is not seen in the male or in genetic females that have been exposed to exogenous testosterone in utero before day 60 of gestation (22). A failure of estrogen to elicit an LH surge was also observed years later in ewes that had become infertile after prolonged grazing on estrogen-containing clover (23). This suggests that exposure to excess exogenous estrogen may permanently impair this mechanism in adult ewes, as it has been reported to do in adult female rats (24).

In the cow, the periovulatory pattern of steroids and gonadotropins has been determined (25, 26). In this species, there were multiple low peaks of estrogen preceding the LH surge rather than a single peak. Progesterone concentration had decreased from the high luteal phase value shortly before estrogen levels began to rise and remained low throughout the period of behavioral estrus and the preovulatory LH surge. There was a second estrogen peak 6–7 days after ovulation in the cow, but this occurred at a time when plasma progesterone concentrations were rising and was not associated with an LH peak. Estrogen concentration increased to about the same extent as in the cycle at about day 30 of pregnancy, but again there was no LH peak and no ovulation unless the normally high progesterone levels had been lowered by the administration of oxytocin. Exogenous estrogen has been shown to induce an LH surge in spayed cows (27). Whether bulls, intact or castrated, respond in the same way does not appear to have been investigated.

Limited data are available for the horse, in which both the estrogen and the LH peaks were prolonged, but a causal relationship was suggested by the finding that maximal estrogen concentrations precede maximal LH levels by about 2 days (28). LH levels remained high for some days after ovulation, and progesterone levels were low throughout this period (29). In the pig, an increase in estrogen concentration preceded the LH surge, and estrogen levels remained high and progesterone levels remained low throughout the period of the LH surge (30). Exogenous estrogen increased plasma LH in mares (31), but the effects on LH secretion do not appear to have been studied in sows.

Primates

The pattern of changes in steroid and gonadotropin levels in the human is well known (7), and the rhesus monkey has been studied in detail (32). The changes in the chimpanzee and in a New World (non-menstruating) monkey, the marmoset, have also been reported (33, 34). The human pattern of a restricted estrogen peak preceding synchronous LH and FSH peaks was seen in the rhesus monkey and the chimpanzee. The marmoset study indicated a cycle length of about 15 days, with a prolonged period (9 days) of high plasma estrogen levels and, displaced so that it began about 2 days later, a similar prolonged period when progesterone levels were high. The relationship to changes in gonadotropin concentrations was, unfortunately, not established, but would be of great interest as both the pattern and the absolute levels (up to 7 ng/ml for estrogen and up to 175 ng/ml for progestins) were most unusual and raised the possibility that unique steroid binding proteins may be present in the plasma of this species. Another New World primate, the squirrel monkey, has previously been reported to have extremely high plasma corticosteroid concentrations but normal transcortin levels (35).

In both the human and the rhesus monkey, the ability of exogenous estrogen to stimulate LH and FSH release has been demonstrated and analyzed in detail (32, 36). Abnormalities have been reported in some cases of amenorrhea in women (37). In the monkey, additional evidence for the importance of estrogen has been provided by the demonstration that active immunization against an estradiol-protein conjugate will block the LH surge and ovulation (38). In sharp contrast to the situation in the hamster (14), rat (15), and sheep (22), there are reports that a facilitatory action of estrogen or of progesterone after estrogen priming can be demonstrated in males after gonadectomy as adults in both the monkey (32) and man (39). This is a major point of difference in view of the emphasis placed on sexual differentiation of the pattern of gonadotropin secretion in the rodent (24). The recent experimental findings in primates are, of course, fully consistent with the failure to suppress ovulation in female rhesus monkeys masculinized by treatment of the mother with testosterone during pregnancy and the occurrence of ovulatory cycles after adequate treatment of girls suffering from congenital adrenal hyperplasia.

Ovulation in mammals is clearly an estrogen-dependent phenomenon, whether the steroid acts as the principal agent to trigger the ovulatory surge of LH in spontaneous ovulators or to render the female sexually receptive so that mating occurs and induces LH release in reflex ovulators. The pattern of secretion and the precise stimulus for FSH release seem to be more variable between species than that of LH although, except in the rabbit and possibly the guinea pig, both gonadotropins are secreted in increased amounts in the immediate preovulatory period. The interesting suggestion that androgens may be involved in the control of FSH in the female rat may indicate a profitable line of investigation (40). Whether the mammalian pattern extends to other vertebrates is also a question that requires much further investigation. In the domestic hen, there is a rise in plasma estradiol shortly preceding the preovulatory rise in LH, but there is little evidence that exogenous estrogen can exert any stimulatory effect on LH release. On the other hand, exogenous progesterone is clearly capable of advancing and accelerating LH release, particularly when injected during the phase of rising plasma estrogen concentration. However, the available evidence suggests that endogenous progesterone increased only very shortly before or coincidentally with the preovulatory surge of LH so that a causal relationship in the normal cycle cannot be taken as established, although it appears possible

(41). Some purely neural stimulus to increasing secretion of hypothalamic releasing factor may be involved; in contrast to mammals, increased pituitary responsiveness to gonadotrophin-releasing hormone (GNRH) at the time of the surge was not observed (42).

Progesterone-Estrogen Interactions

Although progesterone secreted at the time of the ovulatory surge of LH may have an ancillary role in determining the magnitude and duration of LH release, there is no evidence that it plays an obligatory or essential role, even in the rat where presurge increases in peripheral plasma progesterone concentration and the effectiveness of exogenous steroid are best documented (7). Whether it is of greater functional importance for FSH than for LH release remains a possibility. Nonetheless, the facilitatory effects of progesterone raise some very interesting questions. The paradoxical findings of facilitatory effects at one stage and inhibitory effects at another stage of the cycle have been explained along the lines originally suggested by Everett by the demonstration in rats and in man that facilitatory effects are dependent on previous priming of the hypothalamo-pituitary system by estrogen (7, 36). It is not known what the biochemical basis of this estrogen effect might be, although, in the guinea pig uterus and chick oviduct, estrogen can apparently increase the number or availability of progesterone binding sites. There are also considerable differences between species. It is not obvious why, after estrogen-priming, progesterone should facilitate gonadotropin release and increase sexual receptivity in the female rat, but retain its inhibitory effect on both parameters in the ewe. Whether this is a trivial matter of variation in experimental design or a fundamental difference is not clear.

The converse interaction, blockade of the facilitatory action of estrogen or its priming effect by concomitant exposure to progesterone, may be of greater importance. In ovariectomized animals, it has been shown that progesterone given with a dose of estrogen that is normally effective will block the effects on LH release in the hamster (43) and the ewe (44) and the priming effect in the rat (45). In the intact monkey, plasma progesterone levels equivalent to those of the normal luteal phase prevented exogenous estrogen from exerting its usual stimulatory effect on LH and FSH secretion (32) and exogenous estrogen was shown to be ineffective during the luteal phase of the cycle in the ewe (46). The physiological significance of this interaction may be similar to the situation in early pregnancy during which, in several species (7, 25, 33), plasma estrogen concentrations are high at the time when the next ovulation would have been expected to occur, but no LH or FSH surges are seen. It seems likely that the ineffectiveness of estrogen at this time is a consequence of the high progesterone levels.

MECHANISMS INVOLVED IN OVULATORY SURGE

The occurrence of increasing estrogen levels preceding the ovulatory surge of gonadotropin, the ability of antisera or chemical antagonists of estrogen action to block the surge, and the demonstrated facilitatory effects of exogenous estrogen on LH release in the female mammal make an overwhelming case for the central role of estrogen in the control of ovulation. The crucial question, then, is where and how is the estrogen acting? Until recently, the virtually universal answer would have been that estrogen acted on the brain, probably the hypothalamus, although possibly also on the preoptic area and the parts of the limbic system that were also known to contain estrogen-retaining neurons (47), to provoke a greatly increased discharge of hypothalamic gonadotropin-releasing factor(s) at the appropriate time, which in turn stimulated gonadotropin release. All the indirect evidence obtained from earlier studies was compatible with this view. The failure to demonstrate increased levels of releasing factors in hypophysial portal vessel blood at the expected times was ascribed to technical difficulties. Serious consideration of alternative possibilities was stimulated by scattered reports that the response of the pituitary to exogenous releasing factor was greatest at the time of behavioral estrus or on the expected day of proestrus. Initially, the findings were interpreted simply as a minor ancillary action of estrogen on the anterior pituitary, and early attempts to demonstrate such an action of exogenous estrogen were not entirely convincing, especially when reliance was placed on the response to a single injection of releasing factor (6). Systematic studies have now shown that the situation is more complex than was initially anticipated and that the magnitude of the changes at the pituitary level are far greater. In discussing these studies, the abbreviation GNRH (gonadotropin-releasing hormone) will be used both for the synthetic decapeptide corresponding in structure to the compound originally isolated from ovine and porcine hypothalami and shown to have the properties of both an LH- and an FSH-releasing factor (6) and which has subsequently been detected in hypophysial portal vessel blood (48) and for the agent presumed to be secreted in vivo. Whether this is the only or even the major compound with gonadotropin-releasing properties is not known. The general term "pituitary responsiveness" will be used in a broad sense to indicate the degree of enhanced secretion of LH or FSH achieved by GNRH administration under different experimental conditions, acknowledging that both sensitivity and capacity determine the magnitude of the response and should properly be considered separately (36).

Studies on Rats

A series of investigations have been carried out by Fink and his co-workers (49). In the first instance, they showed that in both conscious and sodium pentobarbitone-anesthetized rats there was a 10-fold increase in the pituitary response to a single intravenous injection of a standard dose of GNRH between the early afternoon of diestrus and the morning of the day of proestrus. There was a further very marked increase (50-fold) on the afternoon of proestrus, maximal values being observed at about 1700 hr. If sodium pentobarbitone was given before the onset of the spontaneous increase in plasma LH, the second phase of increased responsiveness was not observed. Subsequently, the responsiveness fell to low levels that were maintained throughout the rest of the cycle. The initial phase of increased responsiveness corresponded to the known period of high plasma estrogen concentration (7). Ovariectomy before this stage of the cycle prevented the initial increase in responsiveness, as did the administration of an antiestrogen, but exogenous estrogen restored it. Estrogen alone did not restore the late increase in responsiveness, but progesterone at 1300 hr, in addition to estrogen, increased, although not to normal, the responsiveness at 1700 hr. The degree of restoration of responsiveness at 1700 hr in these experiments was comparable to the degree of restoration of the secretion of LH at this time in the absence of exogenous GNRH in similar experiments on the same colony of rats (45).

A positive correlation between the response to a single injection of GNRH and the preinjection level of LH both during the normal proestrous surge and in the experiments on ovariectomized rats suggested that exposure to endogenous GNRH might have resulted in an increased responsiveness to further endogenous or exogenous GNRH. In a further study, this was shown to be so. Rats anesthetized with sodium pentobarbitone in the afternoon were given two injections of GNRH 60 min apart, and the responses to the first and second dose were compared. The LH response to the second dose was essentially no different from the first at the estrous stage of the cycle, somewhat greater at metestrus or diestrus, and very much greater at proestrus. Ovariectomy at diestrus reduced the response to both doses the next afternoon (expected day of proestrus), and estrogen replacement increased both responses. With or without estrogen, the second response was greater than the first. The removal of either the ovaries or the adrenals or both before the injection of the first dose of GNRH did not prevent the increased response to the second dose in rats tested at proestrus, indicating that acute changes in steroid secretion were not involved in the priming effect of GNRH. In a further study (50), it was demonstrated that electrical stimulation of the preoptic area in rats at proestrus, which was shown to induce LH release and to increase the release of GNRH and possibly other active peptides into the hypophysial portal vessels (48), also exerted a priming effect on the pituitary. A continuous infusion of a low level of GNRH over 45 min or more, but not for a shorter period, or repeated injections of small amounts of GNRH at 15-min intervals also had a priming effect in proestrous rats. The same dose of GNRH by constant infusion did not have a priming effect in female rats at diestrus, indicating a possible need for estrogen for the effect to occur. The gonadotropin response to electrical stimulation of the medial preoptic area in the female rat was shown to vary throughout the estrous cycle in the same way as the response to exogenous GNRH, and it was not necessary to postulate any increase in the amount of GNRH released to account for the increasing effect on LH release during proestrus (49). Direct measurement of the GNRH content of portal vessel blood after stimulation showed that release was higher on the morning of proestrus than on the afternoon of diestrus, but actually decreased in the afternoon of proestrus, although this was the time when maximal gonadotropin responses were seen (48). The early increase in sensitivity is probably an effect of estrogen, and the later decrease may be due to progesterone (51).

The experiments on intact or acutely ovariectomized rats did not provide an unequivocal answer to the question of whether previous exposure to estrogen was necessary for the priming effect of GNRH to be demonstrable. Retention of estrogen or persistence of its effects on the pituitary may have confused the picture, and only limited studies were made on males. In experiments on rats gonadectomized 1–2 months before (52), it was shown that an enhanced response to a second dose of GNRH was only observed in estrogen-treated females and that estrogen treatment of castrated males increased the response to the first dose of GNRH and allowed a greater second response to be demonstrated, although the effect was less marked than in the female. Progesterone after estrogen pretreatment resulted in a further marked increase in responsiveness in the female and some increase in the male. The additional enhancement of the response in the female could be negated by sodium pentobarbitone given with the progesterone, but the lesser enhancement produced by progesterone in the male or in genetic females which had received androgen in the neonatal period and were anovulatory was not affected by the barbiturate. The effects of progesterone after estrogen pretreatment may be 2-fold, a minor action on the pituitary seen in both sexes and an additional action seen only in normal females that may depend on stimulation of endogenous GNRH release. In the absence of exogenous GNRH, only normal females show an increase in gonadotropin secretion in response to progesterone under the conditions of these experiments (15). Many other relevant studies on rats are cited in the papers referred to above.

Other Studies

There have been no systematic studies of other laboratory animals or farm animals comparable in extent to those carried out on the rat. The human, however, has been studied in detail by Yen and his collaborators and their work has recently been reviewed (36). The parallelism between the human and the rat is striking.

In women, estrogen in the follicular phase facilitated gonadotropin release, and progesterone, after estrogen priming, was also effective. The responsiveness to GNRH was found to be greatest at mid-cycle and was reduced following the administration of estrogen antagonists. Despite some initial difficulties, the responsiveness was shown to be increased by exogenous estrogen. Estrogen followed by progesterone in the early follicular phase resulted in a further increase in responsiveness (53). The responses to successive pulse injections of GNRH were progressively greater at the mid-follicular stage of the cycle in contrast to the findings in intact men. Again, a constant infusion of a low dose of GNRH in the late, but not the early, follicular phase resulted in a dramatic increase in the response after about an hour. The findings in the human are very similar to those in the rat: an increased pituitary responsiveness produced by estrogen or progesterone after estrogen pretreatment, and, in the estrogen-primed state, increasing responses to repeated injections of GNRH at short intervals, and a dramatic increase in the response to a constant infusion of GNRH after a short lag period. The effect of estrogen on the pituitary was evident in males as well as females (54).

To the question of how estrogen induces the ovulatory surge of gonado-tropin the answer must now be that a very important part of the mechanism is an action on the pituitary gland itself (55), possibly preceded by a period of GNRH withdrawal and reduced responsiveness due to a direct action on the hypothalamus (56). This not only increases the immediate responsiveness to GNRH, but also alters the functional state of the gland in such a way that GNRH itself can exert a priming effect on the gland, so that successive pulses or constant low levels of GNRH are progressively more effective in stimulating LH release. It is debatable whether anything more than a transient increase or a slight rise in the tonic level of release of GNRH at the onset of the ovulatory surge need be postulated in view of the dramatic changes in responsiveness that can be demonstrated at this time. This possibility and the biochemical basis of the observed changes in responsiveness have been discussed in detail (36, 49, 57), as has the direct evidence for and against an increase in GNRH secretion (48, 58, 59).

Evidence of increased GNRH secretion at mid-cycle in the rhesus monkey has been presented (60). Very recently a significant increase in the concentration of immunoassayable GNRH in pituitary stalk blood during the period of the pro-estrous LH surge has been observed in rats anesthetized with "Althesin" (a steroidal anesthetic), rather than with urethane as in the earlier studies referred to above (G. Fink, personal communication, 1976). This suggests that both increased GNRH release and altered sensitivity are part of the surge mechanism in this species.

Onset and Termination of Surge

The acute pharmacological blockade of ovulation in the rat could be due to either inhibition of tonic secretion or suppression of an initial transient surge of GNRH secretion or both. The possibility that a small initial transient increase in GNRH release may be sufficient to trigger the ovulatory surge of gonadotropin raises some interesting questions as to the possible source of this GNRH in the rat. There is a low but detectable amount of GNRH in the suprachiasmatic nucleus (61), and lesions of this nucleus have been reported to reduce the GNRH content of the median eminence (62). The anatomical evidence indicates a minor but definite projection from this nucleus to the median eminence (63, 64). Lesions of the nucleus specifically block spontaneous ovulation in the rat and affect many other hormonal and behavioral circadian rhythms that are linked to the light-dark cycle (65). It seems possible that these neurons might supply a small pulse of GNRH just at the onset of the critical period, which would so affect pituitary sensitivity that continued unchanged tonic discharge of GNRH from arcuate neurons would be sufficient to maintain the surge. In gonadotropin-primed immature female rats, it has been shown that sodium pentobarbitone injected into the preoptic-suprachiasmatic area ceases to be effective in blocking ovulation some time before injection in the arcuate-median eminence becomes ineffective (66). In the adult female at proestrus, a cut in the region of the suprachiasmatic nucleus ceases to be effective in blocking ovulation some time before hypophysectomy becomes ineffective (67).

An additional problem, if the ovulatory surge is not initiated by a sustained increase and terminated by a decrease in GNRH secretion, is why the surge should have a finite and relatively constant duration in different species. The data for the human is scanty, but there are indications that plasma LH levels were beginning to fall despite continued infusion or repeated injections of GNRH in women at mid-cycle (36). In anestrous or ovariectomized sheep, prolonged infusions of GNRH failed to maintain elevated plasma levels of LH, although pituitary LH content was still quite high (68). A similar result was obtained in cyclic sheep although not commented on (69). Detailed studies have been carried out on rats anesthetized with phenobarbitone. Continued infusions of GNRH, begun at 1300 hr on the day of proestrus, resulted in LH peaks of greater magnitude but with a similar time course from onset to peak as the naturally occurring surge. Despite continued infusion, plasma LH levels had fallen to baseline values by the usual time, about 1900 hr (70). Similar effects were observed in ovariectomized rats, with or without estrogen pretreatment (71). Very similar results in intact rats at proestrus have been obtained by Blake (72), who has also shown that depletion of pituitary LH cannot explain the decrease in response and that continued infusion of GNRH will cause a release of FSH very similar to that seen in the normal cycle (73). The biochemical changes involved may be the opposite to those involved in the priming effect, and it is possible to devise explanations in terms of a shrinking pool of available LH but less easy to devise experiments to test them.

Other Changes in Pituitary Responsiveness

There are other physiological and clinical situations in which variations in pituitary responsiveness to GNRH may be highly significant. In the sheep, a very marked decrease in the pituitary responsiveness to GNRH during pregnancy was demonstrated for both LH and FSH, which persisted up to the time of parturition, despite a striking rise in plasma estradiol at that time, and into the postpartum period (74). As in other species, high doses of progesterone were found to reduce pituitary responsiveness and LH content, but rather long-term treatment was required to produce this effect in the sheep (75). The pregnant woman has been shown to have a reduced responsiveness to GNRH with respect to both LH and FSH, which persists into the postpartum period (76, 77). During the recovery period, the FSH response was initially greater than the LH response. This is the opposite of that which is observed in nonpregnant eugonadal women, but it is similar to the pattern observed in prepubertal girls (78).

There is a second very interesting situation in which the prepubertal type of response is seen in late adolescent or young adult females and that is in cases of anorexia nervosa, a disease characterized by extreme weight loss and amenorrhea induced by a voluntary but pathological reduction in food intake. When such patients were grossly underweight, both the LH and FSH responses to GNRH were reduced, particularly the LH response. As body weight increased under treatment, the FSH response improved first and increased linearly as normal or near normal body weight was achieved. The LH response did not increase initially, so that for a time a prepubertal pattern of response was observed. The LH response increased abruptly at about 85% of normal body weight. These changes in response could not, however, be related to changes in plasma estrogen concentration (79, 80). The mechanism involved in the body weight-pituitary response relationship is quite obscure, but may involve changes produced by long-term GNRH deprivation. An association between a critical body weight and the onset of menstruation in normal pubertal girls has been described, and there are reports of similar findings in rats (78). The recent clinical studies should stimulate a reexamination of these problems in experimental animals.

A change in the pattern of the LH and FSH response to GNRH has also been described in the human male. Patients with corrected cryptorchidism who had normal plasma testosterone and LH concentrations but elevated basal FSH levels associated with evidence of tubular dysfunction showed a normal LH but an exaggerated FSH response to GNRH. It was suggested that inhibin may normally act at the pituitary level to modulate the FSH response to GNRH (81).

NEGATIVE FEEDBACK

If the gonads of an adult are removed, secretion of gonadotropin increases; if excess exogenous steroid is administered, gonadotropin secretion can be completely suppressed in either intact or gonadectomized animals. The findings have been explained by postulating the existence of a negative feedback system operating to maintain some constant optimal level of plasma total or, more probably, free (i.e., nonprotein-bound) steroid concentration (2, 3). Although not as explicitly formulated as in the case of the pituitary-thyroid or pituitary-adrenal cortex systems (82), it has been generally assumed that this tonic control of gonadotropin secretion was probably a type of proportional control, the rectifying change (increase or decrease in gonadotropin secretion) being proportional to but inversely related to the divergence of (free) plasma steroid hormone from the "set" values. The further assumption that gonadotropin output can be smoothly and continuously varied is probably necessary if a system conceptualized in this way is in fact to maintain a relatively stable plasma steroid concentration on an hour-to-hour basis. The evidence now available, however, indicates that a considerable revision of ideas about negative feedback control of gonadotropins is necessary.

Gonadectomized Animals

Plasma gonadotropin levels rise rapidly (hours or days) after gonadectomy of adult animals. There are initial differences in the female in relation to the stage of the cycle at which the operation is performed, and in males the rise in FSH may be rather slower than that of LH and slower than in the female, possibly because of the effects of inhibin. Of greater interest than the minutiae of the early response is the chronic pattern of secretion that is established and the effect of exogenous steroids. Recent reports have been unanimous in indicating that the gonadectomized animal does not maintain a steady high level of secretion. On the contrary, both sexes showed a series of more or less regularly occurring episodes of rapidly rising LH levels, followed by a decline with a half-time of anything between 1 and 3 times the known half-life of the hormone to lower interpeak levels which, however, were still above nonsurge values for the intact animal. FSH levels, although elevated, showed much less variation, without clear episodes of greatly increased secretion. This pattern has been demonstrated in man, monkey, sheep, rat, and guinea pig (32, 83–85). The secretory episodes could be acutely suppressed by exogenous steroids, the result being a constant although still elevated plasma LH level. More prolonged treatment reduced the tonic LH and FSH levels to normal. The episodes of secretion are thought to be due to intermittent GNRH release and to originate in the central nervous system, although there is no direct proof of this. They can be suppressed by some anesthetics in the rat (86), although not in the monkey, whereas α-adrenergic but not β-adrenergic blocking drugs were effective in the latter species (32). Deafferentation experiments suggested that an isolated intact mediobasal hypothalamus was sufficient to sustain episodic secretion (32, 87).

The widely fluctuating gonadotropin levels complicate the assessment of the mean plasma concentration, but it is important to determine the relationship between plasma steroid and gonadotropin levels over days or weeks. Two groups have carried out such experiments, both using multiple implants of silastic tubing containing crystalline steroids which produce a steady blood level that can be manipulated by adding or removing implants. In ovariectomized rhesus monkeys (88), it was shown that normal levels of LH occurred in association with quite widely different plasma estrogen levels in different animals, and in general these levels were rather higher than those seen in intact animals other than at mid-cycle. Progesterone acted synergistically with estrogen (if estrogen levels were greater than about 30 pg/ml) at plasma concentrations that were ineffective in the luteal phase in intact animals with comparable estrogen concentrations. The difference in an individual animal between an estrogen level that maintained near normal gonadotropin concentrations and one that was associated with very high levels was quite small, about 20–30 pg/ml. This finding could indicate either a very narrow range of operation of a proportional control system or an all-or-none type of control.

The second study (89) was on testosterone–LH interaction in castrated male rats. Again, the plasma steroid concentration required to maintain normal LH levels varied considerably between animals. In an individual rat, the change in plasma testosterone levels between different stages of the experiment was about 50 ng/100 ml. It was found that a change of this magnitude would allow LH levels to increase from normal to castrate values, again suggesting almost an all-or-none type of control. In contrast to the monkey study, sustained plasma steroid levels that were somewhat below, rather than above, the mean for intact animals were adequate to maintain normal plasma LH concentrations.

There is evidence that negative feedback effects can occur at both the hypothalamic and the pituitary level (2, 55), and in the long term, it seems likely that both levels will be involved. Whether one or the other site of action appears to be more important may be more a function of the particular experimental design than an indication of fundamental differences between sexes or species.

Intact Animals

Frequent measurements of plasma hormone levels (the appropriate interval is at least shorter than the half-life of the hormone of interest) have shown that there are major episodic increases in plasma LH and testosterone levels in normal intact adult males of many species, including man, sheep, rat, mouse, rabbit, and the domestic fowl, which may or may not be associated with changes in FSH (90, 91). The changes are in general smaller and less predictable than in gonadectomized animals, and this, plus the fact that the stress of frequent bleeding in small animals tends to suppress the secretory episodes (author's unpublished observations), has handicapped the analysis of possible causal relationships. Some generalizations can be made, however, from inspection of the published data. There is no indication that individual episodes of increased LH secretion are preceded and triggered by decreases in testosterone concentration. Conversely, although major LH spikes may be accompanied or followed by increases in circulating testosterone, such an association is by no means universal for the low level spikes seen in adult males. As in the castrated male, the secretory episodes can be suppressed by high doses of exogenous steroids. Pharmacological studies have not been extensive; no effect of α-adrenergic blocking agents was detected in normal men (92). Episodes of LH secretion begin mainly in a low and end in a high range of plasma LH concentration, which suggests that some form of short loop feedback may be involved, but the evidence for this in the case of LH is minimal (93).

The situation in the intact female is less clear cut. There is evidence of episodic secretion in women in both the follicular and the luteal phase of the cycle, and variations in the pattern suggest that there may be some physiological control by steroids (92). Secretion is clearly episodic in the young ewe (94). As in the male, it seems very unlikely that these fluctuations represent the active operation of a negative feedback mechanism although only limited simultaneous studies of both steroid and gonadotropin levels have been performed. The intact females of other species do not appear to show episodic release. It has not been detected in the monkey (32), the laying hen (91), or the rat. Possibly, in these species, the levels of estrogen in the tissues or the persistence of estrogen effects are such that gonadotropin secretion is relatively heavily suppressed throughout the cycle.

Episodic secretion is not explicable in terms of the minute-to-minute operation of a negative feedback control, but rather seems to be superimposed on a system that may be sensitive only to sustained changes in steroid levels, operate on a longer time scale, and possibly on an all-or-none basis. The fluctuations seen in adults may be of little physiological significance, as discussed later.

Developing Animals

Steroid hormone control of gonadotropin secretion in relation to puberty was initially envisaged in terms of decreasing sensitivity of a neural "gonadostat" so that gonadotropin secretion and circulating steroid levels gradually increased during development until adult values were achieved. While such a scheme can account for some of the findings in some species at certain stages of development, recent studies have shown that there are many anomalies. Much of the available information has been collected in a recent book on the control of the onset of puberty (78) and in a symposium on perinatal sexual endocrinology (95).

In man and in the rhesus monkey, there is an early transient rise in fetal plasma testosterone concentration at about the 12th week of gestation in the male, but not the female; fetal plasma and pituitary gonadotropin levels are low at this time, and it seems likely that testicular steroidogenesis is being driven by gonadotropin of placental origin. Subsequently, there is a period during which, despite high plasma estrogen levels in both sexes, the female, but not the male, shows high plasma LH and FSH values that approach adult castrate values. The absence of these changes in the male cannot be accounted for by higher plasma testosterone concentrations. Toward term, gonadotropin levels are low, and they remain low for a few days after birth, but increase in both sexes as the placental steroids rapidly disappear from the circulation. In the male, both LH and FSH concentrations increase, and there is a considerable rise in plasma testosterone (and also estradiol, but not estrone, values) which persists for some months, but typical low prepubertal values for both gonadotropins and steroids are seen from about 6 months of age. Alterations in sex hormone binding cannot account for these findings. In the female, the rise in LH is less striking than in the male, but the increase in FSH is greater and persists for up to 1 year. As in the male, plasma estradiol (but not estrone) levels are at early pubertal values during this period and fall later. These findings were not predicted by the older theories and, to take account of them, it is necessary to propose that a high sensitivity of the "gonadostat" is a late postnatal development in man. The fact that cases of gonadal dysgenesis, in which steroid feedback is absent, also show a postnatal rise and then a subsequent fall in circulating gonadotropin concentrations suggests that the period of prepubertal quiescence in both sexes from about 3–10 years of age is neural in origin, rather than a consequence of day-to-day operation of a negative feedback system. Pubertal development from about 10 years of age is associated with demonstrable changes in the responses to estrogen or estrogen antagonist administration, but the normal pattern and the changes in cases of gonadal dysgenesis at the same age show that a simple change of feedback sensitivity does not provide an adequate explanation. There is abundant evidence (96) that the onset of puberty is associated with the occurrence of striking episodes of nocturnal LH and FSH secretion, at first closely linked to periods of slow wave sleep but gradually losing this association and also appearing during waking hours as maturation proceeds over a period of years. In the male, the LH pulses are soon associated with major increases in testosterone levels. It is even clearer than in the adult that the LH spikes are not triggered by a fall in testosterone concentration, nor does a progressive increase in testosterone over a period of hours suppress the spiking. Episodic nocturnal gonadotropin secretion precedes the menarche in pubertal girls, and menarche in turn precedes the onset of ovulatory cycles. A facilitatory action of estrogen and evidence of ovulation is not seen until some months at least after the menarche and, although limited estrogen output may also be involved, is probably the last endocrine mechanism necessary for potential fertility to develop in the human female (97). The low LH response and the disproportionately high FSH response to exogenous GNRH in prepubertal children has been mentioned previously. The change to the adult pattern occurs during puberty, but whether this is a consequence of the (presumed) nocturnal periods of exposure to endogenous GNRH or to a gradually changing steroid environment is not clear. The persistence of a high FSH:LH ratio in cases of gonadal dysgenesis is in favor of the latter view (83).

Some of the changes in the human also occur in other primates. High postnatal LH values have been reported in the rhesus monkey although whether the immunoreactive material is the same as adult simian LH has been questioned (95), and high FSH values over the first 2 years were observed in female (but not in male) chimpanzees (98). A prepubertal quiescent period during which gonadotropin levels do not increase in response to gonadectomy has been reported in the rhesus monkey as has the inability to demonstrate the facilitatory action of estrogen until some time after the menarche (78). Whether sleep-linked episodic gonadotropin secretion occurs during puberty in lower primates does not seem to have been established.

The sheep, which is born relatively mature, has been studied by Foster and his co-workers (94). Plasma LH was detectable and steady at about 130 days of gestation, and both sexes could respond to exogenous GNRH. Levels of LH and FSH were low at birth and for a few days thereafter, possibly because of the influence of steroids of placental origin. Subsequently mean levels were lower in the first 2 weeks in the male than in the female. After gonadectomy early in postnatal life, LH levels were higher in castrated males than in controls, but no difference was seen in females up to 14 days of age. After about 8 weeks of age, however, episodic secretion of LH (but not FSH) was detected in males, each LH spike being followed by an increase in plasma testosterone concentration. The magnitude and frequency of spiking appear to decrease rather than increase during development, perhaps because of a progressive rise in mean testosterone levels. The data are not easily explicable in terms of decreasing "gonadostat" sensitivity.

In the female, there may be a short quiescent period, probably overlapping with the period of suppression by placental steroids, but LH values for animals ovariectomized at 2 weeks of age were greater than controls after about 6 weeks of age. At about this stage, the pattern of secretion in the intact female lamb also changed (99). Frequent major spikes of LH secretion were observed, peak values reaching the adult castrate range. Mean FSH values were in the normal adult range, and episodes of secretion were not seen although mean levels might increase somewhat during periods of frequent LH release. These studies covered a period from birth to 9 weeks of age and might have indicated a transient period of postnatal hypersecretion that was later to be inhibited, the degree of suppression gradually lessening as the animals approached puberty. Subsequent studies showed that this was not so (94). The pattern of intermittent LH secretion persisted through the first breeding season and possibly into the second. A rather higher and more sustained LH peak representing the ovulatory surge was now detected before each period of behavioral estrus and predictable differences in frequency (lower in the luteal phase) were found once cycles were established. As in the male, this pattern is incompatible with a progressive decline in "gonadostat" sensitivity. The occurrence of the ovulatory surge type of secretion is probably determined by the ability of the ovary to secrete sufficient estrogen, as the capacity to respond to exogenous estrogen in this way is seen as early as 7 weeks of age and the response reaches adult levels by 27 weeks, although the first cycle is generally seen between 30 and 40 weeks of age (100).

The other species that has been studied in detail is the rat, and there are both similarities to and differences from both the human or the ovine pattern. The rat is born in a relatively immature state, and testicular secretions masculinize the brain in the first few postnatal days. In the male, plasma testosterone was about 50 ng/100 ml at this time, slightly lower at weaning (22 days) and began to increase linearly after 30 days of age, reaching a peak at about 80 days and falling slightly thereafter to adult values. There was considerable variation between individuals at all ages and from day to day in adults. Plasma LH was low or undetectable in the early postnatal period, but FSH was at about 50% of adult values (101). From about day 5, LH was detectable and quite variable, whereas FSH levels were fairly constant. Both gonadotropins were under some form of negative feedback control; castration resulted in a prompt increase and exogenous steroids reduced plasma concentrations. From day 22, plasma FSH concentration increased to a value greater than that in the intact adult, although not to castrate levels, but LH increased very little, if at all. This is difficult to reconcile with a simple decrease in "gonadostat" sensitivity. Direct evidence (78) indicated that the steroidogenic response to LH of the maturing rat testis increased following fairly prolonged exposure to FSH and that this could explain the increasing testosterone levels at a time when LH was only increasing slowly. There is evidence from the work of McCann and associates (78) that, when androgen replacement was begun at the time of castration, the sensitivity at 58 or 88 days of age was less than at 15 or 28 days. This could account for the late (60–90 days) rise in LH to adult values in the face of peak plasma testosterone concentrations. The fall in FSH to adult levels occurred slightly earlier during the phase of rising testosterone concentration, but, as this was also the stage at which the first cycle of spermatogenesis was being completed, it is possible that inhibin may be involved.

In the female, LH and FSH were detectable on the day of birth. LH values were very variable, means varying from the nonsurge level in adult females to several times this value being reported up to about 30 days of age. FSH values were more consistent; values several times the adult nonsurge value were seen in the first few days after birth and increased to very high levels, as great or greater than the long-term ovariectomized adult value at around 15 days. Subsequently, there was a fall to adult nonsurge levels at around 30 days. A negative feedback mechanism could be shown to be operative in the neonatal period as high doses of exogenous steriods reduced plasma gonadotropin levels as in the male (24), although because of the already high levels, effects of ovariectomy were more difficult to demonstrate. There are very considerable differences in the published literature in mean LH levels for both male and female rats between about 5 and 25 days of age. Various studies have reported peaks or nadirs on one or more days, especially when group numbers have been small. The reported time of occurrence varied between studies, and no clear association with any critical developmental change was apparent (95). In earlier studies in this laboratory, pooled samples were generally assayed, and reasonably smooth curves for changes in LH during development were obtained (24). But in view of the conflicting reports, further studies using only samples from individual rats were carried out. The distribution of LH values from individual rats in supposedly homogeneous groups was repeatedly found to be quite abnormal. For example, eight female rats might have LH values of around 1 ng/NIH S-13/ml and the two others values of 15 and 22. Statistically nonsignificant differences in the inci- dence of high values between matched control groups frequently resulted in striking differences when group mean values were compared. Similar though less marked effects were seen in males. In contrast, FSH values for the same samples showed a normal distribution and low coefficient of variation. It seems likely that episodic secretion of LH is occurring in these young rats. A systematic attempt to demonstrate this was frustrated by the finding that the population of high LH values was not seen in samples from conscious rats with implanted cardiac catheters. It seems probable that the secretory episodes were suppressed by nonspecific stress associated with the procedure because rats handled 60 min before decapitation or anesthetized rats, except for a few animals bled under ether anesthesia, also had a very low incidence of high values (102). It is obviously difficult to determine the mean level of LH to which these young animals are normally exposed, and great caution is needed in the interpretation of apparent transient differences between different stages of development.

The pattern of LH and FSH secretion seen in the young females is similar to that of the ovariectomized adult, despite the confusing fact that high levels of estradiol (possibly in part of adrenal origin) are found in the plasma of both female and male rats up to about 22 days of age (103). A specific estrogen-binding protein is also present which greatly reduces the biological effectiveness of the natural estrogens. This protein, which is found only in late pregnancy and in early postnatal life, is the α-fetoprotein of the rat and disappears after about 22 days of age. At this time, even the lower total estrogen, because less is protein bound, may provide a more effective feedback signal and thus account for the low gonadotropin levels at around 30 days. A major increase in the number of specific nuclear binding sites for estradiol in the hypothalamus and limbic system has also been reported to occur at about 25 days of age (104). There is also a transient rise in plasma progesterone at about this time (105).

Between this age and the first ovulation at around 40–45 days, the gonado-tropin system of the female rat seems to be in a dormant state. The adjective "dormant" is used because ovariectomy still leads to a prompt increase in gonadotropin secretion, in contrast to the absence of a response during the quiescent period of prepubertal development in the human and the monkey. The cause of the dormant period in the female rat is not known (although the increase in receptor sites may be involved (104)) nor the determinant of the time of first ovulation. Facilitatory effects of exogenous steroids can be demonstrated as early as 24 days of age. A recent study (106) failed to show any gradual increase in LH, FSH, or prolactin levels, suggestive of a decrease in "gonadostat" sensitivity until immediately before the first ovulatory surge, which was pre-ceded by the first detectable increase in estradiol. The latter observation is of particular importance, because the effects of repeated small doses of exogenous estrogen in inducing precocious ovulation have led to speculation that such a mechanism might be operating under normal conditions. It should be noted, however, that the study was based on single daily samples taken at 1500 hr, and increasing frequency of intermittent sleep-linked episodic secretion cannot be excluded. Evidence suggesting that ring A-reduced C-19 steroids of ovarian origin may be involved in the onset of ovulation has recently been reviewed (107), and the negative findings with respect to estrogen strengthens the case in favor of a possible role for these other steriods.

SUMMARY AND CONCLUSIONS

Three aspects of the physiological interactions of steroid hormones and gonado-tropins have been considered, i.e., the role of estrogen in initiating the ovulatory surge of gonadotropin, the mechanism of this action, and the negative feedback system in adult and developing animals.

Use of estrogen as a direct triggering mechanism seems to be the rule in mammals. A major action of estrogen is on the pituitary gland itself, both to alter the initial responsiveness to GNRH and, probably, to allow GNRH to exert a sensitizing effect of its own on the gland. While there are major physiological problems remaining, in particular the question of whether any sharp increase in the rate of secretion of GNRH occurs, the broad outline of the origin of the ovulatory surge appears to have been established. This particular steroid-gonado-tropin interaction should be an area in which detailed investigation of the biochemical basis of the effects is both possible and profitable. The methodolo-gies available and current interest in their application were discussed at a recent symposium on subcellular mechanisms in reproductive neuroendocrinology (108).

It is not possible to provide a succinct statement about the current status of physiological studies on the negative feedback control of gonadotropins. It seems clear that in all adult males and in the females of some species there is a pattern of episodic release of LH and, to a much lesser extent, if at all, of FSH superimposed upon a tonic level of secretion. Both the episodic release and the tonic release are thought to be driven by GNRH, are enhanced by gonadectomy, and can be suppressed by high doses of exogenous steroids. The balance of present evidence is against the individual episodes of secretion being initiated by falling steriod levels. In view of the long-term persistence of androgens and estrogens in their target tissues, coupled with the existence in many species of steroid binding proteins in the plasma (109) with a partial buffering function, the consequent transient increases in plasma steroid levels probably produce only minor fluctuations in the biological effectiveness at the level of the target tissues. Episodic secretion does not appear to add anything to the efficiency of the negative feedback regulation, but neither does it seem likely to materially detract from it.

Perhaps the persistence of episodic secretion of gonadotropins is the price that has to be paid for the more general availability of a rapidly acting neural drive to pituitary function capable of responding to changes in the external environment and of integrating pituitary function with the sleep-waking cycle and sleep stages. Comparison with nongonadotropic hormones may provide some support for this concept. The adrenal cortex may be required to respond maximally in emergencies and shows a circadian rhythm of activity related to the sleep-activity cycle. Episodic secretion was first demonstrated for this gland and intermittent adrenocorticotropic hormone (ACTH) secretion begins late in sleep at a time when episodes of rapid eye movement (REM) sleep (paradoxical sleep) are frequent, although the association is not a precise one (110). In man and monkey, growth hormone is secreted episodically during slow wave sleep (SWS) early in the sleep cycle, and this secretion is independent of normal metabolic control. The secretion is tightly linked to sleep and rapidly follows enforced changes in the timing of sleep, in contrast to ACTH secretion, for which the rhythm may take days to adapt to an altered sleep pattern. The regulation of TSH secretion appears to be mainly at the pituitary level; at least, in the rat, increased thyroid-stimulating hormone-releasing factor (TRF) secretion seems to be specifically associated only with acute exposure to a reduced environmental temperature (111).

Under what conditions could a neurally driven acute increase in gonadotropin secretion be of physiological importance and of survival value to the species? Clearly, in an induced ovulator such as the rabbit, if the population under natural conditions is scattered, the ability to ovulate and conceive after brief coital stimulation from the male has survival value. Again, in seasonal breeders, particularly if, as now appears to be the case for both mammals and birds, the photosensitive period is restricted to a limited portion of the 24-hr day (112), a rapid response at this critical time may be essential. The human is characterized by a delayed onset of puberty, and recent findings suggest that this process is normally dependent on the attainment of a required degree of somatic and neurological maturity. The demonstration that the early endocrine change is the onset of major nocturnal gonadotropin secretory episodes, initially closely linked to early SWS stages, but progressively spreading through the sleep period and then into waking hours is a very important finding indeed (96). It suggests to this author that the benefits of this pattern of neural control of the onset of puberty may be such that any possible disadvantages of the persistence of episodic secretion into adult life are quite outweighed.

The exploration of the biochemical, neurophysiological, and neuroanatomical factors involved in the initiation and maintenance of episodic secretion is already under way and seems likely to produce many new and exciting findings, particularly in relation to the role of brain amines. The emphasis in this chapter has inevitably shifted, in view of current findings, from an emphasis on humoral control in relation to the ovulatory surge to emphasis on neural control in relation to episodic secretion and puberty. To redress the balance, it should perhaps be pointed out that most attention has naturally been given in the first instance to the major biologically active compounds (testosterone, estradiol, progesterone) when plasma steroid concentrations have been determined and the relationship to gonadotropin secretion examined. These are not the only steroids secreted by the gonads nor are their metabolic products, whether secreted by the gonads or produced in target or nontarget tissues, devoid of important biological activities, and they are present in appreciable quantities in plasma. In situations in which anomalous steroid-gonadotropin relationships appear to exist, it may be necessary to consider the whole spectrum of plasma steroids and their concentrations before concluding that the classical concept of negative feedback control cannot provide an adequate explanation for the findings.

ACKNOWLEDGMENTS

Many scientists generously allowed me access to accounts of their studies before publication. Whether cited or not, their papers were of immense help in preparing this review. The cooperation of Drs. C. A. Blake, J. M. Davidson, B. T.

Donovan, G. Fink, P. C. B. MacKinnon, J. D. Neill, G. A. Schuiling, S. C. Yen, and their co-workers is gratefully acknowledged. I also wish to thank Mrs. Sarah Pertwee for her assistance and Dr. B. A. Cross for his tolerance of the prolonged use of laboratory space for literary purposes.

REFERENCES

1. Harris, G. W. (1955). Neural Control of the Pituitary Gland. Arnold, London.
2. Davidson, J. M. (1969). Feedback control of gonadotropin secretion. *In* W. F. Ganong and L. Martini (eds.), Frontiers in Neuroendocrinology 1969, pp. 343–388. Oxford University Press, New York.
3. Schwartz, N. B. (1969). A model for the regulation of ovulation in the rat. Recent Prog. Horm. Res. 25:1.
4. Bogdanove, E. M. (1964). The role of the brain in the regulation of pituitary gonadotropin secretion. Vitam. Horm. 22:205.
5. Odell, W. D., and Daughaday, W. H. (1971). Principles of Competitive Protein-Binding Assays. J. B. Lippincott, Philadelphia.
6. Gual, C., and Rosemberg, E. (1973). Hypothalamic Hypophysiotropic Hormones. Excerpta Medica, Amsterdam.
7. Brown-Grant, K. (1971). The role of steroid hormones in the control of gonadotropin secretion in adult female mammals. *In* C. H. Sawyer and R. A. Gorski (eds.), Steroid Hormones and Brain Function, pp. 269–288. University of California Press, Los Angeles.
8. Coutts, J. R. T., and Govan, A. D. T. (1975). Symposium Report No. 7. Functional morphology of the ovary. J. Reprod. Fertil. 45:557.
9. Neill, J. D., and Smith, M. S. (1974). Pituitary-ovarian interrelationships in the rat. *In* V. H. T. James and L. Martini (eds.), Current Topics in Experimental Endocrinology. Vol. II, pp. 73–106. Academic Press, New York.
10. Baranczuk, R., and Greenwald, G. S. (1973). Peripheral levels of estrogen in the cyclic hamster. Endocrinology 92:805.
11. Bosley, C. G., and Leavitt, W. W. (1972). Dependence of preovulatory progesterone on critical period in the cyclic hamster. Am. J. Physiol. 222:129.
12. Bast, J. D., and Greenwald, G. S. (1974). Serum profiles of follicle-stimulating hormone, luteinizing hormone and prolactin during the estrous cycle of the hamster. Endocrinology 94:1295.
13. Greenwald, G. S. (1975). Proestrous hormone surges dissociated from ovulation in the estrogen-treated hamster, Endocrinology 97:878.
14. Norman, R. L., and Spies, H. G. (1974). Neural control of the estrogen-dependent twenty-four-hour periodicity of LH release in the golden hamster. Endocrinology 95:1367.
15. Brown-Grant, K. (1974). Steroid hormone administration and gonadotrophin secretion in the gonadectomized rat. J. Endocrinol. 62:319.
16. Blatchley, F. R., Donovan, B. T., and ter Haar, M. B. (1976). Plasma progesterone and gonadotropin levels during the estrous cycle of the guinea pig. Biol. Reprod. 15:29.
17. Dufy-Barbe, L., Franchimont, P., and Faure, J. M. A. (1973). Time-courses of LH and FSH release after mating in the female rabbit. Endocrinology 92:1318.
18. Cumming, I. A. (1975). The ovine and bovine oestrous cycle. J. Reprod. Fertil. 43:583.
19. Jackson, G. L. (1975). Blockage of estrogen-induced release of luteinizing hormone by reserpine and potentiation of synthetic gonadotropin-releasing hormone-induced release of luteinizing hormone by estrogen in the ovariectomized ewe. Endocrinology 97:1300.
20. Jackson, G. L., Thurmon, J., and Nelson, D. (1975). Estrogen-induced release of LH in the ovariectomized ewe: independence of time of day. Biol. Reprod. 13:358.
21. Colombo, J. A., Baldwin, D. M., and Sawyer, C. H. (1974). Timing of the estrogen-induced release of LH in ovariectomized rats under an altered lighting schedule. Proc. Soc. Exp. Biol. Med. 145:1125.

22. Karsch, F. J., and Foster, D. L. (1975). Sexual differentiation of the mechanism controlling the preovulatory discharge of luteinizing hormone in sheep. Endocrinology 97:373.
23. Findlay, J. K., Buckmaster, J. M., Chamley, W. A., Cumming, I. A., Hearnshaw, H., and Goding, J. R. (1973). Release of luteinizing hormone by oestradiol-17β and a gonadotrophin-releasing hormone in ewes affected with clover disease. Neuroendocrinology 11:57.
24. Brown-Grant, K. (1974). On 'critical periods' during the post-natal development of the rat. In M. G. Forest and J. Bertrand (eds.), INSERM Symposium 32, Endocrinologie sexuelle de la période périnatale, pp. 357–376. INSERM, Paris.
25. Glencross, R. G., Munro, I. B., Senior, B. E., and Pope, G. S. (1973). Concentrations of oestradiol-17β, oestrone and progesterone in jugular venous plasma of cows during the oestrous cycle and in early pregnancy. Acta Endocrinol. (Kbh) 73:374.
26. Lemon, M., Pelletier, J., Saumande, J., and Signoret, J. P. (1975). Peripheral plasma concentrations of progesterone, oestradiol-17β and luteinizing hormone around oestrus in the cow. J. Reprod. Fertil. 42:137.
27. Short, R. E., Howland, B. E., Randel, R. D., Christensen, D. S., and Bellows, R. A. (1973). Induced LH release in spayed cows. J. Anim. Sci. 37:551.
28. Pattison, M. L., Chen, C. L., Kelley, S. T., and Brandt, G. W. (1974). Luteinizing hormone and estradiol in peripheral blood of mares during the estrous cycle. Biol. Reprod. 11:245.
29. Plotka, E. D., Foley, C. W., Witherspoon, D. M., Schmoller, G. C., and Goetsch, D. D. (1975). Periovulatory changes in peripheral plasma progesterone and estrogen concentrations in the mare. Am. J. Vet. Res. 36:1359.
30. Henricks, D. M., Guthrie, M. D., and Handlin, D. L. (1972). Plasma estrogen, progesterone and luteinizing hormone levels during the estrous cycle in pigs. Biol. Reprod. 6:210.
31. Garcier, M. C., and Ginther, O. J. (1975). Plasma luteinizing hormone concentration in mares treated with gonadotropin-releasing hormone and estradiol. Am. J. Vet. Res. 136:1581.
32. Knobil, E. (1974). On the control of gonadotropin secretion in the rhesus monkey. Recent Prog. Horm. Res. 30:1.
33. Reyes, F. I., Winter, J. S. D., Faiman, C., and Hobson, W. C. (1975). Serial serum levels of gonadotropins, prolactin and sex steroids in the non-pregnant and pregnant chimpanzee. Endocrinology 96:1447.
34. Preslock, J. P., Hampton, S. H., and Hampton, J. K., Jr. (1973). Cyclic variations of serum progestins and immunoreactive estrogens in marmosets. Endocrinology 93:1096.
35. Brown, G. M., Grota, L. J., Penney, D. P., and Reichlin, S. (1970). Pituitary-adrenal function in the squirrel monkey. Endocrinology 86:519.
36. Yen, S. S. C., Lasley, B. L., Wang, C. F., Leblanc, H., and Siler, T. M. (1975). The operating characteristics of the hypothalamic-pituitary system during the menstrual cycle and observations of biological action of somatostatin. Recent Prog. Horm. Res. 31:321.
37. Glass, M. R., Shaw, R. W., Butt, W. R., Edwards, R. L., and London, D. R. (1975). An abnormality of oestrogen feedback in amenorrhoea-galactorrhoea. Br. Med. J. 3:274.
38. Ferin, M., Dyrenfurth, I., Cowchock, S., Warren, M., and Vande Wiele, R. L. (1975). Active immunization to 17β-estradiol and its effects upon the reproductive cycle of the rhesus monkey. Endocrinology 94:765.
39. Stearns, E. L., Winter, J. S. D., and Faiman, C. (1973). Positive feedback effect of progestin upon serum gonadotropins in estrogen-primed castrate men. J. Clin. Endocrinol. Metab. 37:635.
40. Gay, V. L., and Tomaccari, R. L. (1974). Follicle-stimulating hormone secretion in the female rat: cyclic release is dependent on circulating androgen. Science 184:75.
41. Wilson, S. C., and Sharp, P. J. (1975). Changes in plasma concentrations of luteinizing hormone after injection of progesterone at various times during the ovulatory cycle of the domestic hen (Gallus domesticus). J. Endocrinol. 67:59.
42. Bonney, R. C., Cunningham, F. J., and Furr, B. J. A. (1974). Effect of synthetic

luteinizing hormone releasing hormone on plasma luteinizing hormone in the female domestic fowl, *Gallus domesticus*. J. Endocrinol. 63:539.

43. Norman, R. L., Blake, C. A., and Sawyer, C. H. (1973). Evidence for neural sites of action of phenobarbital and progesterone on LH release in the hamster. Biol. Reprod. 8:83.

44. Scaramuzzi, R. J., Tillson, S. A., Thorneycroft, I. H., and Caldwell, B. V. (1971). Action of exogenous progesterone and estrogen on behavioral estrus and luteinizing hormone levels in the ovariectomized ewe. Endocrinology 88:1184.

45. Tapper, C. M., Greig, F., and Brown-Grant, K. (1974). Effects of steroid hormones on gonadotrophin secretion in female rats after ovariectomy during the oestrous cycle. J. Endocrinol. 62:511.

46. Howland, B. E., Akbar, A. M., and Stormshak, F. (1971). Serum LH levels and luteal weight in ewes following a single injection of estradiol. Biol. Reprod. 5:25.

47. Stumpf, W. E. (1970). Estrogen neurons and estrogen-neuron systems in the periventricular brain. Am. J. Anat. 129:207.

48. Fink, G., and Jamieson, M. G. (1976). Immunoreactive luteinizing hormone releasing factor in rat pituitary stalk blood: effects of electrical stimulation of the medial preoptic area. J. Endocrinol. 68:71.

49. Fink, G., Aiyer, M. S., Jamieson, M. G., and Chiappa, S. A. (1975). Factors modulating the responsiveness of the anterior pituitary gland in the rat with special reference to gonadotrophin releasing hormone (GnRH). *In* M. Motta, P. G. Crosignani, and L. Martini (eds.), Hypothalamic Hormones, pp. 139–160. Academic Press, London.

50. Fink, G., Chiappa, S. A., and Aiyer, M. S. (1976). Priming effect of luteinizing hormone releasing factor elicited by preoptic stimulation and intravenous infusion and multiple injections of the synthetic decapeptide. J. Endocrinol. 69:359.

51. Sherwood, N. M., Chiappa, S. A., and Fink, G. (1976). Immunoreactive luteinizing hormone releasing factor in pituitary stalk blood from female rats: sex steroid modulation of response to electrical stimulation of preoptic area or median eminence. J. Endocrinol. 70:501.

52. Aiyer, M. S., Sood, M. C., and Brown-Grant, K. (1976). The pituitary response to exogenous luteinizing hormone releasing factor in steroid-treated gonadectomized rats. J. Endocrinol. 69:255.

53. Lasley, B. L., Wang, C. F., and Yen, S. S. C. (1975). The effects of estrogen and progesterone on the functional capacity of the gonadotrophs. J. Clin. Endocrinol. Metab. 41:820.

54. Wang, C. F., Lasley, B. L., and Yen, S. S. C. (1975). The role of estrogen in the modulation of pituitary sensitivity to LRF (luteinizing hormone-releasing factor) in men. J. Clin. Endocrinol. Metab. 41:41.

55. Greeley, G. H., Jr., Allen, M. B., Jr., and Mahesh, V. B. (1975). Potentiation of luteinizing hormone release by estradiol at the level of the pituitary. Neuroendocrinology 18:233.

56. Smith, E. R., and Davidson, J. M. (1974). Location of feedback receptors: effects of intracranially implanted steroids on plasma LH and LRF response. Endocrinology 95:1566.

57. Pickering, A. J. M. C., and Fink, G. (1976). Priming effect of luteinizing hormone releasing factor: *in vitro* and *in vivo* evidence consistent with its dependence upon protein and RNA synthesis. J. Endocrinol. 69:373.

58. Eskay, R. L., Oliver, C., Ben-Jonathan, N., and Porter, J. C. (1975). Hypothalamic hormones in portal and systemic blood. *In* M. Motta, P. G. Crosignani, and L. Martini (eds.), Hypothalamic Hormones, pp. 125–137. Academic Press, London.

59. Nett, T. M., Akbar, A. M., and Niswender, G. D. (1974). Serum levels of luteinizing hormone and gonadotropin-releasing hormone in cycling, castrated and anestrous ewes. Endocrinology 94:713.

60. Neill, J. D., Patton, J. M., and Tindall, G. T. (1975). LHRH in the rhesus monkey: its presence in hypophysial portal blood and its distribution in the hypothalamus. *In* Proceedings of the 57th Meeting of the Endocrine Society, p. 298 (Abstr.).

61. Palkovits, M., Arimura, A., Brownstein, M., Schally, A. V., and Saavedra, J. M.

(1974). Luteinizing hormone-releasing hormone (LH-RH) content of the hypothalamic nuclei in the rat. Endocrinology 96:554.

62. Schneider, M. P. G., Crichton, D. B., and McCann, S. M. (1969). Suprachiasmatic LH-releasing factor. Neuroendocrinology 5:271.

63. Réthelyi, M., and Halász, B. (1970). Origin of the nerve endings in the surface zone of the median eminence of the rat hypothalamus. Exp. Brain Res. 11:145.

64. Swanson, L. W., and Cowan, W. M. (1975). The efferent connections of the suprachiasmatic nucleus of the hypothalamus. J. Comp. Neurol. 160:1.

65. Raisman, G. (1975). Anatomical and functional studies on the preoptic area and its involvement in the initiation of ovulation in the female rat. Adv. Biosci. 15:25.

66. Hagino, N. (1969). The hypothalamic time sequence controlling gonadotrophin release in the immature female rat. Neuroendocrinology 5:1.

67. van Rees, G. P., Wildschut, J., and Schuiling, G. A. (1971). Time factors involved in pre-ovulatory gonadotropin secretion. Acta Endocrinol. (Kbh). Suppl. 155, 42 (Abstr.).

68. Chakraborty, P. B., Adams, T. E., Tarnavsky, G. K., and Reeves, J. J. (1974). Serum and pituitary LH concentration in ewes infested with LH-RH/FSH-RH. J. Anim. Sci. 39:1150.

69. Hooley, R. D., Baxter, R. W., Chamley, W. A., Cumming, I. A., Jonas, H. A., and Findlay, J. K. (1974). FSH and LH response to gonadotropin-releasing hormone during the ovine estrous cycle and following progesterone administration. Endocrinology 95:937.

70. Schuiling, G. A., De Koning, J., and Zurcher, A. F. (1976). On differences between the pre-ovulatory luteinizing hormone surges of 4- and 5-day cyclic rats. J. Endocrinol. 70:373.

71. Schuiling, G. A., and Gnodde, H. P. (1976). The secretion of luteinizing hormone caused by continuous infusions of luteinizing hormone releasing hormone in the long-term ovariectomized rat: effect of oestrogen pretreatment. J. Endocrinol. 71:1.

72. Blake, C. A. (1976). Simulation of the proestrous luteinizing hormone (LH) surge after infusion of LH-releasing hormone in phenobarbital-blocked rats. Endocrinology 98:451.

73. Blake, C. A. (1976). Simulation of the early phase of the proestrous follicle stimulating hormone rise after infusion of luteinizing hormone-releasing hormone in phenobarbital-blocked rats. Endocrinology 98:461.

74. Chamley, W. A., Findlay, J. K., Jonas, H., Cumming, I. A., and Goding, J. R. (1974). Effect of pregnancy on the FSH response to synthetic gonadotrophin-releasing hormone in ewes. J. Reprod. Fertil. 37:109.

75. Plant, H. C., and Ward, W. R. (1973). Effect of progesterone on the pituitary responsiveness to luteinizing hormone releasing hormone (LHRH) in intact anoestrous ewes. J. Physiol. (Lond.) 232:45P (Abstr.).

76. Jeppson, S., Rannevik, G., and Kullander, S. (1974). Studies on the decreased gonadotropin response after administration of LH/FSH releasing hormone during pregnancy and the puerperium. Am. J. Obstet. Gynecol. 120:1029.

77. LeMaire, W. J., Shapiro, A. G., Rigall, F., and Yang, N. S. T. (1974). Temporary pituitary insensitivity to stimulation by synthetic LRF during the postpartum period. J. Clin. Endocrinol. Metab. 38:916.

78. Grumbach, M. M., Grave, G. D., and Mayer, F. E. (1974). The Control of the Onset of Puberty. John Wiley and Sons, New York.

79. Warren, M. P., Jewelewicz, R., Dyrenfurth, I., Ans, R., Khalaf, S., and Vande Wiele, R. L. (1975). The significance of weight loss in the evaluation of pituitary response to LH-RH in women with secondary amenorrhea. J. Clin. Endocrinol. Metab. 40:601.

80. Sherman, B. M., Halmi, K. A., and Zamudio, R. (1975). LH and FSH response to gonadotropin-releasing hormone in anorexia nervosa: effect of nutritional rehabilitation. J. Clin. Endocrinol. Metab. 41:135.

81. Bramble, F. J., Houghton, A. L., Eccles, S. S., Murray, M. A. F., and Jacobs, H. S. (1975). Specific control of follicle stimulating hormone in the male: postulated site of action of inhibin. Clin. Endocrinol. 4:443.

82. Stear, E. B., and Kadish, A. M. (1969). Hormonal control systems. Mathematical Biosci. Suppl. 1.

83. Yen, S. S. C., Tsai, C. C., Vandenberg, G., and Rebar, R. (1972). Gonadotropin dynamics in patients with gonadal dysgenesis: a model for the study of gonadotropin regulation. J. Clin. Endocrinol. Metab. 35:897.
84. Gay, V. L., and Sheth, N. A. (1972). Evidence for a periodic release of LH in castrated male and female rats. Endocrinology 90:158.
85. Donovan, B. T., ter Haar, M. B., and Rosemberg, L. E. (1974). Pulsatile gonadotrophin release in the adult guinea pig. J. Endocrinol. 65:3P (Abstr.).
86. Blake, C. A. (1974). Localization of the inhibitory actions of ovulation-blocking drugs on release of luteinizing hormone in ovariectomized rats. Endocrinology 95:999.
87. Blake, C. A., and Sawyer, C. H. (1974). Effects of hypothalamic deafferentation on the pulsatile rhythm in plasma concentrations of luteinizing hormone in ovariectomized rats. Endocrinology 94:730.
88. Karsch, F. J., Weick, R. F., Hotchkiss, J., Dierschke, D. J., and Knobil, E. (1973). An analysis of the negative feedback control of gonadotropin secretion utilizing chronic implantation of ovarian steroids in ovariectomized rhesus monkeys. Endocrinology 93:478.
89. Damassa, D. A., Kobashigawa, D., Smith, E. R., and Davidson, J. M. (1976). Negative feedback control of LH by testosterone: a quantitative study in male rats. Endocrinology 99:736.
90. Baker, H. W. G., Santen, R. J., Burger, H. G., de Kretser, D. M., Hudson, B., Pepperell, R. J., and Bardin, C. W. (1975). Rhythms in the secretion of gonadotropins and gonadal steroids. J. Steroid Biochem. 6:793.
91. Wilson, S. C., and Sharp, P. J. (1975). Episodic release of luteinizing hormone in the domestic fowl. J. Endocrinol. 64:77.
92. Santen, R. J., and Bardin, C. W. (1973). Episodic luteinizing hormone secretion in man: pulse analysis, clinical interpretation, physiologic mechanisms. J. Clin. Invest. 52:2617.
93. Boyar, R., Perlow, M., Hellman, L., Kapen, S., and Weitzman, E. (1972). Twenty-four hour pattern of luteinizing hormone secretion in normal men with sleep stage recording. J. Clin. Endocrinol. Metab. 35:73.
94. Foster, D. L., Lemons, J. A., Jaffe, R. B., and Niswender, G. D. (1975). Sequential patterns of circulating luteinizing hormone and follicle-stimulating hormone in female sheep from early postnatal life through the first estrous cycles. Endocrinology 97:985.
95. Forest, M. G., and Bertrand, J. (eds.) (1974). INSERM Symposium 32, Endocrinologie sexuelle de la période périnatale. INSERM, Paris.
96. Boyar, R. M., Rosenfeld, R. S., Finkelstein, J. W., Kapen, S., Roffwarg, H. P., Weitzman, E. D., and Hellman, L. (1975). Ontogeny of luteinizing hormone and testosterone secretion. J. Steroid Biochem. 6:803.
97. Winter, J. S. D., and Faiman, C. (1973). The development of cyclic pituitary-gonadal function in adolescent females. J. Clin. Endocrinol. Metab. 37:714.
98. Winter, J. S. D., Faiman, C., Hobson, W. C., Prasad, A. V., and Reyes, F. I. (1975). Pituitary-gonadal relations in infancy. I. Patterns of serum gonadotropin concentrations from birth to four years of age in man and chimpanzee. J. Clin. Endocrinol. Metab. 40:545.
99. Foster, D. L., Jaffe, R. B., and Niswender, G. D. (1975). Sequential patterns of circulating LH and FSH in female sheep during the early postnatal period: effect of gonadectomy. Endocrinology 96:15.
100. Foster, D. L., and Karsch, F. J. (1975). Development of the mechanism regulating the preovulatory surge of luteinizing hormone in sheep. Endocrinology 97:1205.
101. Brown-Grant, K., Fink, G., Grieg, F., and Murray, M. A. F. (1975). Altered sexual development in male rats after oestrogen administration during the neonatal period. J. Reprod. Fertil. 44:25.
102. MacKinnon, P. C. B., Mattock, J. M., and ter Haar, M. B. (1976). Serum gonadotrophin levels during development in male, female and androgenized female rats and the effect of general disturbance on high luteinizing hormone levels. J. Endocrinol. 70:361.

103. Ojeda, S. R., Kalra, P. S., and McCann, S. M. (1975). Further studies on the maturation of the estrogen negative feedback on gonadotropin release in the female rat. Neuroendocrinology 18:242.
104. Plapinger, L., and McEwen, B. S. (1973). Ontogeny of estradiol-binding sites in rat brain. I. Appearance of presumptive adult receptors in cytosol and nuclei. Endocrinology 93:1119.
105. Meijs-Roelofs, H. M. A., de Greef, W. J., and Uilenbroek, J. T. J. (1975). Plasma progesterone and its relationship to serum gonadotrophins in immature female rats. J. Endocrinol. 64:329.
106. Meijs-Roelofs, H. M. A., Uilenbroek, J. T. J., de Greef, W. J., De Jong, F. H., and Kramer, P. (1975). Gonadotrophin and steroid levels around the time of first ovulation in the rat. J. Endocrinol. 67:275.
107. Eckstein, B. (1975). Studies on the mechanism of the onset of puberty in the female rat. J. Steroid Biochem. 6:873.
108. Naftolin, F., Davies, I. J., and Ryan, K. J. (eds.) (1976). Subcellular Mechanisms in Reproductive Neuroendocrinology. Elsevier, Amsterdam.
109. Corvol, P., and Bardin, C. W. (1973). Species distribution of testosterone-binding globulin. Biol. Reprod. 8:277.
110. Rubin, R. T. (1975). Sleep-endocrinology studies in man. Prog. Brain Res. 42:73.
111. Montoya, E., Seibel, M. J., and Wilber, J. F. (1975). Thyrotropin-releasing hormone secretory physiology: studies by radioimmunoassay and affinity chromatography. Endocrinology 96:1413.
112. Stetson, M. H., Elliott, J. A., and Menaker, M. (1975). Photoperiodic regulation of hamster testis: circadian sensitivity to the effects of light. Biol. Reprod. 13:329.

International Review of Physiology
Reproductive Physiology II, Volume 13
Edited by Roy O. Greep
Copyright 1977 University Park Press Baltimore

4
Gonadotropin Interactions with the Gonad as Assessed by Receptor Binding and Adenylyl Cyclase Activity

R. J. RYAN, L. BIRNBAUMER, C. Y. LEE, and M. HUNZICKER-DUNN

Department of Molecular Medicine, Mayo Medical School, Rochester, Minnesota;
Department of Cell Biology, Baylor College of Medicine, Houston, Texas; and
Department of Biochemistry, Northwestern University School of Medicine,
Chicago, Illinois

Parts of this research were funded by Grants HD 9140, HD 6513, and HD 9581 from the National Institutes of Health, United States Public Health Service, and funds from the Mayo Foundation.

This review will be concerned predominantly with luteinizing hormone (LH) and human chorionic gonadotropin (hCG) as they affect the ovary. Follicle-stimulating hormone (FSH), prolactin, prostaglandins, other hormones, and the testis will be considered in less detail.

The chemistry (1, 2) and physiology (3) of the gonadotropins have been reviewed recently, and only certain aspects will be outlined in this introduction.

FSH, LH, and hCG, as well as thyroid-stimulating hormone, are glycoproteins containing sialic acid and having two noncovalently linked subunits. One subunit, termed α, is common to all, while the β subunits are hormone unique. Their primary amino acid sequences are known. The β subunits of LH (of pituitary origin) and hCG (of placental origin) share large regions of homology, as would be expected from their similar biological properties. In addition, there are restricted regions of homology common to all of the β subunits, as would be expected from their combination with a common α subunit. There is a high degree of cross-linking, by disulfide bridges, within the α and β subunits, but the assignment of these bridges has not yet been made. Available data indicate that biological functions occur only when the α and β subunits are combined to form the native molecule.

The concentrations of LH in serum, which range from approximately 0.3–30 ng/ml or 10^{-11} to 10^{-9} M, show variations between the sexes, within the day, within the life cycle, and within the menstrual or estrus cycle. Concentrations are low ($\sim 10^{-11}$ M) and fairly constant prior to puberty, aside from somewhat higher values and rudimentary cyclicity during infancy (4). At puberty, concentrations rise (high 10^{-11} M to low 10^{-10} M) and low amplitude, short frequency (30–90 min) pulsatile variations appear (5–7). In the male, this situation persists without other dramatic variation. In the female, marked variations occur with the menstrual or estrus cycle (8, 9). Concentrations are low but gradually rising in the preovulatory phase, rise markedly (high 10^{-10} to low 10^{-9} M) 12–24 hr prior to ovulation, and drop to lowest concentrations in the luteal phase (10^{-11}–10^{-10} M). Serum LH rises to the high 10^{-10} or low 10^{-9} M concentrations in the human female at the menopause or in either sex following castration. Under these circumstances, the short frequency pulsatile variations increase in amplitude.

Serum FSH is generally similar to LH in concentration (10^{-11}–10^{-9} M) and variation, with several exceptions. The short frequency pulsatile variations in serum FSH tend to be of lower amplitude than the variations of LH. During the menstrual cycle, serum FSH shows two peaks, one during menses and a second coincident with the mid-cycle peak of LH. Following the menopause or castration, serum FSH tends to show a proportionately greater increase than serum LH.

HCG is normally absent from the circulation except during pregnancy. It can be detected (10^{-11} M) about the time of the first missed menses or shortly following implantation, and reaches a peak serum concentration (10^{-7} M) at the end of the first trimester. Thereafter it declines, but is still measurable throughout pregnancy. Various tumors in the male or nonpregnant female are known to produce hCG or an hCG-like material.

In the male, LH is responsible for the proliferation and differentiation of the Leydig or interstitial cell and the production and secretion of testosterone by this cell. In the female, it is required for the differentiation of the theca and interstitial cells of the ovary and the secretion of androgen and, indirectly, estrogen. The action of LH on steroidogenesis is believed to be concerned with the rate-limiting step, the mitochondrial conversion of cholesterol to pregneno-lone. LH is primarily responsible for the phase of rapid growth and ovulation of suitably matured follicles. During the process of ovulation, LH is responsible for the luteinization of the granulosa and the structural formation of the corpus luteum. In most species, it is also primarily responsible for the function of the corpus luteum, namely the secretion of progesterone. In the rodent, prolactin is also required for the function of the corpus luteum. In the rabbit, estrogen is currently believed to be the major luteotropin (10–14). It has also been observed that large doses of LH are luteolytic (15–17).

LH has a variety of other effects on the ovary. It increases ovarian blood flow (18) and the activity of the enzyme ornithine decarboxylase (19). It induces a decrease in the concentrations of ascorbic acid (20) and cholesterol (21). It has one known extragonadal effect, namely a change in the coloration of the breast feathers of the weaver finch (22). hCG has biological effects that are similar to LH, but they tend to be more pronounced because hCG has a longer half-life in the circulation, which relates to its higher content of sialic acid (23).

FSH, in the male, is concerned with development of the germinal epithelium. How this effect was mediated was an enigma, because most, if not all, of the effects of FSH on spermatogenesis and spermiogenesis could be mimicked by androgens. Present data suggest that FSH acts on the Sertoli cell to induce the formation of an androgen-binding protein which then facilitates the action of androgens on the germinal epithelium (24).

In the female, FSH, like LH, is required for the production of estrogen. The dual dependency of estrogen production on both LH and FSH has been a mystery. Recent data suggest an elegant solution (25). LH, in stimulating the conversion of cholesterol to pregnenolone, leads to the production of androgen by the theca cell; this androgen then diffuses across the limiting membrane of the follicle and is taken up by the granulosa cell. FSH induces the aromatizing enzyme system in the granulosa cell which allows the conversion of theca-produced androgen to estrogen.

FSH acts on the follicle to induce antrum formation and proliferation of the granulosa. This action is enhanced by estrogen. These effects of FSH (and estrogen) precede and may be responsible for the induction of LH receptors and LH-stimulatable adenylyl cyclase in the granulosa. It is interesting to speculate that FSH may be responsible for induction of an estrogen receptor in the granulosa, analogous to the androgen-binding protein in the Sertoli cell.

FSH can also induce the ovulatory process, but under physiological circum-stances, this appears to be only a facilitating action.

There are also data indicating that the process of follicular atresia in the ovary is under pituitary regulation. Hypophysectomy leads to a decrease in the number of follicles undergoing atresia (26), but the details of this action remain to be elucidated.

The initial step in the action of protein hormones is generally believed to be binding to a membrane-bound receptor, which then leads to activation of a membrane-bound cyclase and subsequently a biological effect. These processes will be examined in greater detail.

BINDING OF GONADOTROPINS TO THEIR TARGET TISSUE RECEPTORS

General Properties of Gonadotropin Receptors

The receptor concept is old in terms of drug action, and a number of theories have been developed to relate receptor occupancy to pharmacological effects (27–29). The concept has been applied to the action of steroid hormones for more than a decade. Although the applicability of receptor theory to the mode of action of protein and polypeptide hormones has been recognized for many years, direct assessment of receptor binding did not occur until about 1970.

Simply stated, a receptor is a molecule that serves to recognize and combine with stimulatory or inhibitory substances delivered to the cell and translate this recognition function into a biological action. In the case of protein and polypeptide hormones, this receptor molecule is thought to be expressed predominantly, if not exclusively, on the outer surface of the plasma membrane.

In order to fulfill the definition of a receptor, certain criteria need to be met.

Specificity This implies a unique interaction between the hormone of interest and its receptor. Other substances, without agonist or antagonist effects of that particular hormone, do not bind to the same receptor, nor do they compete for binding. Furthermore, the concept of specificity has been extended to the target tissue(s) and cell(s) and the specific portions of the cell that are accessible to the hormone and are related to the primary biological events.

Affinity The receptor hormone interaction must have an association constant (K_a) high enough or a dissociation constant (K_d) low enough to be applicable to the lower range of concentrations of the hormone in blood under physiological conditions.

Limited Number of Binding Sites The hormone receptor interaction must be a saturable process, and this implies a limited number of binding sites per unit of cell or tissue. The appraisal of saturability is generally a functional estimate and does not truly indicate the number of receptor molecules that are present, unless all are functionally active and the number of moles of hormone bound per mole of receptor is known. In line with the criterion of specificity, the number, or at least concentration, of binding sites in target tissues must be in excess of that found in nontarget tissues. This, of course, begs the question of what is a target tissue.

A major question with respect to this criterion is what is the limit and what does it mean? The general feeling is that the limited number of binding sites is a few. From a physiological standpoint, a few would be that number that could be saturated by the highest physiological concentrations of the hormone. However, there may be pharmacological effects that would require a larger number, and there may even be excess or spare receptors.

Relationship Between Receptor Occupancy and Biological Actions of Hormone This criterion is the most difficult to establish. At present, we must be content with correlations between binding data and biological effects. The correlations are phenomenological, temporal, and dose related. Of course, the more correlations that can be made, the greater the degree of confidence that can be placed in the relationship. Ultimate fulfillment, however, must await detailed knowledge of the molecular mechanisms involved in the translation of the binding event into a biological action.

Involved in this criterion are the questions of whether the relationship between receptor occupancy and biological effect is concerned with the number of receptors that are occupied (either in a graded (27, 28), quantal or threshold fashion (30)) or with the rate of occupancy (29).

One of the greatest handicaps in the study of receptors for protein and large polypeptide hormones is the lack of analogues with agonist and particularly antagonist properties. These have proved particularly useful in establishing criteria for receptors of small molecules.

Efforts to study gonadotropin receptors first appeared in 1971, with reports concerned with the binding of LH or hCG to rat testes by deKrester, Catt, and Paulsen (31); Leydig cell tumors by Moudgal, Moyle, and Greep (32); and pregnant mare's serum gonadotropin (PMSG)-hCG primed rat ovaries by Lee and Ryan (33). Subsequently, there have been a host of reports concerned with LH-hCG receptors in ovarian tissue of the rat (34–37), cow (38–40), and human (41); granulosa cells of the pig (42, 43); and Leydig cells of the rat (44–47). FSH receptors have been studied in the testis of the rat (48–50) and bull (51).

Midgley (52, 53) has done an elegant study of the localization of LH, FSH, and prolactin receptors in the rat ovary using autoradiographic techniques. FSH receptors were localized to the granulosa cells. LH-hCG receptors were found on theca cells, interstitial cells, granulosa cells of large follicles, and corpora lutea. Prolactin receptors were found on luteal and interstitial cells. In subsequent studies using cell fractionation techniques (54) and electron microscopic (EM) autoradiography (55), it has been shown that LH-hCG receptors in the rat corpus luteum are predominantly, if not exclusively, localized to the plasma membrane. The possibility that some LH (hCG) enters the cells cannot be excluded, however, since some localization within the cytoplasm (55) and nucleus (W. Anderson and R. J. Ryan, unpublished observations) of the rat luteal cell has been observed by EM autoradiography and hCG effects on nuclear functions have been reported (56). It is difficult to prove, however, that the silver grains represent intact, biologically active hormone and that the nuclear events are directly mediated by the gonadotropin.

A variety of approaches have been used to study in vitro binding of gonadotropins. These have involved tissue slices (33, 35), dispersed cells (31, 32, 42, 43, 46, 49), homogenates (34, 35), crude particulate fractions (36), and plasma membrane preparations in various stages of purity (38–40). On the other hand, the number of radiolabeled hormones employed has been limited to human LH and hCG, ovine LH, and human FSH. This may relate to the availability of highly purified materials for radiolabeling. There are data, however, in the rat testis system indicating differences in affinity of LH preparations from various species, with the human hormones having the highest affinities (47). Thus, hormones from other species may be more difficult to study.

Factors concerned with in vitro binding of LH-hCG have been studied in greater detail than binding of FSH. Binding is dependent upon pH, the optimal being at pH 7.5 (34). Both molarity and specific salts affect binding. In contrast to the adenylyl cyclase system, divalent ions such as Ca^{2+}, Mg^{2+}, or Mn^{2+} do not appear to be required (34, 37). In fact, Ca^{2+} and Mg^{2+} in concentrations in excess of 10 mM inhibit binding of LH-hCG to rat ovarian tissue (34, 37). Sodium at 50 mM and potassium at 100 mM also inhibit binding (34, 37). These salt and molarity effects are mediated, at least in part, by slowing the initial association rate constant (57).

Time and temperature variables are important and, as expected, equilibrium is achieved sooner at 37°C than at lower temperatures (34–36, 45). However, there has not been a systematic study to determine the free energy of these reactions.

Evidence has been presented to indicate that unbound hormone (36) and unoccupied receptors (37, 58) are damaged during incubation conditions. These processes, like binding, are time- and temperature-dependent (36, 37, 58) and generally necessitate that binding assays be carried out at temperatures less than 25°C. Damage, at least to free hormone, is greater with crude homogenates than with washed particles and is least with whole cell preparations (36). Damage is also affected by the hormone employed (hCG less than human LH (hLH)), the concentration of hormone (proportionately greater with low concentrations), and the quantity of tissue (36). Corrections for incubation damage may be made either empirically (36) or by a computer curve-fitting procedure, for accurate assessment of binding kinetics (58). These phenomena have not been studied with respect to FSH binding.

In addition to damage during incubation, there are data to indicate the existence of inhibitors. A small molecular weight (3800 daltons) inhibitor has been partially purified and characterized from the soluble fraction of rat ovarian homogenates (59).

Although the role of nucleotides in adenylyl cyclase activity (see below) and in glucagon-receptor interactions (60) has been studied extensively, there are no published systematic studies with respect to nucleotides and gonadotropin binding. In the glucagon-rat liver cell membrane system, nucleotides accelerate the dissociation rate constant, with GTP and GDP being most potent (60). These effects are independent of temperature and the presence or absence of cold hormone (60). Preliminary data in the LH-rat ovarian system (61) suggest that ATP is more effective than GTP and particularly slows the association rate constant (R. J. Ryan, unpublished observations). Several nucleotides inhibited total binding in the bovine corpus luteum membrane (40).

Assay Methods

Radioiodination   The chloramine-T method of Greenwood et al. (62) has been widely used for the iodination (^{125} I) of hCG or LH for binding studies. In order to minimize damage to the hormone, some modification is required (36, 45, 49). The procedure of Lee and Ryan (36) follows. Ten μg of gonadotropin in 10 μl was added to 25 μl of 0.5 M phosphate buffer, pH 7.5, containing 1.0 mCi of ^{125} I. The reaction was initiated by addition of 10 μl of chloramine-T (1.25 mg/ml in 0.5 M phosphate buffer). After 40–60 s in an ice bath, the reaction was terminated by addition of 250 μl of sodium metabisulfite at a concentration of 0.1 mg/ml and followed by 100 μl of 1% KI. Labeled hormone was purified by gel chromatography on Bio-Gel P10 but other methods have been employed (37, 45, 47). The labeled gonadotropin had a specific activity of 60–70 μCi/μg, which corresponds to approximately 1 atom of ^{125} I per molecule of hormone.

For quantitative binding studies, it is necessary to know the mass of biologically active tracer. Mass can be assessed by radioimmunoassay procedures (36). The mass of biologically active tracer can be assessed by an in vitro radioreceptor assay using a differently (^{131} I) labeled hormone as tracer and a direct comparison with purified unlabeled hormone (36). This procedure is preferable to an in vivo bioassay because it requires less hormone and has greater precision and can be justified by the excellent agreement between the in vitro receptor assay and the in vivo bioassay (33). Radioiodinated hLH and hCG, when prepared as described above and assayed by receptor and radioimmunoassays, showed retention of 90–100% and 80–90% of biological activity, respectively (36).

The lactoperoxidase method (63, 64) has also been used to radioiodinate ovine LH (38) and human FSH (hFSH) for binding studies. With the use of this method, hFSH could be iodinated with substantial retention of biological activity, where the chloramine-T method failed (H. Takahashi and C. Y. Lee, unpublished observations). No noticeable damage to ovine LH was observed using the lactoperoxidase method for iodination (38).

Preparation of Receptors

Ovarian Slices   Heavily luteinized rat ovaries previously induced by PMSG and hCG priming were employed. Female Holtzman rats were primed with 50 I.U. of PMSG at 25 days of age and 50 I.U. of hCG (33) or 125 μg of ovine LH (64) (NIH-S17) 56 hr later, which causes an increase in both ovarian weight and hCG receptor sites per mg of weight. Animals were used between 8 and 15 days after PMSG priming. Ovaries were removed and trimmed and either used immediately or frozen on dry ice and stores at −60°C for up to 1 year without loss of activity. Thin sections were made with a Stadie-Rigg microtome on the fresh or thawed ovaries, discarding the first and last slices.

Porcine Granulosa Cells Pig ovaries were collected and dissected at the slaughterhouse, placed in normal saline solution, packed in ice, and transported to the laboratory. This process required 4–6 hr. Cells were aspirated from small (1–2 mm), medium (3–5 mm), and large follicles (6–12 mm), pooled by follicle size, centrifuged at 300 $\times$ *g*, and the cells were resuspended in buffer as described by Channing and Ledwitz-Rigby (66). The cells could either be used immediately or stored overnight in an ice bucket.

Particulate Fractions A simple and rapid method for preparation of particulate fraction from luteinized rat ovaries has been described before (36). In brief, rat ovaries were sliced, as above, and homogenized with a motor-driven Teflon-glass homogenizer in 10 volumes (v/w) of 40 mM Tris buffer (pH 7.4) containing 5 mM $MgCl_2$. Homogenates were centrifuged at 2,000 $\times$ *g* for 15 min. The pellets were washed twice and resuspended in the same Tris buffer. The washed pellets can be resuspended in 10% sucrose and stored at −60°C for months without loss of binding activity.

Plasma membrane preparations obtained from bovine corpora lutea have been used to study ovine LH-receptor interaction. The method for purifying plasma membrane from bovine corpus luteum by sucrose density centrifugation has been described (39).

Soluble Receptors Solubilization of membrane-bound hCG-LH receptors has been achieved by the use of nonionic detergents (37, 67, 68). Particulate fractions (2,000 $\times$ *g* pellets) were prepared from luteinized rat ovaries and extracted with 10 volumes of 0.25% Triton X-100 and centrifuged at 110,000 $\times$ *g* for 1 hr, as described before (67). The soluble receptors in the supernatant were shown to retain binding activity (67). Solubilization of hCG receptors from rat testes has also been reported (37, 68).

Assay of Binding Activity

Ovarian Slices Labeled hLH was incubated with ovarian slices in modified Krebs-Ringer phosphate buffer according to the method described previously (33, 65). At the end of incubation, ovarian slices were removed and washed with cold buffer. Radioactivity in both slices and media was measured and binding was expressed as a T:M ratio (cpm per g of tissue to cpm per ml of medium). Nonspecific binding was measured by control incubations containing a large excess of unlabeled hCG (200 I.U./ml) and constituted only 3–4% of total binding.

Porcine Granulosa Cells Cells were incubated with labeled hCG in medium 199 plus Earle's salts and 15% pig serum (66). The binding reaction was terminated by centrifugation. After washing twice, the cell pellets were counted for bound [125]-hCG. Nonspecific binding was measured in the presence of a large excess of unlabeled hCG and represented 1–10% of total binding.

Particulate Fraction A typical assay mixture contained 2 ng of labeled hCG and 2–5-mg equivalents of ovarian weight, as 2,000 $\times$ *g* pellet of ovarian homogenate, in a final volume of 1 ml of 40 mM Tris buffer containing 0.1%

bovine serum albumin, pH 7.4. The incubation was at $25°C$ for 5–16 hr. Millipore membrane filtration was employed to separate bound from free hormone (36). Separation of bound from free hormone can also be achieved by centrifugation. Nonspecific binding represented only 1–2% of total binding.

Soluble Receptors The incubation conditions for the binding assay of soluble receptors has been reported before (67). Bound and free hormone were separated by using polyethylene glycol (MW 6,000–7,500) which precipitates bound hormone. Nonspecific binding represented about 5% of total binding and could be further reduced by resolubilization and reprecipitation of the hormone receptor complex.

Incubation Damage Experimental evidences indicated that labeled gonadotropin was inactivated during incubation. The degree of inactivation was influenced by time, temperature, the concentration of hormone, and the receptor preparation (36). The most sensitive method available for the assessment of inactivation is based on the ability of unbound hormone to bind to fresh receptor. For the control, labeled gonadotropin is incubated under the same condition in the absence of receptor preparation (36, 69).

Damage of the hCG-LH receptor during incubation has also been demonstrated (37, 58). The degree of receptor degradation is measured by preincubation of the receptor preparation in the assay buffer. At the end of the preincubation period, ^{125}-hCG is added to the incubation tube for the second incubation, and binding activity is determined. For the control, a receptor preparation preincubated in ice is incubated under the same conditions as the experimental samples.

Specificity

Binding of gonadotropin to its target tissues (ovary and testis) is highly specific for both hormone and tissue. Figure 1 illustrates that unlabeled hCG, hLH, and

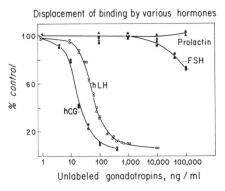

Figure 1. Hormonal specificity of ^{125}I-hCG binding to a particulate fraction of PMSG-hCG-primed rat ovaries. Control tubes contained no unlabeled hormone and had an absolute specific binding of 30% (34).

ovine LH readily compete with ^{125}I-hLH for binding to ovarian receptors and that FSH and prolactin are without effect (34). The results also indicate that hCG and LH share the same receptor. The good correspondence of LH potency estimates of several preparations of pituitary gonadotropin by the radioreceptor assay and the ovarian ascorbic acid depletion bioassay is further evidence of specificity (33).

Target tissue specificity of gonadotropin binding has also been demonstrated. The in vivo administration of radioactive labeled hCG resulted in an active concentration of hormone by the ovaries of the immature mouse (70) and pseudopregnant rat (71). A relatively low uptake of labeled hormone by liver and kidney was noted and may reflect hormone metabolism, excretion, or as yet undefined actions. In an in vitro system, labeled hLH bound significantly only to rat ovary and not to muscle, liver, spleen, or kidney (33).

The cellular and subcellular localization of gonadotropin binding has been discussed above. LH receptors are primarily located on the plasma membranes of luteal and thecal cells. Because it has recently been demonstrated (72) that porcine and rat granulosa cells develop microvillous processes with follicular maturation, it can be asked whether the distribution of receptor sites on the plasma membrane is random or organized. The possibility of intracellular translocation of hormone has been discussed above.

Kinetics and Equilibrium Studies

The binding of ^{125}I-hLH to ovarian receptors is dependent on pH, temperature, and salt concentration (34, 46). Binding of LH in vitro to particulate fractions of ovarian homogenates reached a plateau in 30 min at $37°C$, approximately 4 hr at $25°C$, and 48 hr at $4°C$ with shaking. Maximal binding activity occurred at pH 7.5. A nonlinear relationship exists between the amount of homogenate used and the amount of labeled hLH bound. The reduced binding at the higher concentration of ovarian tissue is probably due to the inactivation of both labeled hLH and receptors. Several nucleotides, including ATP, AMP, cyclic adenosine $3':5'$-monophosphate (cAMP), CTP, and cyclic guanosine $3':5'$-monophosphate (cGMP) at concentrations of $1–2$ mM significantly inhibited total binding of labeled hCG to bovine corpus luteum membrane. CTP and cyclic GMP were most effective (40).

The quantitative aspect of gonadotropin-receptor interaction has been investigated (36). In equilibrium experiments, thrice washed fractions (2,000 $\times$ g) of luteinized rat ovaries were incubated with increasing amounts of labeled gonadotropin until equilibrium was achieved. A saturable binding process was observed (Figure 2). The specific binding data may be analyzed by a Scatchard plot (73) as illustrated in Figure 3. The slope of the curve gives the reciprocal of the equilibrium dissociation constant (K_d) and the intercept at the abscissa yields the number of binding sites. The relationship used is $[Bound]/[Free] = 1/K_d$ $([Binding sites] - [Bound])$. The K_d value for hLH binding was calculated to be 6.7×10^{-11} M. When correction for the inactivation of free hLH was made, the

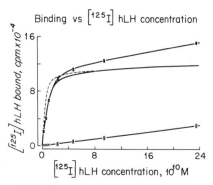

Figure 2. Saturability of ^{125}I-hLH binding to a fixed, limited quantity of particulate fraction from PMSG-hCG-primed rat ovaries. ●———● indicates total binding; ×———× indicates nonspecific binding; ——— indicates specific binding; – – –, specific binding corrected for inhibition of binding of free hormone during incubation (36).

value of K_d decreased by a factor of approximately 2. However, correction for the inactivation did not alter the number of binding sites significantly (36) (Figure 3). Degradation of receptors has also been taken into account for analysis of equilibrium binding data in the rat testis (58). Furthermore, the concentration of biologically active hormone must also be considered as discussed above.

A Hill plot for gonadotropin binding can be made from specific binding data (Figure 4). Values of 0.9 and 1.1 were obtained for the Hill coefficient of hLH and hCG binding, respectively. The analyses from Scatchard and Hill plots indicate the absence of cooperativity for hCG (LH)-receptor interaction (36).

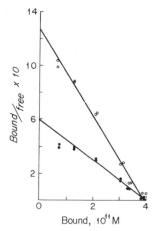

Figure 3. Scatchard plot of the binding data illustrated in Figure 2. ●———●, uncorrected specific binding; ○———○, specific binding corrected for inhibition of binding of free hormone during incubation (36).

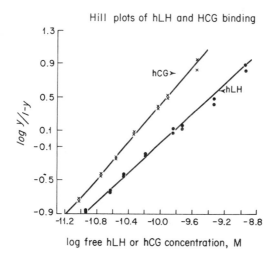

Figure 4. Hill plots of hLH and hCG binding to particulate fractions of PMSG-hCG-primed rat ovaries. The Hill coefficients of approximately 1 (see text) give no indication of positive or negative cooperativity among binding sites (36).

Table 1 lists the dissociation constants of gonadotropin binding to target tissues reported from different laboratories. The values of dissociation constants for hCG, hLH, and hFSH binding are in the range of $10^{-10}-10^{-11}$ M regardless of the method of receptor preparation and the source of target tissues. These values are in accordance with the general concept of high affinity of gonadotropin binding to target tissue and are applicable to the concentrations of hormone that exist in serum. The variation in the reported values could be explained on the basis of the assay conditions, the degree of iodination and incubation damage, and the receptor and gonadotropin preparations.

Kinetic studies of gonadotropin-receptor interaction have been performed and the association and dissociation rate constants for hLH and hCG have been reported. The association rate constant was derived from the assumption that gonadotropin-receptor interaction behaves as a second order reaction. The equation is $k_{+1} = 2.303/t(H - R)$ [log $R(H - X)/H(R - X)$], where k_{+1} is the association rate constant at a particular temperature, H is the hormone concentration, R is the number of binding sites of receptor as determined from the equilibrium study, and X is the concentration of hormone bound at time t. The association rate constant is independent of both hormone and receptor concentrations, because varying hCG and receptor concentrations, separately, yielded similar values (36). The association rate constant for hCG at 25°C was estimated to be 5.7×10^6 M^{-1} s^{-1}. The value for hLH under the same conditions was 3.2×10^6 M^{-1} s^{-1}.

The dissociation of labeled hormone-receptor complex as a function of time was determined by addition of a large excess of nonlabeled gonadotropin to

Table 1. Dissociation constants for hCG-LH and FSH binding to ovarian and testicular receptors

Tissue	Hormone	Dissociation constant	Temperature (°C)	Reference
Rat luteinized ovary	hCG	5.0×10^{-10} M	4	Danzo (74)
		6.2×10^{-10} M	37	
Rat luteinized ovary	hCG	3.8×10^{-11} M	25	Lee and Ryan (36)
	hLH	4.9×10^{-11} M	25	
Bovine corpus luteum	oLH	3.0×10^{-9} M	23	Gospodarowicz (38)
Bovine corpus luteum	hCG	1.5×10^{-10} M	37	Haour and Saxena (75)
	hLH	1.4×10^{-10} M	37	
Bovine corpus luteum	hCG	1.2×10^{-10} M	38	Rao (40)
Rat testis	hCG	2.5×10^{-11} M	24	Ketelslegers et al. (58)
Porcine granulosa cell	hCG	1.7×10^{-10} M	37	Kammerman and Ross (61)
	hLH	1.9×10^{-10} M	37	
Porcine granulosa cell	hCG	2.4×10^{-10} M	37	Lee (43)
Rat testis	hFSH	6.7×10^{-10} M	37	Bhalla and Reichert (49)
Bovine testis	hFSH	9.8×10^{-11} M	25	Cheng (51)
Rat granulosa cell	hFSH	5.0×10^{-10} M	37	Louvet and Vaitukaitis (76)

incubates containing preformed complexes. Dissociation was biphasic. The dissociation rate constant can be obtained from the equation derived from the first order reaction: $k_{-1} = 0.693/t_{1/2}$, where the half-life $(t_{1/2})$ is the time necessary for half the hormone-receptor complex initially present to dissociate. The initial dissociation rate constants were 2.4×10^{-4} s^{-1} and 1.4×10^{-4} s^{-1}, respectively, for hCG and hLH (36). If the gonadotropin-receptor complex was incubated over longer periods of time, a second (k_{-2}) slower dissociation rate constant was observed. The equilibrium dissociation (K_d) constant can be calculated from the ratio of the dissociation rate constant (k_{-1}, k_{-2}) and the association rate constant (k_{+1}). The values of K_d calculated from the ratio $k_{-1}:k_{+1}$ were 4.4×10^{-11} M and 4.2×10^{-11} M, respectively, for hLH and hCG binding, which compare favorably with the constants $(3.8 \times 10^{-11}$ M for hCG, 4.9×10^{-11} M for hLH) obtained from equilibrium data (36). However, K_d calculated from $k_{-2}:k_{+1}$ was in the range of 10^{-12} M. This discrepancy with equilibrium data suggests the possibility that the hormone-receptor interaction involves a number of intermediate reactions. The constant derived from equilibrium studies by definition reflects all of the possible intermediate steps. That derived from the ratio of the rate constants may differ because, under the conditions used experimentally, the rate constants may be biased by certain rate-limiting intermediate reactions. It should also be pointed out that the dissociate rate constants, at least, are temperature-dependent (36). Systematic study of the effects of temperature on the rate constants has not been reported.

Assessment of Number of Binding Sites

The Scatchard method of analysis is usually used. The intercept on the X axis (concentration of bound hormone) of the Scatchard plot indicates the number of binding sites. It is difficult to compare the estimated number of binding sites reported in the literature, because they generally are based on protein content of membrane preparations which vary in degree of purification. Some of the discrepancies also may result from the different receptor and hormone preparations and various assay conditions which reflect different degrees of incubation damage of receptor and hormone. The receptor sites occupied by endogenous hormone also must be considered when the concentration of circulating hormone is high. The number of binding sites also depend on the physiological condition. It has been shown that the number of hLH binding sites in the immature rat ovary increase from 2.8×10^{-15} mol/mg of weight to 18×10^{-15} mol/mg of weight following PMSG-hCG priming (36). This induction phenomenon will be discussed below.

Relationship of Receptor Binding to Biological Effects

It is generally believed that the primary step of gonadotropin action is binding to specific receptors in target tissues and subsequent activation of adenylyl cyclase and stimulation of steroidogenesis. In Table 2, phenomenological relationships

Table 2. Phenomenological relationships between hCG binding and hormonal responses

| Tissue | hCG binding (sites/cell or/mg of weight) | Hormonal response | | References |
		hCG stimulable adenylate cyclase (fold increase)	Progesterone (concentration)	
Porcine granulos cell				
Small follicle	300–400	2	1.2–1.8 ng/follicle	Lee (43)
Medium follicle	1,000–1,600	4	3.8–5.8 ng/follicle	Chang et al. (78)
Large follicle	8,000–10,000	11	25–42 ng/follicle	
Rat ovaries				
Immature	Low	Low	Low	Lee (unpublished)
10 days after PMSG-hCG	High	High	High	Lee et al. (65)
18 days after PMSG-hCG	Low	Low	Low	Lee and Ryan (79)
Estrogen-treated (18 days after PMSG-hCG)	High	High	High	Lee and Ryan (79)

between hCG binding and hormone responses under a variety of physiological conditions are listed.

It has been reported that porcine granulosa cells harvested from large preovulatory follicles have a greater ability to bind labeled hCG than cells from smaller follicles (43, 74–77). Increase in hCG binding of cells from large follicles is also consistent with increased cAMP and progesterone formation by LH stimulation in these cells compared with cells from medium and small follicles (43, 66). The increased binding ability of large follicle cells was found to be due to a larger number of binding sites and not to an increased binding affinity (43, 61). The increased number of hCG binding sites per cell is roughly proportional to the degree of adenylyl cyclase activation by hCG (43) and the progesterone content per follicle (Table 2) (78).

The dependence of labeled hCG binding to the rat ovary on the functional state of the ovary also has been reported (65). Increased binding activity by PMSG-ovine LH (oLH) priming was shown to correlate with an increase in ovarian progesterone content. Binding activity and progesterone levels were maintained at maximal for 9 days and subsequently underwent a sharp decline. The decrease in binding activity was also closely associated with luteolysis as evidenced by diminished ovarian progesterone. The change of hCG-mediated adenylyl cyclase activity also followed a pattern similar to that of binding activity. If corpus luteal function of the pseudopregnant rat ovary was maintained over a longer period of time by estrogen administration, as evidenced by a high ovarian progesterone concentration, the high levels of hCG binding and adenylyl cyclase activity also were maintained (79).

If the sequence of events is hormone binding, adenylyl cyclase activation, and steroidogenesis (or other biological effect), then this same order should be followed in studies of the time frame of these reactions. Unfortunately, no meaningful data are available for time correlations of the gonadotropin-gonadal interaction. Time correlations are available for other systems, as will be discussed subsequently, and are complicated by the role of nucleotides on the reactions. Dose-effect relationship also will be discussed separately.

Induction of Receptors and Adenylyl Cyclase in Ovarian Tissue

Induction of LH-hCG receptors and their sensitive adenylyl cyclase has been examined in some detail in the PMSG-oLH-primed immature rat (C. Y. Lee, unpublished observations). Ovine LH was chosen because of its relatively short half-life in vivo. The results were generally similar, in terms of binding data, to previous reports in which PMSG-hCG (41) and PMSG-oLH (65) priming were used. The number of binding sites of unprimed 27-day-old rats was 2×10^{-15} moles of hCG/mg of weight. The number of binding sites decreased slightly 4 days after PMSG (40 hr after oLH), possibly due to receptor occupancy by the exogenous gonadotropin. On the 5th day following PMSG, the number of receptor sites increased progressively until a maximum (26×10^{-15} mole of hCG/mg of weight) was reached on the 8th day. During this time, ovarian weight increased approximately 10-fold; thus, there was a 120- to 130-fold increase in the total receptor sites per ovary. During this induction of receptors, there was no significant change in dissociation constants (mean of 2.3×10^{-11} M with a range $1.0-3.2 \times 10^{-11}$ M).

Maximally stimulable hCG-sensitive adenylyl cyclase activity correlated temporarily with the number of hCG receptors during this induction process in the PMSG-oLH–primed rat (C. Y. Lee, unpublished observations). Thus, hCG-stimulable cyclase 4 days after PMSG was only 1.6-fold relative to basal, but thereafter increased progressively to reach a maximum (5-fold) 8 days after PMSG.

The porcine granulosa cell system provides a more physiological setting for examining receptor induction. Data presented in Table 2 and elsewhere (43, 77) indicate a progressive increase in the number of LH-hCG binding sites per granulosa cell and a parallel increase in the maximal LH-hCG-stimulable adenylyl cyclase during follicular maturation. When the cyclase data are calculated as molecules of cAMP produced/s/cell, then the production rate for cells from small follicles (240 molecules/s/cell) increases 29-fold with maturation to a large follicle (7,000 molecules/s/cell). This corresponds to a 26-fold increase in binding sites.

Maximal FSH-stimulable adenylyl cyclase in the porcine follicle follows a different course (C. Y. Lee, unpublished observations; M. Hunzicker-Dunn and L. Birnbaumer, unpublished observations). Greatest stimulation (approximately 11-fold) is seen in small (1–2 mm) follicles and decreases to approximately 1-fold in large (6–12 mm) follicles. This raises several interesting questions. Are there two populations of granulosa cells, one containing FSH receptors that decrease in number with follicular maturation and a second with LH-hCG receptors that increases in number with maturation? Or, is there one population of cells that lose one receptor while gaining the other? Furthermore, there is indication (78) that the concentration and content of cAMP in pig follicular fluid progressively increases with follicular maturation in parallel with the increase in number of LH binding sites and stimulable cyclase. Why isn't cAMP high in the fluid of the small follicle when FSH-stimulable cyclase is high? Does this imply that FSH and LH differentially affect cAMP secretion?

Although the phenomenon of receptor (and cyclase) induction appears established, little is known concerning the mechanisms involved or the processes of physiological control. It is not known whether the increased number of binding sites represents newly synthesized molecules, unmasking or activation of pre-existing molecules, or slower turnover of previously existing sites. Recently, data have appeared indicating that FSH is responsible for induction of the LH receptor and estrogen for the induction of the FSH receptor (80).

Physicochemical Properties of Gonadotropin Receptor

Treatment of crude homogenates or particulate fractions from ovaries or testes with proteolytic enzymes (trypsin, chymotrypsin, pronase) and phospholipases (C and D) reduce LH-hCG binding activity (35). DNase was without effect and

RNase had only a small effect, possibly due to contamination with proteases (35). Neuraminidase treatment enhanced binding to a small but reproducible degree (35). It is tempting to infer from these studies that the receptor is a lipoprotein, perhaps partially masked in the membrane by a glycoprotein. Caution must be exercised, however, because these effects on binding could be mediated by a general change in membrane structure without specific alteration of the receptor.

Chemical modification of membranes has suggested that tyrosine residues (40) and SH groups (81) are involved in the LH-hCG binding site. These data are subject to the same criticism as the enzymatic treatment of membranes. Furthermore, LH and hCG contain no free SH groups (1), and the hormone-receptor complex is dissociable (albeit slowly), making it unlikely that disulfide bridging is involved in receptor binding. A second study of SH group involvement was negative (40).

Efforts to solubilize gonadal LH receptors with common salt solutions or ionic detergents have been unsuccessful or only minimally successful (35). Solubilization has been achieved with a variety of nonionic detergents (Triton X-100, Luberol PX, and Emulphogen) (35, 46, 67, 68, 75, 82) with HLB (Hydrophilic, Lipophilic Balance) numbers in the range of 13−14. The soluble receptors have retained their hormonal specificity (67) with minimal (67) or no loss of affinity (68).

Physical characterizations of solubilized rat testicular and ovarian receptors have been made by Dufau et al. (82), using techniques that are not dependent upon chemical purity. These data indicate similarity of the hCG-receptor complex from both tissues with respect to sedimentation coefficient (7.5 S), Stokes radius (60−64 Å), and molecular weight (228,000).

Further purification of Triton-solubilized receptors from rat ovaries with the use of hCG-Sepharose 4B affinity chromatography has been reported (67). This procedure yielded a protein that was homogeneous by 1) disc gel electrophoresis under several circumstances and 2) production of a single population of antibody following immunization of rabbits and goats. The antisera were capable of minimal, but significant, reduction of ovarian weight and hCG binding to ovarian homogenate, when administered in vivo to PMSG-hCG-primed rats (R. J. Ryan, unpublished observations). When used in vitro, the antisera completely inhibited hCG binding to ovarian particulate fractions (R. J. Ryan, unpublished observations). However, the yield of this protein was 100-fold greater than anticipated from binding studies on the particulate starting material. Furthermore, the antisera reacted, with identity, to a protein contained in Triton extracts of nontarget tissues such as liver, spleen, and kidney (67). If this protein is the LH-hCG receptor, which certainly is not proven, then it requires a reconsideration of notions concerned with target organ specificity and what controls the activity of masked or functionally inactive receptor molecules.

A 30% ethanol extract of testicular homogenates has produced a heat-stable factor which prevents the precipitation of ^{125}I-hLH and ^{125}I-hFSH by Carbowax 400 (50). This material alters the circular dichroic spectra of LH and FSH.

It was also found in extracts of rat liver and kidney. The physiological significance of these observations is unclear.

Very little has been done concerning the organization of the gonadal plasma membranes and the relationship of one constituent to another. Recent data by T. A. Bramley and R. J. Ryan (unpublished observations) suggest some intriguing aspects. Ovarian homogenates from PMSG-hCG-primed rats were subjected to sucrose gradient centrifugation, and fractions were analyzed for a variety of activities. Two peaks of hCG binding were found, one with a density of 1.14 and the other with a density of 1.18. The first peak was associated with alkaline phosphatase activity and the second with 5'-nucleotidase activity and Mg-dependent ATPase. Na,K-ATPase activity was present in both peaks, but they were differently affected by Ca^{2+} chelating agents. hCG-stimulable adenylate cyclase was found to be associated only with the second peak. These data certainly suggest that the membrane has an organization, if not within a single cell type, at least with respect to different cells in the same organ.

Structural Requirements of Gonadotropin Molecules For Binding to Their Receptors

This subject is, as yet, poorly understood and has been reviewed elsewhere (1, 2). Therefore, only a few aspects will be considered here.

Data are available to indicate that the isolated subunits of hLH, oLH, and hCG do not bind to receptors nor do they compete with the native molecule for binding (35, 83). They do not stimulate steroidogenesis in vitro (83) or exhibit biological activity in the standard in vivo assays.

Why is the juxtaposition of the two subunits required and what is its physiological meaning? Only speculation can be offered, based on a few hints from available data. First, there is evidence from spectroscopic and immunological data that the conformations of the subunits in the native molecule are different from the conformations of the isolated subunits. This suggests that the shape of the native molecule must be so arranged that it fits the receptor site or sites. The binding site(s) on the LH molecule could involve juxtaposed regions on both the α and β subunits. It could involve a region or regions on α or β, or both, that only have the proper shape because of association with the opposite subunit. Data are not available to decide among these possibilities.

There are data to suggest that more than one site on the LH molecule is involved in binding. The biphasic dissociation of LH-hCG from its receptor (36) is compatible with this notion, particularly because the more rapid dissociation phase is reduced by a more prolonged preincubation of hormone with receptor (36). Furthermore, chemical modification data suggest that residues on both subunits are involved in receptor binding, but whether they are part of the same or different binding sites cannot be decided. Lastly, the data of Bishop and Ryan (84) indicate a considerable degree of relative mobility between the subunits of hLH when they are associated in the native molecule and recent preliminary data

(W. H. Bishop, unpublished observations) suggest a change in this relative mobility following receptor binding.

Thus, it is possible that gonadotropins bind to receptors at more than one site. The initial site may be concerned with receptor occupancy and the additional site(s) concerned with activation of adenylyl cyclase. Whether binding occurs at one or more sites, there is probably a considerable energy exchange between the hormone and plasma membrane with the binding event. This energy may contribute to the activation of the cyclase either directly or indirectly.

GONADOTROPINS AND ACTIVATION OF ADENYLYL CYCLASE

It is now recognized that many, if not all, of the effects of gonadotropins are mediated by cAMP. Initial studies leading to this conclusion were carried out by Marsh and Savard, who established the steroidogenic activity of cAMP in ovarian tissues (85–87) and by Sandler and Hall (88), who mimicked in vitro steroidogenic actions of LH in rat testis with cAMP. Since then, LH and FSH have been shown to increase cAMP levels in all ovarian tissues on which they are known to act, i.e., Graafian follicles (89), granulosa cells (90), and corpora lutea (91), as well as Leydig cells and Sertoli cells (92, 93). It is also accepted that these increases in cAMP levels are receptor-mediated events, exhibiting hormonal specificity, and that they are the result of adenylyl cyclase activation by the occupied receptors. It was demonstrated unequivocally in tissue homogenates that LH-dependent increases of cAMP accumulation are not due to inhibition of cyclic nucleotide phosphodiesterase (94, 95). Thus, understanding of the molecular events involved in the initiation of gonadotropin action on their target tissues will require not only an understanding of binding events leading to receptor occupation, as discussed so far in this chapter, but also the enzyme activation event that results from receptor occupation.

In the following sections, the following topics are reviewed: 1) some of the basic properties of adenylyl cyclase systems as discovered and studied in systems other than those sensitive to gonadotropin; 2) the properties of gonadotropin-sensitive adenylyl cyclase systems as seen in homogenates and partially purified membrane particles from various sources, including some of the methodological details; 3) the current knowledge of variations of hormonal responsiveness of ovarian adenylyl cyclases, correlating them with physiological states of the animals; and 4) present knowledge of gonadotropin-induced desensitization of adenylyl cyclase, a less well known, but nonetheless interesting, phenomenon that may be related to differentiation of follicular structures into corpus luteum and to gonadotropin-induced luteolysis.

It is not our primary purpose to discuss whether or not cAMP mediates actions of gonadotropin, and the chapter in this respect is incomplete. An excellent review discussing involvement of cyclic nucleotides in gonadal function has recently been written by Marsh (96), and the reader is referred to that work for more complete information.

General Properties of Adenylyl Cyclase

In all nucleated cells thus far studied, adenylyl cyclases are membrane-bound enzymes that are found in the plasma membrane (97, 98). A few exceptions have been noted (adrenal microsomes (99), for example), but it is not clear to what extent these exceptions are due to artifactual contamination with plasma membranes. Many, if not all, of the adenylyl cyclases in nucleated cells are stimulated by specific hormones and all but two (see below) are stimulated by fluoride ion.

Divalent cations play a crucial, yet not well understood role, in regulation of adenylyl cyclase activity. Sutherland and co-workers (100, 101), discoverers of adenylyl cyclase, were first to point out an absolute requirement for Mg^{2+}. Subsequently, in fat (102) and in cardiac tissue (103), it was established that the true substrate for the enzyme is MgATP and that free ATP is inhibitory to the reaction. Mg adenyl-5'-yl imidodiphosphate (Mg AMP-PNP), a synthetic analogue of ATP that is not hydrolyzed by nucleoside triphosphate phosphohydrolases, also serves as a substrate for the reaction (104).

At concentrations ranging from 0.1–5.0 mM, Ca^{2+} invariably inhibits all expressions of adenylyl cyclase activity; Cu^{2+} and Zn^{2+} act similarly (105, 106). In a few instances (107, 108), submicromolar concentrations of Ca^{2+} have been found necessary for adenylyl cyclase activity. CO^{2+}, on the other hand, partially replaces Mg^{2+} in supporting basal, fluoride-stimulated, and hormone-stimulated activities in fat cell "ghosts" (105).

Although adenylyl cyclases from mammalian tissues all respond well to hormones in the presence of Mg^{2+}, depending on the system studied, Mn^{2+} can replace Mg^{2+} with a variable degree of efficacy. In general, it has been found that basal and fluoride-stimulated activities tend to be either equal or higher in the presence of Mn^{2+} (107). Two adenylyl cyclase systems have been described recently which depend on Mn^{2+} for activity. One is derived from mature spermatozoids (109) and the other from a slime mutant of *Neurospora crassa* (110).

Activation of adenylyl cyclase by hormones is tissue-specific and dependent on the receptor coupled to them. This activation process in most systems appears to be rapidly reversible, and in consequence, the state of activity of any given adenylyl cyclase system is under the moment-to-moment control of the circulating hormone concentration. Evidence to this effect was obtained by Birnbaumer et al. (111), working with the glucagon-stimulated adenylyl cyclase system of liver plasma membranes, in which the effect of glucagon could be abolished or diminished by washing the membranes, by dilution of the incubation medium with hormone-free media, and by addition of a competitive inhibitor (des-His[1]-glucagon). Similar reversibility was demonstrated in the heart membrane adenylyl cyclase system stimulated by catecholamines and other competitive β blockers (112).

Adenylyl cyclases may be stimulated by more than one hormone receptor in a single tissue, as shown in rat adipose tissue. Thus, in ghosts of free fat cells it was shown, using kinetic arguments and selective competitive inhibitors to the action of catecholamines, adrenocorticotropic hormone (ACTH) and glucagon, that each of these hormones interacts with distinct receptor sites, and that all of them stimulate a single adenylyl cyclase system (107, 113). As will be shown below, corpus luteum adenylyl cyclase may be another example of an adenylyl cyclase system activated by more than one hormone.

Recently, it has become clear that the hormonal activation of adenylyl cyclase systems is under regulation by nucleotides and nucleosides. Initially (104), it was noted that under stringent assay conditions (using as substrate either low ATP or AMP-P(NH)P) the liver plasma membrane adenylyl cyclase exhibited an almost total requirement for GTP (or ATP) to show activation by glucagon. Similar findings were soon reported for other adenylyl cyclase systems activated by other hormones (for review see Birnbaumer and Yang (114)) and led to the early conclusion that a nucleotide-dependent step is involved in hormonal activation in an obligatory manner (115).

The type of modulation affected by nucleotides varies with the system studied. Both positive and negative effects on hormonal stimulation have been observed with either nucleoside triphosphates or nucleoside diphosphates. In several systems, nucleoside tri- and diphosphates can be shown to have opposite effects. Thus, it was found that the responsiveness of a beef renal adenylyl cyclase to vasopressin is enhanced by GDP and inhibited by GTP, whereas that of a cat heart adenylyl cyclase to catecholamines is enhanced by GTP and inhibited by GDP (114). Opposite effects of this type appear to be quite common (116). However, not all adenylyl cyclase systems are modulated bidirectionally (114), and not all are regulated by guanyl nucleotides (114, 117–119). A summary of some of the effects of nucleotides on hormonal stimulation is presented in Table 3 (120–123).

Although GTP affects relative stimulation by hormone in an apparently opposite way in tissues such as rat liver and beef renal adenylyl cyclase, the mechanism by which it acts on these and other adenylyl cyclases may be very similar. Thus, in both the liver glucagon-sensitive adenylyl cyclase (116, 124) and the renal vasopressin-sensitive adenylyl cyclase (119), GTP affects basal activity (increase), rate of stimulation by submaximally stimulating concentrations of hormone (increase), final degree of stimulation achieved (increase in the liver system and decrease in the kidney system), and, finally, transient kinetics. In both systems, many if not all of the effects of GTP are mimicked by guanylyl-5'-yl imidodiphosphate (GMP-P(NH)P), which has recently been extensively used in testing for basic kinetic properties of adenylyl cyclases by several research groups (125–127). The effect of GMP-P(NH)P differs, however, from that of GTP, mainly in that it is irreversible, i.e., it cannot be reversed by a washing procedure.

Based on the combined effects of hormones and nucleotides on transient and steady state kinetics of liver, fat, and adrenal adenylyl cyclases, Rodbell and collaborators (116, 128, 129) have recently presented a three state model, in

Table 3. Examples of positive and negative modulation by nucleotides of hormone stimulation (relative to basal) of adenylyl cyclase systems

---

Enhancement
  GTP: Glucagon action in rat liver (104), hamster β cell (121), and rat
        epididymal fat (120).
        Catecholamine action in rat liver (123), cat myocardium (114), and
        rat epididymal fat (120).
        ACTH action in rat adrenal cortex and rat epididymal fat (120).
        Prostaglandin E action in renal medulla (114), human platelets (122),
        and beef thyroid (117).
  ATP: LH action in rabbit corpus luteum (114) and pig Graafian follicles
        (118).
        Neurohypophyseal hormone action in beef renal medulla (119).

  ITP: Thyroid-stimulating hormone action in beef thyroid (117) and horse
        thyroid.
  GDP: Prostaglandin E action in rabbit corpus luteum of
        pseudopregnancy (114).
        Neurohypophyseal hormone action in beef renal medulla (114).
Inhibition
  GDP: Catecholamine action in cat myocardium (114) and rat
        epididymal fat (120).
        ACTH action in rat adrenal cortex.
        Glucagon action in rat liver (116).
  ADP: LH action in rabbit corpus luteum (114).
  GTP: Neurohypophyseal hormone action in beef renal medulla (119).

---

which they propose a mechanism by which nucleotides affect adenylyl cyclase and ways by which hormones exert their stimulatory action. The three state model as described by Rodbell's group is summarized in Table 4. Basic features are 1) the existence of three discrete states of activity (E, E$'$, and E$''$), 2) the central regulatory role of guanyl nucleotides without which no hormonal stimulation can occur, and 3) the absence of an allosteric activating site for magnesium ion. Basal activity depends on the abundance of form E, which has both low V_{max} and high affinity for the inhibitory protonated form of ATP (ATPH^{3-}). E$'$ and E$''$ are two nucleotide-affected states whose existence was suggested by the finding that in liver and fat stimulation of adenylyl cyclase by GTP and GMP-P(NH)P is secondary to the rapid formation of a state exhibiting an activity that is equal to or lower than basal activity. This results in time courses of cAMP accumulation with lags and transient inhibitions. Thus, E$'$ is a state of activity whose existence is transient and precedes the formation of E$''$. Curve fitting experiments of data collected at both varying pH and varying Mg concentration led to the conclusion that in comparison to E, E$'$ has a high V_{max} and a still higher affinity for ATPH^{3-} (hence a resulting low activity). E$''$ has both high V_{max} and low affinity for the inhibitor, thus accounting for increased catalytic activity in the presence of guanyl nucleotide. Because, in the absence of

Table 4. Three activity state model for regulation of adenylyl cyclase as proposed by Rendell et al. (129)

E, E', E'', activity states of adenylate cyclase; N, a guanyl nucleotide.

1. In the presence of GMP-P(NH)P:

$$E + N \rightleftharpoons E'_N \rightleftharpoons E''_N \rightleftharpoons E''_N$$

Rapid   Slow (lag)
↑   ↑

  a) Very slowly reversible   a) Accelerated by hormone
                                    b) Sometimes increased by hormone toward
                                         E''_N so that

$$(E''_N/E'_N) < (E''_N/E'_N)$$

                                    without   with
                                    hormone   hormone

2. In the presence of GTP:

$$E + N \rightleftharpoons E'_N \rightleftharpoons E''_N$$

Rapid   Rapid (no lag or lag very short)
↑   ↑

         Rapidly reversible   Increased by hormone toward E''_N

3. Inhibition by protonated free substrate at catalytic site: E = low; E'_N = high; and E''_N = low.
4. V_{max} of states: E = low; E'_N = high; and E''_N = high.
5. Substrate: Mg·ATP.
6. Allosteric site for Mg^{2+}, whose occupation leads to activation of enzyme: no.
7. Allosteric site for Mg^{2+}, whose occupation may lead to inhibition of enzyme: yes.
8. Allosteric site for nucleotide (GTP or GMP-P(NH)P): specific for *free* nucleotide.
9. Activation mechanism: State transition (induced by nucleotide binding) from an initial state having medium affinity for inhibitory protonated substrate and low V_{max} to a final state having low affinity for the inhibitory protonated substrate and high V_{max}; with the existence of a transient intermediate state having high affinity for protonated free substrate and high V_{max}.

---

hormone, lags are short or nonexistent when GTP is used, but are very pronounced with GMP-P(NH)P, this is taken to indicate that the isomerization reaction is slower with GMP-P(NH)P thaw with GTP. Furthermore, since activities obtained with GMP-P(NH)P alone are higher than with GTP alone, it is assumed that the equilibrium between E' and E'' is shifted more toward E'' when the imidodiphosphate analogue is used than when GTP is used. Finally, since addition of hormone was found to result, in the presence of either GTP or its analogue, in increased activities and in obliteration of transient states of activity (lags or inhibitions), it was concluded that hormones both shift the equilibrium between E' and E'' toward E'' and accelerate the isomerization reaction. Thus, in the three state model, hormones do not affect the activity of any one of the states of activity of this enzyme system, but are assumed to modify only the equilibrium between nucleotide-induced states and the rate at which this equilibrium is established. By assuming that GMP-P(NH)P has a much higher affinity for the E state than GTP and also a much smaller specific rate constant of dissociation, the model accounts for the lack of experimentally observable reversibility of GMP-P(NH)P action of adenylyl cyclases.

This lack of reversibility of GMP-P(NH)P action has led Cuatrecasas et al. (127) to propose an alternate model for the mode of action of nucleotides and hormones on adenylyl cyclase (Table 5). They proposed activation by guanyl nucleotides to be the result of the formation of a covalent enzyme-PP or enzyme-P(NH)P complex, depending upon whether GTP or GMP-P(NH)P was used to activate the system. Based on kinetic data, they suggested that normal stimulation of the enzyme by hormones (in the presence of GTP) is the result of an increased rate of formation (and therefore concentration) of a highly active and highly unstable enzyme-PP complex. In the presence of GMP-P(NH)P, activation of the enzyme would both be slow and irreversible because of low rates of formation of active and highly stable enzyme-P(NH)P complex and because of the impossibility of this complex to decay to enzyme plus P(NH)P. In their model, hormonal stimulation of the enzyme in the presence of GMP-P(NH)P would be exclusively due to increased rates of formation of enzyme-P(NH)P complex and would be observed only at early times of incubation. Higher total stimulated activity in the presence of hormone, as opposed to that seen in the absence of hormone (116, 127), would be the result of GMP-P(NH)P protection of enzyme against heat inactivation when hormone is present.

The model of Cuatrecasas et al. (127) is attractive because, like the three state model of Rodbell and collaborators (128), it does account for experimental findings. Interestingly, both models assume three basic states: a basal state unaffected by nucleotides; a second state affected by nucleotides but not yet active (E' or enzyme-PP complex); and an active state affected by nucleotides. Interesting also is the fact that neither model invokes the formation of a complex between receptor and enzyme to account for hormonal stimulation; rather, it is assumed that the action of receptor is to modify the rate at which a final active state of the enzyme is formed. Both models have in common the

Table 5. Regulation of adenylyl cyclase as proposed by Cuatrescasas et al. (127)

concept that nucleotides play an intrinsic and obligatory role and that it is by regulating the action of the nucleotide that hormones stimulate adenylyl cyclase. While the model of Cuatrecasas et al. (127) accounts for irreversible kinetics using an irreversible step, that of Rodbell and co-workers (129) does so by using a slowly reversible, high affinity interaction between E and GMP-P(NH)P. Rodbell's model, with its demonstrated capability to fit many experimental data on a quantitative basis, offers a mechanistic explanation of the kinetic parameters affected upon nucleotide and hormonal stimulation, which the model of Cuatrecasas et al. does not.

Is there an allosteric site for Mg? Early experiments with the fat cell system (101), as well as experiments by Drummond and collaborators (130) and Perkins and Moore (131) with heart and brain adenylyl cyclases, had suggested, contrary to the assumptions made in the three state model, that adenylyl cyclases may have allosteric sites for Mg whose interaction with the divalent cation leads to enhancement of activity. Thus, in these systems it was postulated that activation by hormones occurred either by an increase of the affinity of the system for activating Mg at the allosteric site or by an increase in V_{max} of the enzyme upon interaction with Mg. More recently (132), the steady state kinetics of the Lubrol-solubilized brain enzyme provided evidence indicating that increases in activity dependent on $MgCl_2$ addition were due to interaction of Mg with an allosteric activating site and not to removal of competitive inhibitor. However, similar analyses testing one hypothesis against the other (133) have not been made for other systems. Thus, Rodbell and colleagues, who based their modeling studies on De Haën's (134) concept that Mg acts via removal of inhibitory free substrate, did not test for existence of direct stimulatory action of Mg ion by exploring the alternative possibility that their data might also fit a model in which the kinetic parameter involved in activation is affinity of an allosteric site for Mg with or without a concomitant V_{max} change.

Do nucleotides interact with more than one type of site? Using antagonism against GMP-P(NH)P action, Lefkowitz (126) established in myocardial membranes that both GTP, which enhances catecholamine action (114) and GDP, which inhibits expression of catecholamine stimulation (114), interact with this system in a manner that is competitive with GMP-P(NH)P. A similar finding was reported for the glucagon-sensitive liver system (116). These findings strongly suggest that all three nucleotides bind to a common regulatory site, regardless of whether their final effect is one of activation or inhibition or whether it is of a reversible or irreversible nature.

In addition to transient lag periods in their progress curves, adenylyl cyclase systems may also exhibit transient burst phases, i.e., time courses of cAMP accumulation in which initial activity is higher than steady state activity. This burst phenomenon has been observed for several adenylyl cyclase activities (120, 121), including fluoride-stimulated activity in rat testis (135) and LH-stimulated activity in membrane particles from rabbit corpus luteum (136) (see below). Several studies, including those with corpus luteum (CL) membranes, revealed that not only lags, but also bursts, are under the influence of guanyl nucleotides. Neither the three state model nor the models proposed by Cuatrecasas et al. or tested by Garbers and Johnson (132) take into consideration or predict the appearance of highly active transient states. Thus, while much has been learned in recent years about in vitro regulation of hormonal stimulation of adenylyl cyclase, the only model that may account for all of the, thus far, described properties is one in which nucleotides *and* occupied hormone receptor each interact with distinct sites of the catalytic unit of adenylyl cyclase and in which transient states of activity such as lags and bursts are the result of slow isomerization rates between the conformation of the system at the moment of ligand binding and establishment of a preferred (more stable) conformation. For a further discussion on possible modes of action of hormone on adenylyl cyclase, see below and Birnbaumer et al. (136).

Gonadotropin-sensitive Adenylyl Cyclase

Marsh et al. (91), after determining that incubation of bovine corpus luteum slices with LH resulted in elevation of cAMP levels, were first to demonstrate that this was due to stimulation of an adenylyl cyclase system by showing LH-stimulated accumulation of cAMP in homogenates incubated with ATP and Mg. Control incubations indicated that cAMP accumulation was not secondary to inhibition of cyclic nucleotide phosphodiesterase. Since then, stimulation of adenylyl cyclase by LH has been shown in a variety of ovarian and testicular tissues, both in intact and broken cell preparations derived from these tissues (Table 6).

Properties and Conditions of Assay Although stimulation of adenylyl cyclase by LH (and FSH) in cell-free systems is unequivocal (94, 95, 135), it has been a characteristic finding that gonadotropin stimulation of cAMP accumulation in intact cells is of a larger relative magnitude, especially if accumulation of labeled cAMP from ^{14}C- or ^{3}H-labeled adenine (89, 137–140) is used as the endpoint.

Reasons for the low relative stimulation, obtained in early studies on LH-activated adenylyl cyclase in cell-free systems derived from reproductive tissues, resided in the conditions of homogenization, composition of storage media, and details of the assay. To obtain highly responsive cell-free systems, it is necessary to carry out gentle homogenizations and to use sucrose-containing media (118). The effect of composition of the homogenization and storage media on absolute and relative LH-stimulated activities of membrane preparations derived from CL of pregnant rabbits is exemplified on Figure 5. Satisfactory homogenates were obtained using hand-operated glass-glass Dounce homogenizers (118), motor-driven Teflon-glass homogenizers at low speeds (36, 43), and Polytron-like homogenizers at low settings (114, 118).

Some basic properties of the LH-sensitive adenylyl cyclase of CL from pregnant rabbits, as reported recently (118) are as follows:

Table 6. Some examples of stimulatory effects of LH-hCG on ovarian and testicular adenylyl cyclase systems

Cell type	Mode of assay	Reference
Ovary		
Follicle	Intact	Marsh et al. (139)
	Intact	Marsh et al. (138)
	Intact	Tsafriri et al. (89)
	Intact	Lamprecht et al. (149)
	Cell-free	Birnbaumer et al. (118)
	Cell-free	Hunzicker-Dunn and Birnbaumer (143)
	Cell-free	Hunzicker-Dunn and Birnbaumer (144)
	Cell-free	Hunzicker-Dunn and Birnbaumer (177)
	Cell-free	Bockaert et al. (141)
Granulosa cells	Intact	Kolena and Channing (137)
	Intact	Kolena and Channing (90)
	Intact	Lee (43)
Corpus luteum	Intact	Marsh et al. (91)
	Intact	Lamprecht et al. (149)
	Cell-free	Marsh (94)
	Cell-free	Anderson et al. (165)
	Cell-free	Birnbaumer et al. (118)
	Cell-free	Hunzicker-Dunn and Birnbaumer (143)
	Cell-free	Hunzicker-Dunn and Birnbaumer (144)
	Cell-free	Hunzicker-Dunn and Birnbaumer (177)
	Cell-free	Bockaert et al. (141)
Interstitial tissue	Cell-free	Dorrington and Baggett (95)
	Cell-free	Hunzicker-Dunn and Birnbaumer (177)
Immature ovary	Intact	Lamprecht et al. (149)
	Cell-free	Hunzicker-Dunn and Birnbaumer (144)
Testis		
Leydig cells	Intact	Dorrington and Fritz (158)
Total testis	Cell-free	Murad et al. (135)
	Intact	Braun and Sepsenwol (140)

1. pH optimum of basal activity is between 8.0 and 8.4, that of LH-stimulated activity is about 7.5.

2. LH-stimulated activity, but not basal (control) activity, is inhibited by addition of various salts (NaCl, KCl, and Tris, bis-Tris-propane, and phosphate buffers). Fifty and 100% inhibition of LH-stimulated activity by NaCl is obtained with 60 and 200 mM NaCl, respectively. NaCl (100 mM, 80% inhibition) also results in a 10-fold increase in the concentration of LH required to stimulate the system by 50%. (These pH and salt effects are similar to those noted for LH binding. See above.)

3. Relative response to LH is low (1.5–2-fold) at 0.1 mM ATP and increases with increasing ATP. GTP increases catalytic efficacy of the system both in the presence and absence of LH, but has no effect on relative stimulation.

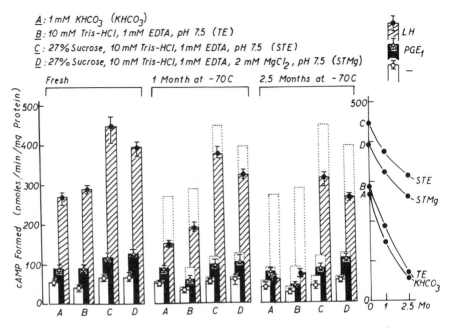

Figure 5. Homogenization conditions: effect on the stability of response of adenylyl cyclase system. Dissected corpora lutea, obtained from rabbits on the 8th day of pregnancy, were homogenized in medium A, B, C, or D. Membrane particles were prepared in the respective mediums and were assayed for adenylyl cyclase activity in the presence of high ATP (3.0 mM) at pH 7.0. Basal activity, *open bars;* 10 μg/ml of LH, *hatched bars;* and PGE$_1$, *solid bars. Dotted bars* indicate responses in fresh tissue. Tissue preparation and additional assay conditions are described elsewhere (118).

4. Optimal relative stimulation by LH is obtained with about a 1 mM excess of added Mg over Mg^{2+}-binding ingredients (routinely ATP and 1.0 mM EDTA).

5. Except for ionic strength, neither pH, nor ATP, GTP, or MgCl$_2$ affects sensitivity of rabbit CL adenylyl cyclase to LH, which is about 0.2 μg/ml of NIH-LH-B8.

Some of the above parameters were determined also for rat CL and pig and rabbit Graafian follicles (118):

1. Like rabbit CL, responsiveness of rat CL adenylyl cyclase to LH is low and labile in particles derived from homogenates prepared in hypotonic media, and is increased by ATP (but not GTP) showing as much as 9-fold stimulation by LH.

2. Like rabbit CL, pig Graafian follicles larger than 6 mm in diameter contain an adenylyl cyclase system that is highly responsive to LH (5-fold, or more, stimulation). LH responsiveness in follicles is enhanced by ATP (but not GTP) and requires for detection both gentle homogenization and addition of sucrose to the homogenization medium.

3. Unlike rabbit and rat CL, however, pig and rabbit Graafian follicles were

found to require only very low (micromolar) concentrations of ATP to show their responsiveness to LH.

4. Lower concentrations of LH are required for half-maximal stimulation of adenylyl cyclase in membrane particles from pig and rabbit follicles which were 0.020 and 0.008 µg/ml of NIH-LH-B8, respectively. From these studies, Birnbaumer et al. (118) concluded that optimal detection of absolute activities, as well as of relative stimulations by LH, requires that incubations be carried out at high (2–3 mM) levels of ATP, low (about 1.0 mM free) concentrations of Mg ion, low ionic strength, and pH values between 7.0 and 7.5.

Regulation by Nucleotides and Transient Kinetics ATP, but not GTP, enhances relative stimulation of ovarian adenylyl cyclase by LH, in spite of evidence indicating that GTP is active as well (Figure 6). GTP was found not to compete with ATP for its effect on LH responsiveness (114, 118), suggesting that the rabbit corpus luteum adenylyl cyclase may contain two distinct regulatory sites for purine nucleotides, one specific for GTP affecting the catalytic unit of the enzyme system, and the other specific for ATP affecting the process through which the gonadotropin receptor alters catalytic activity. ADP was found to exert effects opposite to those of ATP, possibly by interacting with the LH receptor, as suggested by the finding of C. Y. Lee and R. J. Ryan (unpublished observations) that ATP increases the rate of hormone exchange at the receptor site, in a manner that may be similar to that by which GTP enhances exchange of glucagon at its specific binding sites in liver plasma membranes (60).

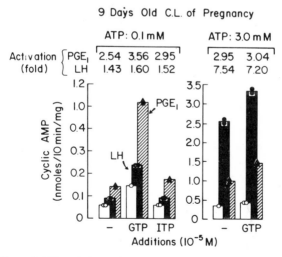

Figure 6. Effect of ATP and GTP on response of rabbit CL adenylyl cyclase to LH and PGE₁. Membrane particles were assayed for adenylyl cyclase activity in the absence (*open bars*) or presence of 10 µg/ml of LH (*solid bars*) and PGE₁ (*hatched bars*). Tissue preparation and additional assay conditions are described elsewhere (114).

Although GTP does not markedly affect relative stimulation by LH of adenylate cyclase either in the rat (L. Birnbaumer, unpublished observations) or rabbit CL or the pig Graafian follicle (141), it does have a profound stimulatory effect on the catalytic capacity of these systems. The pig system is especially interesting because, like the renal and other adenylyl cyclases, it too exhibits transient kinetics with burst and lag phases (at low 0.1 mM ATP) and is not desensitized (see below). Bursts and lags in time courses of cAMP accumulation by pig Graafian follicle adenylyl cyclase and the effect of GTP are illustrated in Figure 7. The finding that gonadotropin-sensitive adenylyl cyclase may exist in several discrete states of activity, thus yielding nonlinear progress curves, should be kept in mind in experiments designed to test for temporal correlation between LH-receptor occupancy and adenylyl cyclase activation. Thus, nonlinearities in the progress curves that are obtained at submaximally stimulating concentrations of LH (or hCG) cannot be directly attributed to formation of hormone-receptor complex, unless preliminary incubations for up to 10 min are carried out under basal adenylyl cyclase assay conditions, allowing for transitions of state to occur before hormone is added to the incubation medium. The mechanism by which GTP exerts its action on any adenylyl cyclase, including ovarian adenylyl cyclase, is not known. It may do so by interacting with the system directly at an allosteric site. However, findings with GMP-P(NH)P raise

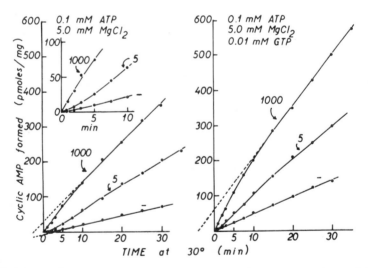

Figure 7. Transient kinetics of pig Graafian follicle adenylyl cyclase: effect of GTP. Adenylyl cyclase activity was determined for the indicated times in membranes prepared from 6–8 mm follicles at 0.1 mM ATP alone (——) or in the presence of either 5 or 1,000 ng/ml of LH and 0.01 mM GTP: – – –, extrapolation of rates of accumulation obtained after 10 min of incubation calculated by least squares regression analysis of the data. Inset, same as large graph, but with expanded scale. Tissue preparation and additional assay conditions are described elsewhere (194).

some doubts about this contention. Thus, although GMP-P(NH)P was found to activate pig Graafian follicle adenylyl cyclase, it does so at very slow rates, taking as much as 4 hr to activate fully (40-fold) (J. Bockaert and L. Birnbaumer, unpublished observations). The possibility has to be considered that guanyl nucleotides may affect adenylyl cyclases indirectly, their effect being secondary to other as yet undefined actions on the membrane in which the systems are embedded. This very slow rate of action of GMP-P(NH)P does not seem to be peculiar to the pig Graafian follicle adenylyl cyclase, because Neer (142), working with a Lubrol-solubilized adenylyl cyclase, reported similarly slow activation rates.

Hormone Specificity of Adenylyl Cyclase Activation in Ovarian Tissue

Response to Prostaglandins Adenylyl cyclase in mature gonadotropin-responsive tissues was stimulated not only by LH, but also by prostaglandins of the E series (PGE_1 and PGE_2) and, to a lesser degree, PGA_1, while prostaglandins of the F series were ineffective (114, 143, 144). Relative activities achieved in homogenates and membrane particles with prostaglandins are discrete (2- to 2.5-fold) when assayed under conditions yielding optimal response to LH. This is in part due to the fact that prostaglandin response of ovarian adenylyl cyclase, especially that of CL, is under positive modulation by GDP (114), a nucleotide whose concentration is minimized when high concentrations of ATP are used as substrate.

Responses to prostaglandins in cell-free systems of gonadal tissues (114, 118, 143, 144) had been predicted by previous studies exploring their effect in intact cells on cAMP levels (90, 145), as well as cAMP turnover using the ATP-prelabeling technique in which flow of radioactive label from ATP into cAMP was determined (91, 145). Using the prelabeling technique, Kuehl et al. (146) observed that LH-stimulated accumulation of label in cAMP is competitively inhibited by a prostaglandin antagonistic (7-oxa-13-prostanoic acid). They suggested that prostaglandins may play a mediating role between occupation of gonadotropin receptor and stimulation of adenylyl cyclase. However, similar results could not be obtained in a cell-free system (L. Birnbaumer and P.-C. Yang, unpublished observations) or confirmed in two other intact cell systems (90, 145). Possibly, the accumulation of labeled cAMP noted by Kuehl et al. (146) was due to a combination of increased adenylyl cyclase activity, decreased phosphodiesterase activity, and altered exit rates of cAMP from the ovarian cells. The prostaglandin analogue may have acted on any one of these parameters. It is important to keep in mind that the prelabeling technique for cyclase activity, although useful, actually determines turnover rates, which also depend on cellular rates of cAMP degradation and on exit of cAMP from the cells.

Both the effects of various prostaglandins on the adenylyl cyclase systems and their role in reproductive physiology remain to be established. Interestingly, prostaglandin synthesis and levels raise dramatically in preovulatory follicles (147, 148) at a time when follicular adenylyl cyclase activity is unresponsive to gonadotropin, but is still sensitive to prostaglandin stimulation (see below).

The site of action of prostaglandins has not been established unequivocally. Kolena and Channing (90) reported that in pig granulosa cells maximally effective doses of LH and PGE_2, as determined by increased intracellular cAMP levels, are additive, suggesting that gonadotropins and prostaglandins affect separate adenylyl cyclase systems. Similar experiments by Lamprecht et al. (149), testing effects on cAMP levels in immature rat ovaries, as well as experiments testing stimulation of adenylyl cyclase in membrane particles of CL from pregnant rabbits and superovulated rats (L. Birnbaumer and P.-C. Yang, unpublished observations), showed effects of gonadotropin and prostaglandin to be only partially additive, suggesting the existence of more than one adenylyl cyclase system, one of them being responsive to both agents.

Response to Catecholamines   Both rat and rabbit CL have been found to contain a gonadotropin-responsive adenylyl cyclase system that is also responsive to β-adrenergic catecholamines (118). Because stimulatory effects of LH and catecholamine were found not to be additive, it was concluded that the same adenylyl cyclase system is responsive to both agents. As with prostaglandins, the role of β-adrenergic catecholamines in corpora lutea is not known. Studies on stimulation of adenylyl cyclase activity in homogenates and membrane particles from pig and rabbit follicles (118) revealed very low stimulations (30–40% over control, as opposed to 500–900% over control with LH), indicating that high catecholamine responsiveness develops with development of the corpus luteum. Absence of effects on the adenylyl cyclase system in nonluteinized tissue (150) and its presence in luteinized tissue of rats (151) has been reported.

Response to Gonadotropins   In regard to specificity, LH has been shown to stimulate adenylyl cyclase activity as determined in either homogenates or membrane particles, in mature (ovulable) Graafian follicles of pigs (141), rabbits (143), and rats (144) and in active (progesterone-secreting) corpora lutea of pregnant cows (91) and pregnant, as well as pseudopregnant, rats (144) and rabbits (143). Whenever tested, these systems also showed high responsiveness to FSH, but with significantly lower affinity; in some of these instances, this can be accounted for by LH contamination of the FSH preparations (L. Birnbaumer and P.-C. Yang, unpublished observations). Thus, it would appear that LH responses seen in adenylyl cyclase assays of these tissues are specific and that these systems are very little, if at all, responsive to FSH. Adenylyl cyclases from the above mentioned sources also respond to hCG, a finding that is in agreement with data indicating that LH and hCG interact with a common receptor (34, 36, 42). In fact, hCG (highly purified, 10,000 I.U./mg) and LH (highly purified, 2.0 NIH-S18 units/mg) have been found to stimulate pig Graafian follicle adenylyl cyclase with equal potency (L. Birnbaumer, unpublished observations). Although adenylyl cyclase in ovulable Graafian follicles (141–144) and in granulosa cells of ovulable follicles (43, 152) is responsive almost exclusively to LH, responses specific for FSH (not obtainable with LH or hCG) have also been found, but in immature follicles (152), lending support to the hypothesis that this hormone plays a key role in follicle maturation.

As for potency and efficacy, cAMP appears to be the second messenger for LH action in reproductive tissues, because alone or in combination with cyclic nucleotide phosphodiesterase inhibitors it mimicks many of the effects of LH (87, 89, 153, 154). Three lines of research indicate that not all the factors intervening in LH action are known and that one or more of the effects of the hormone may not be mediated by cAMP. Thus, cAMP has not yet been shown to induce ovulation. Furthermore, quantitative in vitro estimations of LH (or hCG) binding, cAMP production, and steroidogenesis (46, 155) indicate that, while LH addition to testis or Leydig cells does lead to both cAMP and testosterone production, the latter occurs at LH concentrations at which it has been impossible to detect changes in cAMP metabolism with the techniques at hand. Finally, recent studies on the mechanism by which ovulatory doses of LH induce desensitization of the follicular LH-sensitive adenylyl cyclase system (see below) suggest that cAMP is not the mediator of this effect.

The question may, therefore, be asked whether studies on LH-stimulated adenylyl cyclase provide information that relates to the mechanism of LH action in a general way (all of the actions of the hormone mediated by cAMP) or only in a limited way (only a few responses of target tissues such as follicles, Leydig cells, and corpora lutea, dependent on adenylyl cyclase stimulation). If some responses were independent of cAMP, then it is necessary to ask if they were triggered by receptors of different properties. Furthermore, it would be necessary to restrict and qualify the assumption (156) that adenylyl cyclase systems are useful models for studying coupled hormone receptor activity. An indirect approach to this problem was presented by Birnbaumer et al. (118), who explored potencies and efficacies of various LH fractions on three LH-sensitive adenylyl cyclase systems and compared these results with potencies determined by in vivo bioassay. Dose-response curves for 10 LH fractions on LH-sensitive adenylyl cyclase activity in membrane particles from pig Graafian follicles, rat CL of superovulation, and rabbit CL of pregnancy were obtained. These were linearized over the proportional range with the use of Hill transformations. Absolute potencies, as well as potencies relative to NIH-LH-S18, were calculated. Figure 8 presents actual adenylyl cyclase data obtained with the pig and rat systems for 5 of 10 fractions tested, and Figure 9 presents in graphical form the relation of biological potency of the tested LH fractions as seen by the in vivo ovarian ascorbic acid depletion test to the relative potencies of these same fractions in each of the three adenylyl cyclase systems. Several findings emerged from these studies:

1. The relative affinity of the rat CL adenylyl cyclase system for the various LH fractions (ranging from 0.041–2.0 NIH-LH-S18 units/mg and of bovine, ovine, and human origin) correlates relatively well with biological potency, a characteristic expected if the receptors coupled to adenylyl cyclase and those responsible for initiation of ascorbic acid depletion in rat ovaries were similar or the same. (It is also of interest to note that the time after PMSG-hCG priming for

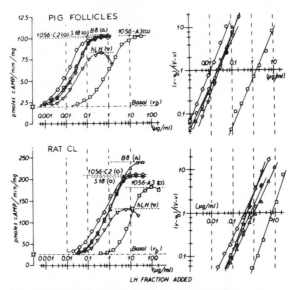

Figure 8. Response of two adenylyl cyclase systems to various LH fractions. Membrane particles prepared either from 6–8 mm pig Graafian follicles or from CL of PMSG-hCG-superovulated rats were assayed for adenylyl cyclase activity in the presence of the indicated concentrations of LH-containing fractions. *Lower* and *upper dashed lines* represent basal (v_b) and maximally stimulated (V) activities. The *right-hand panels* represent Hill transformations of the dose-response data shown on the left. Tissue preparation and additional assay conditions are described elsewhere (141).

maximal LH binding, LH-stimulable adenylyl cyclase and LH-sensitive ascorbic acid depletion all coincide. Furthermore, all these effects are altered similarly by estrogen treatment of the PMSG-hCG-primed rat.) The correlation curve, however, does not have the theoretical slope of 1.0, as expected if the only factor involved is receptor affinity. Rather, it was found to be 0.74 ± 0.12 (mean ± S.D. of the regression). One explanation for this deviation from ideal is that the bioassay data depend not only on receptor activity but also on hormone half-life, for which no corrections were made. From these data, it was concluded that the correlation of rat cyclase data with rat bioassay data is consistent with the assumption that rat cyclase activation may provide information about LH-receptor interaction in general and not only in very limited circumstances. Lee and Ryan (33) provided similar potency data comparing LH binding in rat CL and in vivo bioassays.

2. Receptors coupled to rat, pig, and rabit adenylyl cyclase appear to recognize common features of all LH molecules, because, except for one of the fractions, all other LH fractions tested tend to align themselves along a common line. This lends further indirect support for the assumption that information gained about LH action by using adenylyl cyclase systems may be of general applicability in

these animals. Although one or more of the effects of LH may not be mediated by LH, it is likely that the discriminatory characteristics of the LH receptor triggering these effects are similar to those of the LH receptor triggering adenylyl cyclase activation.

3. The apparent affinities of the analyzed adenylyl cyclase systems for LH fractions vary both with the species and tissue of origin. Both CL preparations tested were found to be considerably less sensitive to low concentrations of LH than the pig or rabbit follicle systems. About 20 times higher concentrations of bovine LH were needed to half-maximally stimulate rabbit CL adenylyl cyclase

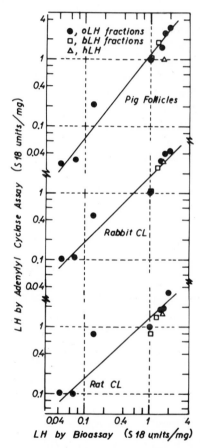

Figure 9. Relationship between biological potency of the tested LH fractions as seen in in vivo ovarian ascorbic acid depletion test and potency of these same LH fractions in two separate adenylyl cyclase systems. The proportional ranges of the dose-response curves (Figure 8) were linearized (Hill plots), and the potency (estimates of intercepts of regressions with the abscissa) and 95% confidence limits relative to NIH-LH-S18 were calculated by least squares regression analysis (141).

than to stimulate rabbit follicle adenylyl cyclase (160 ng/ml versus 7.5 ng/ml). Furthermore, the rat CL system was found to be still less sensitive to LH (by a factor of approximately 5) than the rabbit CL system. This finding is in sharp contrast with absolute affinities of receptor sites for hCG and LH, which were found by Lee and Ryan (36) and Lee (43) to be 0.1–0.5 ng/ml, and about equivalent in pig granulosa cells and rat CL of superovulation. Reasons for this discrepancy may lie in the use of different assay conditions for determination of binding affinity and apparent affinity of adenylyl cyclase activation, the most important one being temperature. It is clear that future studies attempting to study mechanism of LH action in cell free systems will have to determine binding and enzyme activation under identical conditions.

4. Contrary to results obtained in bioassays, in which all LHs elicit the same maximal response, the degree of stimulation that can be obtained in the various adenylyl cyclase systems varies with the origin of LH. Thus, the adenylyl cyclase system in rat CL is stimulated to a large extent by bovine LH, with various ovine LH fractions and human LH (LER 960) giving about 80 and 50% as much stimulation, respectively. Differences of this type, although present, were found to be less pronounced in rabbit CL. In other studies (L. Birnbaumer, unpublished observations), it was found that pig FSH, whose activity on adenylyl cyclase of estrous pig follicles can be accounted for by contaminating pig LH (stimulation to 50% of maximum was obtained with 6,300 μg/ml of NIH-FSH-Pl) stimulates this system about 25% more than ovine LH. The finding that varying degrees of activation are obtained with LHs of different origin is of interest, because, if coupled to specific binding studies, it may afford a tool to investigate quantitative relationships between binding and adenylyl cyclase activation. Eventually, knowledge of the tertiary and quarternary structure of bovine, ovine, and human LH may shed some light on which molecular characteristics and amino acid residues of LH are involved in enzyme activation.

Hormonal Specificity of Adenylyl Cyclases in Testicular Tissues Adenylyl cyclase activity specific for FSH has been demonstrated in homogenates of immature rat testis (48, 157) and in FSH-responsive Sertoli cell-enriched preparations (92). A testicular LH-responsive adenylyl cyclase activity has been described in Leydig cells, as seen by increased cAMP levels in intact cells (46, 158). In contrast to the receptor-gonadotropin interactions in testicular tissues, receptor cyclase interactions have received, as yet, little attention.

In Vivo Regulation of LH-sensitive Adenylyl Cyclase Activity

Neither adenylyl cyclase activity nor its responsiveness to gonadotropins are set, immutable parameters of reproductive tissues. Thus, studies on levels of enzymatic activity and responsiveness to LH have revealed large variations, especially in the latter, that depend on the physiological state of the animals at the time of observation. Our knowledge of the in vivo regulation of adenylyl cyclase activity in reproductive tissues of rats, rabbits, and pigs is reviewed below. Common, as well as differential features, are pointed out.

Rat

Female Although newborn rat ovaries contain adenylyl cyclase activity, it is unresponsive to gonadotropins, when either the prelabeling technique (149) or direct measurement of enzyme activity (144) is used. Both assessments of activity showed that LH responsiveness of adenylyl cyclase appears between days 9 and 10 after birth, after which time relative stimulation and absolute activities in whole homogenates remain relatively constant until puberty and the first ovulation. Appearance of responsiveness by day 10 is coincident with the reported appearance of increased ovarian estrogen secretion and stimulation of interstitial cell size in response to gonadotropin administration. FSH-stimulated activity (obtained in the presence of 10 μg/ml of NIH-FSH-Pl) did not differ significantly from LH-stimulated activity, in consistency with the contention that an activity specific for LH was being determined (144). A similar experiment (159) was carried out measuring LH- and FSH-stimulated adenylyl cyclase activity in homogenates from ovaries of rats between 14 and 31 days of age. Rather than finding relatively constant and equal activities with both hormones, LH- and FSH-stimulated activities underwent transient increases, peaking by day 18 (approximately 4-fold over basal) for LH stimulation and by day 21 (approximately 5-fold over basal) for FSH stimulation. Stimulation by LH on day 21 was only 2-fold over basal, indicating that activity seen in the presence of FSH on that day is specific and cannot be accounted for by contaminating LH in the FSH preparation. No explanation for the discrepancy of results between these research groups is currently available.

Studies that measured activities in both whole ovaries and dissected structures such as follicles and CL revealed that activities in these structures vary independently of that of interstitial tissue containing both preantral and immature follicles, as well as remains of regressed corpora lutea (corpora albicans). LH-stimulated adenylyl cyclase activities determined in follicles dissected on each day from rats with 4-day estrous cycles have shown that immature follicles are poorly responsive to LH (about 2-fold) on metestrus and diestrus, but acquire high responsiveness between 1,000 hr of diestrus and 1,000 hr of proestrus, coincident with acquisition of the capacity to ovulate in response to an ovulatory dose of LH or hCG. After the endogenous LH surge, and with the nearing of ovulation, follicle adenylyl cyclase loses LH-responsiveness at the same time as its basal activity increases, so that by the time of ovulation the system is desensitized by approximately 50% to stimulation by LH. Desensitization proceeds throughout the night of proestrus to estrus, so that by the morning of estrus the adenylyl cyclase of the newly formed CL is totally desensitized to LH stimulation. These findings are illustrated in Figure 10.

Adenylyl cyclase activity and its responsiveness to LH in CL of the rat have been found to vary with the type of CL examined. Thus, CL of the cycle, which never become very active in terms of progesterone secretion, have an adenylyl cyclase system that is unresponsive to LH on day 1 (estrus) and acquire some responsiveness (2- to 2.5-fold) by the morning of day 2. Both activity and

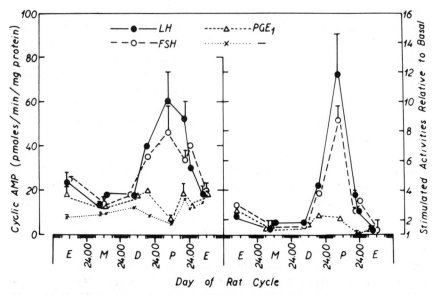

Figure 10. Variation of an LH-responsive adenylyl cyclase system in follicles of cycling rats. Absolute (*left panel*) and relative (*right panel*) activities were measured in homogenates of dissected ovarian follicles obtained from rats with a 4-day estrus cycle. When present, LH, FSH, and PGE_1 were 10 μg/ml. Assay conditions and tissue preparation are described elsewhere (144).

responsiveness are maintained throughout metestrus, but are rapidly lost on the next day (diestrus), coincident with their regression. CL of pregnancy, on the other hand, rather than showing declining responsiveness to LH on day 3, increase their LH-sensitive activity steadily over the next days until day 9, showing a small (not always significant) decrease on days 10 and 11, followed by a sharp rise to maximal activity on days 15 and 16. Thereafter, LH-stimulated activity was found to decline as parturition approached. Activity in CL of pseudopregnancy increased until day 11, declining thereafter, coincident with termination of pseudopregnancy and CL regression. The patterns of LH-sensitive adenylyl cyclase activity in CL of pregnant and pseudopregnant rats are illustrated in Figure 11 (*continuous lines*). The pattern of LH-sensitive adenylyl cyclase in PMSG-oLH primed rats has been described above (see under "Relationship of Receptor Binding to Biological Effects").

Other studies (144) showed that blockage of the endogenous LH surge with Nembutal results in blockage both of ovulation and of the desensitization seen between the afternoon of proestrus and ovulation. Injections of prolactin (PRL) into cycling animals from metestrus through estrus (mimicking PRL surges observed in pseudopregnant rats) (160) resulted both in maintenance of the CL and in maintenance of active LH-sensitive adenylyl cyclase. Similarly estrogen administration to rats on metestrus, known to prolong CL lifespan (161), was found to result in maintenance of levels of LH-sensitive adenylyl cyclase.

Not only "normal" maturation of follicles, as occurs in the cycling rat between diestrus and proestrus, but also PMSG- and hCG-induced maturation of antral follicles in immature 24–26-day-old rats is associated with appearance of highly responsive adenylyl cyclase activity (144). Induction of maturation of follicles in immature rats with PMSG or FSH has been shown to be associated with an increase of LH-specific binding sites in their granulosa cells (80). A study of the time course of appearance of receptor sites and LH sensitivity that would determine whether one precedes the other or whether they occur simultaneously has not yet been done in the rat.

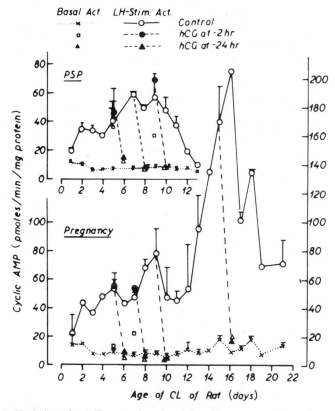

Figure 11. Variation of an LH-responsive adenylyl cyclase system in CL throughout pseudopregnancy and pregnancy of rats: effect of a luteolytic dose of hCG. Adenylyl cyclase activities were measured in the absence (basal) or presence of 10 μg/ml of LH in homogenates of dissected ovarian CL obtained 1–13 days after cervical stimulation (*upper panel*) and 1–21 days after mating (*lower panel*). For determining the effect of a desensitizing hCG injection on LH-stimulated adenylyl cyclase activity in CL, rats were killed 2 or 24 hr after hCG injection. Assay conditions and tissue preparation are described elsewhere (177).

Male Responsiveness of rat testis FSH-stimulated adenylyl cyclase activity has also been found to vary with age and development. Thus, presence of FSH-specific enzyme in testes of immature rats, as seen by determination of FSH action on cAMP-dependent protein kinase activity (92, 162), on intracellular cAMP levels (93, 163), on accumulation of labeled cAMP from prelabeled pools of ATP (140, 157), or on adenylyl cyclase activity determined directly in unfractionated homogenates (157) is undetectable at birth, increases to maximal activity by days 6–10, and then declines steadily, so that no effect of FSH in any of the above mentioned assays can be detected. The reasons for loss of these variations are not known. Final loss of responsiveness does not appear to be due to loss of FSH receptors, for addition of methylisobutylxanthine to testis incubates, restores the effectiveness of FSH as seen by stimulation of protein kinase (162) or accumulation of cAMP (140). To our knowledge, regulatory events such as observed in rat ovaries with respect to LH-stimulated adenylyl cyclase (144) and in rat testes with respect to FSH-stimulated adenylyl cyclase (157) have not been investigated in testicular interstitial cells with respect to LH-sensitive adenylyl cyclase activity.

Rabbit As in the rat, mature estrous follicles of the rabbit contain an active, highly responsive LH-sensitive adenylyl cyclase system. It too undergoes desensitization after exposure to ovulatory doses of LH or hCG. This was first suggested (138, 139) with the use of the ATP prelabeling technique to demonstrate accumulation of cAMP in response to in vitro LH exposure and has recently been confirmed by direct measurement of adenylyl cyclase in homogenates (143). As in the rat, small, immature follicles (which in the rabbit are less than 1 mm in diameter) were found to respond only minimally to LH, in spite of the presence of active basal activity (143). While the degrees of responsiveness to LH of the adenylyl cyclases in mature follicles of rats and rabbits is about the same, they differ in that the rabbit system becomes desensitized much more rapidly (see below). As in the rat, newly formed rabbit CL (day 1) do not contain a gonadotropin-responsive adenylyl cyclase, but acquire it within the first 3 days of their lifespan. Thus, LH-stimulated adenylyl cyclase activity in CL of pregnant rabbits was found to develop under normal conditions in three stages (illustrated in Figure 12), characterized by three maxima (on days 3, 12, and 27) and two minima (on days 9 and 21), and to decay to levels found in interstitial tissue by the time of parturition, coincident with CL regression. Activity in CL of pseudopregnant rabbits, known to have a normal lifespan of about 15 to 16 days, was found to persist for the duration of pseudopregnancy and to follow a pattern similar to the pattern in CL of pregnancy during their first 14 days of life, but decaying by day 16 (143).

Hysterectomy of rabbits prior to induction of pseudopregnancy, known to prolong CL lifespan (164), lead to maintenance of active, LH-responsive adenylyl cyclase beyond that of regular CL of pseudopregnancy (143). These findings are consistent with the possibility that active CL are associated and perhaps dependent on an active, LH-dependent adenylyl cyclase system.

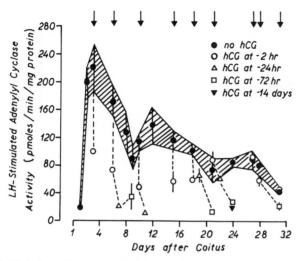

Figure 12. Variation of an LH-responsive adenylyl cyclase system in CL throughout pregnancy of rabbits: effect of a luteolytic dose of hCG. The main curve (*hatched area*) illustrates LH-stimulated adenylyl cyclase activities (•) ± S.E.M. in homogenates of dissected rabbit ovarian CL obtained 1–31 days after mating each female to two fertile bucks in succession. LH was 10 μg/ml. For determining the effect of a desensitizing hCG injection on LH-stimulated adenylyl cyclase activity in CL, rabbits were injected on the days indicated by the *arrows* and were killed 2 (o), 24 (△), and 72 (□) hr after the hCG injection. In one instance, rabbits were killed 14 days (▲) after hCG-induced desensitization. – – –, the time course for desensitization of the original CL. Details are given elsewhere (177).

Pig Follicle Although reports have appeared indicating that CL of pigs contain an LH-sensitive adenylyl cyclase system (165), thus suggesting a role of LH in regulation of the pig corpus luteum, the effects reported were minimal (20 to 30% over basal) and did not allow precise estimates of variation during various physiologic states. Assay conditions such as developed by Birnbaumer et al. (118) did not allow for detection of better LH responses.

As in rats and rabbits, mature, but not immature, follicles from pig ovaries contain an adenylyl cyclase system that is highly responsive to LH (141). Lee (43) has shown that at least part of that activity is localized in granulosa cells of mature follicles. Whereas granulosa cells from large, ovulable follicles were found to contain an active (11-fold) LH-responsive adenylyl cyclase, cells obtained from small, immature follicles were found to respond to LH with only a 2-fold stimulation, but to FSH with an 11-fold stimulation. Thus, follicular maturation in this species is associated not only with appearance of LH receptor (determined by binding) and LH-sensitive adenylyl cyclase (43, 141), but also with a decline of FSH-specific adenylyl cyclase. Although follicular adenylyl cyclase has been shown to be localized in granulosa cells, its existence in theca cells has not been excluded. Because autoradiographic observations in the rat, demonstrating the localization of hormone specific sites within follicles have indicated that receptors for LH occur in theca cells as well as in granulosa cells (52, 53), it is likely that LH-responsive adenylyl cyclase activity in mature porcine follicles is distributed in both granulosa and these cells.

Collapsed, postovulatory pig follicles have been found (166) to have a desensitized cyclase system. Thus, the plasma membranes of the pig, like the rat and rabbit, contain the necessary machinery to promote desensitization of the adenylyl cyclase system to LH stimulation when exposed to high levels of LH.

Some Physiological Implications of Variations in LH-sensitive Adenylyl Cyclase in Ovarian Tissues   Serum progesterone levels in pregnant rabbits (167) and pregnant rats (168) were compared to patterns of LH-stimulated adenylyl cyclase activity in CL of these animals (143, 144). In the rabbit, striking correlation between these two parameters was found after implantation (Figure 13), strongly suggesting a regulatory role for LH and its effector system in the direct control of steroid production by the rabbit. This contradicts the role assigned to estrogen as the primary luteotrophic agent in the rabbit (10–14) and the assumption that luteotrophic effects of LH in the intact rabbit are secondary to its estrogen-producing effect on follicles. However, evidence showing that LH stimulates adenylyl cyclase activity in homogenates and membrane particles from rabbit CL and that hCG-induced desensitization of CL adenylyl cyclase can be obtained independently of estrogen variations in the ovary, (see below) strongly argues in favor of LH being capable of acting in vivo directly on the CL of the rabbit. This, coupled to the above-mentioned correlation of LH-sensitive adenylyl cyclase activity in CL and progesterone levels in serum of pregnant

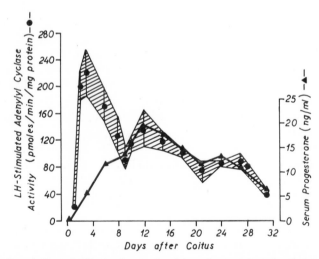

Figure 13. Correlation of LH-stimulated adenylyl cyclase activity (*hatched area*) in CL of pregnant rabbits with serum progesterone levels (▲——▲). Adenylyl cyclase activities were measured in the presence of 10 μg/ml of LH in homogenates of dissected ovarian CL obtained 1–31 days after coitus. Adenylyl cyclase assay conditions and tissue preparation are described elsewhere (143). Progesterone data were taken from Challis et al. (167).

rabbits, suggests that the adenylyl cyclase system may be, under normal physiological conditions, a regulatory step in steroidogenesis of the rabbit CL. This is not to deny a crucial role for estrogen in this structure. Perhaps, rather than regulating pogesterone output on a moment-to-moment basis, the role of estrogen is to maintain—by stimulation of protein synthesis—adequate levels of all enzymes involved, including those affected by cAMP.

Working through this mechanism, estrogen might be capable of compensating, in the absence of LH (as in the hypophysectomized rabbit), for the lack of stimulatory gonadotropin and cAMP, by inducing higher levels of steroidogenic enzymes, thus relieving the CL from its dependence on LH. Clearly, further experiments along these lines, determining levels of the various enzymes involved in regulation of progesterone secretion in the normal and the estrogen-supported CL, need to be carried out to understand the fine tuning of this gland in rabbits and the role of LH in its regulation.

Whereas a good correlation between serum progesterone levels and LH-stimulated adenylyl cyclase was found in the pregnant rabbit after implantation, in the rat this correlation appears to be only qualitative (Figure 14), suggesting that

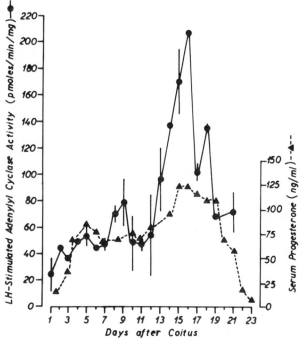

Figure 14. Correlation of LH-stimulated adenylyl cyclase activity (●———●) in CL of pregnant rats with serum progesterone levels (▲———▲). Adenylyl cyclase activities were measured in the presence of 10 μg/ml of LH in homogenates of dissected ovarian CL obtained 1–21 days after fertile mating. Adenylyl cyclase assay conditions and tissue preparation are described elsewhere (82). Serum progesterone levels were taken from Morishige et al. (168).

factors other than LH levels and LH-sensitive adenylyl cyclase play important roles in regulating progesterone secretion. There are data (144, 160, 169) to indicate a good agreement between the patterns of progesterone in serum and LH-stimulated adenylyl cyclase in CL throughout the estrous cycle and pseudo-pregnancy. However, the increase in LH-stimulated adenylyl cyclase activity in CL of pregnancy after day 12, coincident with lowering of serum LH levels (168) and establishment of placental dependence of the CL (170), is not accompanied by a similarly marked increase of progesterone production. Further investigation is needed to unravel the role of LH and cAMP, if any, during the later stages of pregnancy of the rat.

In general, however, the following lines of evidence (144) suggest that the functional capacity of ovarian structures in the rat is associated with and may require adequate levels of LH-stimulated adenylyl cyclase activity: 1) the pre-pubertal rat ovary becomes responsive to gonadotropins when it acquires a gonadotropin-responsive adenylyl cyclase system; 2) follicles from PMSG- or hCG-primed prepubertal rat ovaries rapidly acquire a highly responsive adenylyl cyclase system prior to ovulation; 3) follicles dissected from cycling rat ovaries on the day of proestrus have acquired an adenylyl cyclase system which is highly responsive to LH stimulation; 4) CL from pseudopregnant and pregnant rats contain an LH-stimulated adenylyl cyclase system which persists throughout the known life span of these CL and declines with their regression; and 5) exogenous treatments which prolong CL life span, such as PRL and estrogen injection or cervical stimulation resulted in maintenance of CL with an active LH-responsive adenylyl cyclase system.

Thus, studies on ovarian tissues have revealed that the LH-sensitive adenylyl cyclase system in follicles differs from that in corpora lutea (Table 7) and is under complex humoral regulation. It would appear that LH-stimulated adenylyl cyclase activity may be a useful marker for the study of the final stages of follicular maturation (ovulability) and the regulation of corpus luteum life-span. Studies on which factors and hormones (e.g., prolactin, estrogens, etc.) are operative in providing an adequate effector system for the receptor-LH inter-action and correlations of the state of responsiveness of the adenylyl cyclase

Table 7. Comparison of follicular and corpora lutea adenylyl cyclase systems

Variable studied	Follicles	CL
Stimulation by LH (responsiveness)	Large	Large
Sensitivity for LH (affinity)	High	Small
Minimal effective LH concentration	Low	High
LH-stimulated adenylyl cyclase activity (absolute)	Low	High
Desensitization by LH (or hCG)	Yes	Yes
Rate of desensitization by LH	Fast	Slow
ATP dependence of LH response	Low (μM)	High (mM)
Nucleoside triphosphate pyrophosphohydrolase	Absent	Present

system with functional activity of follicles and corpora lutea will surely provide new insights into mechanisms involved in maturation of follicles and regulation of corpus luteum function.

Desensitization of Adenylyl Cyclase in Ovarian Tissues

Follicular Desensitization in Vivo Exposure of the highly responsive adenylyl cyclase in mature, ovulable follicles to LH, such as occurs in cycling rats, in mated rabbits, and cycling pigs, as well as to exogenous gonadotropin, results, within approximately the time span necessary for ovulation to occur, in a loss of the capacity of adenylyl cyclase to be stimulated by LH, i.e., in gonadotropin-induced desensitization of the adenylyl cyclase systems to further gonadotropin stimulation. In ovarian tissues, this phenomenon was originally shown by Marsh et al. (139), who found rabbit preovulatory follicles to lose their ability to respond to in vitro stimulation by LH, and by Lamprecht et al. (149), who found that preliminary incubation of dissected rat follicles with LH (18 hr) resulted in establishment of a state of refractoriness of the follicles to accumulate cAMP in response to a renewed LH stimulus. These findings, made with the prelabeling technique, have been confirmed and expanded by direct measurement of LH-stimulated activity by Hunzicker-Dunn and Birnbaumer (143, 144) as mentioned earlier in this chapter. Thus, in separate experiments on desensitization of the rabbit follicular adenylyl cyclase (143), it was established that this effect of gonadotropin was specific for LH or hCG and not mimicked by FSH or prolactin. It could be elicited either by exogenous administration of gonadotropin or by endogenous LH as obtained after mating. It required ovulatory doses of gonadotropin and could be obtained after indomethacin treatment that blocked prostaglandin synthesis to the extent of blocking ovulation. Finally, it was shown that, as in rat follicles, adenylyl cyclase activity highly responsive to LH, present in dissected estrous follicles of the rabbit, can be desensitized in vitro if incubated for 2 hr in Krebs-Ringer buffer with 1 –10 μg/ml of LH. Rabbit follicular desensitization of adenylyl cyclase differed from desensitization of rat follicular adenylyl cyclase only in that it was significantly faster (143, 144), with 50% desensitization occurring 1 hr after gonadotropin administration in the rabbit, and after about 6–8 hr in the rat.

The physiological role and implication of desensitization in follicles is not clear, especially since the process is slower in rats than rabbits.

There is considerable evidence indicating that the steroidogenic effect of LH or hCG on ovarian tissues is mediated by cAMP (85–87, 171–173). Thus, an ovulatory dose of LH or hCG would be expected to stimulate adenylyl cyclase, producing large increases in cAMP and in steroid output. This sequence of events has been shown to occur in estrous rabbit follicles, because, when an ovulatory dose of hCG initially stimulates adenylyl cyclase, there is an increase in follicular cAMP (174) and a rapid dramatic increase in steroidogenesis (173, 174, 175). Large (mg) quantities of steroids, including estrogens, progestins—especially 20α-hydroxy-4-ene-3-one—and androgens are poured into follicular fluid and the ovarian vein. However, in the rabbit follicle this increase in steroid production is short-lived. At the time of ovulation (10 hr after administration of hCG), steroid *levels* in both the ovarian vein and follicular fluid have dropped to unmeasurable quantities, indicating an earlier time at which steroid synthesis was terminated. Once steroidogenesis in the follicle is terminated, it cannot be reinstituted by exogenous gonadotropins (176), and steroid output is not resumed until some 38 hr after follicle rupture, at which time progestin levels generally increase (176). It would seem, therefore, that desensitization of the adenylyl cyclase system is responsible for this loss of steroidogenic activity in the preovulatory follicle.

However, although the adenylyl cyclase system in rabbit follicles was found to be unresponsive to LH long before ovulation (139, 143), in rat follicles it was found to decline only about 50% when tested between 8 and 10 hr after the endogenous surge (144). It seems reasonable to assume, therefore, that complete desensitization may not be a necessary prerequisite for ovulation to occur. It remains to be seen in individual follicles, whether desensitization is complete at ovulation and whether incomplete desensitization would interfere with the proper development of this process. Studies on variation of steroidogenesis and possible appearance of steroidogenic quiescence have not been carried out in the preovulatory rat follicle. Although desensitization may play an important role in the ovulatory process, it is not always followed by ovulation, as seen from the following two examples (M. Hunzicker-Dunn and L. Birnbaumer, unpublished observations): 1) large (2–3 mm), unovulated (atretic?) follicles dissected from rabbit ovaries 16 hr after an hCG injection were found to contain an unresponsive adenylyl cyclase system, and 2) unovulated follicles dissected 24 hr after an hCG injection in which ovulation was blocked by indomethacin treatment (70 mg of indomethacin, subcutaneously with hCG, and again 8 hr after hCG) were also found to contain an adenylyl cyclase system that was desensitized. Clearly, further work is needed to establish the physiological role of follicular desensitization.

Follicular desensitization of adenylyl cyclase can be interpreted either as only a temporary effect, with adenylyl cyclase in the newly formed CL merely recovering its responsiveness to gonadotropin, or as a permanent effect, with adenylyl cyclase in the differentiated, developing CL being a "new" enzyme or having acquired a "new" LH receptor and not just a system that has recovered a temporarily lost activity. The adenylyl cyclase system in CL exhibits characteristics which are different from those in follicles: 1) the adenylyl cyclase in CL becomes responsive to catecholamines (118); 2) the response of the adenylyl cyclase to LH in CL becomes dependent upon high ATP concentration (118); 3) the adenylyl cyclase system becomes less sensitive to LH in CL (118); 4) the adenylyl cyclase system stimulated by LH is more active in active CL (143, 144); 5) the LH response of the follicular adenylyl cyclase system is desensitized rapidly under influence of LH or hCG, whereas that in CL is desensitized more slowly (see below).

In addition to these differences in the adenylyl cyclase system (see Table 7), luteinization in the rabbit is associated with the appearance of a nucleoside triphosphate pyrophosphohydrolase (118) and the disappearance of a 19-hydroxylase-aromatase, the enzyme responsible for estrogen formation from androgens. These observations suggest that differentiation of granulosa and theca cells into luteal cells includes the differentiation of various enzyme systems, including adenylyl cyclase. Whether or not follicular (or granulosa cell) LH receptors are lost upon desensitization and the new response involved synthesis of a new receptor site has not been explored.

Luteal Desensitization in Vivo   Not only follicular adenylyl cyclase but also corpus luteum adenylyl cyclase is desensitized to LH stimulation by LH and hCG. This was recently shown in pregnant and pseudopregnant rats and rabbits (143, 177). As in follicles, the doses of gonadotropin needed are high, of the type that promotes both new ovulations and regression of the CL in which the doses trigger desensitization of the adenylyl cyclase system. The luteolytic effect of ovulatory doses of hCG and LH in several species has been well documented (15–17).

Desensitization in CL was studied by injecting animals with ovulatory doses of hCG and then testing the state of activity of the LH-responsive adenylyl cyclase in homogenates of CL obtained from these animals at subsequent times. It was found (177) that administration of hCG or LH, but not FSH or prolactin, induced loss of LH-stimulated activity in all types of CL studied, but at rates that varied, depending on the species (rat or rabbit) and the physiological state (early or late pregnancy or pseudopregnancy). Thus, in the rabbit, a decline of at least 50% in the LH-stimulated activity was demonstrable within 2 hr of hCG, both during pseudopregnancy and during the first 18 days of pregnancy. However, after day 21 of pregnancy, adenylyl cyclase was found to be unaltered at 2 or 24 hr after hCG injection, requiring as much as 72 hr before becoming desensitized (see Figure 12). It seems from these studies, that adenylyl cyclase from CL of the last third of pregnancy are afforded partial protection from the desensitizing effects of hCG. This effect was not noted in follicles contained in ovaries of rabbits in their last third of pregnancy or in hCG-induced 3-day-old CL in 24-day pregnant rabbit ovaries. This indicates that the process responsible for the partial resistance to desensitization of the CL of pregnancy is absent in follicles and requires in CL more than 3 days to become effective. Desensitization of adenylyl cyclase induced by hCG appears to be a persistent effect, because no reversal has yet been demonstrated. Furthermore, activity in a 14-day-old hCG-induced corpus albicans dissected from a 24-day pregnant rabbit contained a desensitized adenylyl cyclase system (177).

Desensitization of the rabbit CL adenylyl cyclase by gonadotropin appears to be a direct effect on the CL and was found not to be secondary to the estrogen deprivation that occurs after desensitization of follicular adenylyl cyclase by the same desensitizing dose of gonadotropin. This was established (177) by 1) demonstrating that desensitization is not affected by prior removal (cauterization) of estrogen-secreting antral follicles, a treatment which was found to be itself without deleterious effect on LH-stimulated adenylyl cyclase in CL, and 2) demonstrating that administration of 17β-estradiol will neither prevent the desensitizing effect of hCG seen at 1 hr nor reverse this effect as seen 3 days after hCG injection (Table 8). The regime of estrogen administration used in this experiment has been shown to be luteotropic and to prevent functional gonado-tropin-induced luteolysis in rabbits, as evidenced by maintenance of weight and progesterone secretion (10–14).

In rats, the injection of an ovulatory dose of hCG (or LH) was shown to induce desensitization of adenylyl cyclase systems in a CL of superovulated, pseudopregnant, and pregnant rats (177). The rate at which desensitization became established, although slower than in the rabbit, appeared to be the same regardless of the physiological state of the CL-bearing animal. Effects of hCG injection on CL adenylyl cyclase in pseudopregnant rats are shown in Figure 11.

Desensitization of hormone-stimulated adenylyl cyclase to stimulation by hormones, induced by the stimulatory hormone, has been demonstrated in several nonreproductive systems (178, 179), and this phenomenon is not, there-fore, unique to gonadotropin-sensitive adenylyl cyclase systems. In no instance is the physiological role or significance of this process known, except for the fact that it is quite generalized. The existence of desensitization in the CL is of interest, because it appears to be an effect of the gonadotropin that precedes luteolysis. Thus, luteolysis can be induced during pseudopregnancy and preg-nancy in both rats and rabbits by the injection of an ovulatory does of LH or hCG. Subovulatory doses of either gonadotropin do not promote luteal regres-sion. Functional luteolysis in the pseudopregnant rabbit is characterized by a decline in progesterone synthesis, demonstrable 16 hr, but not 9 hr, after an acute injection of LH. Thereafter, morphological luteolysis follows, charac-terized by a decline in CL weight, demonstrable 36 hr after an LH injection (180). Desensitization of the adenylyl cyclase system in CL is also induced by the injection of an ovulatory dose of hCG into pseudopregnant and pregnant rats and rabbits, subovulatory doses of hCG being without effect. Complete desensi-tization of the luteal adenylyl cyclase system appears to precede morphological luteolysis in pregnant rabbits (before day 21) and probably in pseudopregnant rabbits, because in both cases the extent of desensitization 2 hr after an hCG injection indicates a rapid decline in responsiveness of the adenylyl cyclase system. Whereas complete desensitization appears to precede morphological luteolysis, partial desensitization ($\sim$ 50% reduction of the LH-stimulated adenyl-yl cyclase), as observed in studies reported above, precedes functional luteolysis. Thus, there would seem to be a close association between hCG-induced desensi-tization of the adenylyl cyclase system to LH and hCG-induction of luteolysis. It is not known whether desensitization is a prerequisite for hCG-induced luteo-lysis. To establish such a cause-effect relationship, it will be necessary to show not only that progesterone synthesis declines in a manner parallel to the rate of loss of the adenylyl cyclase system, but also that desensitization of the cyclase system induced by other means also leads to luteolysis.

Table 8. Desensitization and persistence of desensitization induced by hCG in corpora lutea of pseudopregnant rabbits receiving estrogen

Time of Pseudopregnancy (days)	Treatments	Age of CL tested (days)	Weight of CL tested (mg)	No. of CL tested	Adenylyl cyclase activities (mean ± S.E.M.)		No. of activities tested
					Basal (pmol/min/mg of protein)	LH-stimulated	
6	None	6	11.3	16	38.0 ± 8.0	158.5 ± 7.5	2
	hCG^a	6	9.1	18	63.5 ± 8.5	82.0 ± 4.0	2
	hCG^a plus E_2 ^{b}	6	11.3	13	51.5 ± 7.5	71.0 ± 1.0	2
9	None	9	13.2	24	25.0 ± 6.0	124.0 ± 16.0	2
	hCG^c	9	9.3	22	22.0 ± 3.0	67.0 ± 2.5	2
	hCG^c	3	4.9	14	20.0 ± 2.0	153.5 ± 0.5	2
	hCG^c plus E_2 ^{d}	9	19.3	38	34.0 ± 4.0	58.0 ± 3.0	4
	hCG^c plus E_2 ^{d}	3	3.3	25	30.0 ± 7.5	133.0 ± 18.0	3

[a] 75 I.U./3.5–4.5 kg of rabbit, administered I.V. at 0900 hr of day 6 of pseudopregnancy. Animals were killed at 1100 hr of the same day.

[b] Treatment with 17β-estradiol (E_2) consisted of three subcutaneous injections of 1.5 μg each (in 0.1 ml of peanut oil), administered with 12-hr intervals and started at 1800 hr of day 5 of pseudopregnancy.

[c] 75 I.U./3.5–4.5 kg of rabbit, administered I.V. at 0900 hr of day 6 of pseudopregnancy. Animals were killed at 0900 hr of day 9 of pseudopregnancy. Both sets of CL, the original 9-day-old one and the new 3-day-old one (identified by fresh ovulation points and small size), were tested separately for their activities.

[d] Treatment with 17β-estradiol (E_2) consisted of nine subcutaneous injections of 1.5 μg each (in 0.1 ml of peanut oil), administered with 12-hr intervals and started at 0800 hr of day 5 of pseudopregnancy.

Even if hCG-induced desensitization is an initial step in hCG-induced luteo-lysis, the physiological means by which CL normally regress does not appear to be associated with surges of LH and/or desensitization of the adenylyl cyclase system. In rabbits, luteal regression at the end of pregnancy or pseudopregnancy does not seem to be induced by an endogenous surge of LH (181), and in rats, even though parturition is associated with a surge of LH that causes postpartum ovulation (182), the loss of progesterone synthesis and hence functional luteo-lysis appears to precede this event. On the other hand, ovulatory levels of LH are not always lytic in rats, because an endogenous surge of LH induced on diestrus by the injection of estradiol benzoate on metestrus promotes ovulation (early in proestrus) but does not promote luteolysis (183). Thus, although desensitization may be closely associated with hCG- and LH-induced luteolysis, it clearly is not the physiological means of CL lysis. Whether it is an obligatory response to a surge of LH is not known.

Desensitization in Isolated Membranes Experiments with isolated membranes from mature porcine Graafian follicles indicated a very stable adenylyl cyclase activity, which in the absence of hormonal additions, yielded linear accumulations of cAMP for up to 35 min when tested at 1–2 mM ATP. Addition of LH to these incubations resulted in time courses that were stimulated but nonlinear, showing a constant decay, until, after 35–40 min of incubation, the rates of accumulation in the presence of LH approached those seen in its absence. Figure 15, taken from work by Bockaert et al. (141), illustrates this finding. It is possible that, in part, this lack of linearity is due to a burst phenomenon, such as is seen at lower ATP concentrations in the presence of GTP. From studies at low ATP (see Figure 7), it can be estimated that such a burst would account for an up to 50% loss of activity within the first 8 min of incubation and should have led to linear time courses thereafter. However, it was consistently found that reductions in activity during the first 10 min of incubation were less than 50% (25% in the experiment shown in Figure 15) and that activity after 10 min of incubation continued to decay until it nearly equaled basal activity if incubated for long enough times (usually 35–40 min). This suggested that in addition to showing state transitions at low ATP concentrations, these membranes may contain another process, which in the presence of high ATP and LH leads to total loss of responsiveness of the system to LH stimulation.

This possibility was investigated (142) by studying the requirements for loss of LH-stimulated activities in short 5-min incubations. This protocol has the advantage that state transitions have already occurred and reductions in activity detected are likely to be due to other reasons. It was found that membranes of pig Graafian follicles contain a process dependent on Mg *and* ATP that, upon incubation of membranes at 30°C, causes the adenylyl cyclase system to lose, in the absence or the presence of LH, its susceptibility to respond to LH. High concentrations of Mg (above 10 mM) were required to detect the process in the absence of LH, the hormone being necessary for detection of loss of LH-stimu-

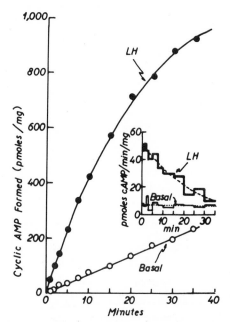

Figure 15. Time course of cAMP accumulation in the absence and presence of LH by membrane particles of porcine Graafian follicles. Membrane particles were prepared from 6–8 mm follicles and assayed for activity at 1.5 mM ATP in the absence (basal) and presence of 10 μg/ml of LH. *Inset,* time course for the loss of LH-stimulated adenylyl cyclase activity. Tissue preparation and additional assay conditions are described elsewhere (141).

lated activity at low (5 mM or less) concentrations of Mg. Loss of LH-stimulated activity was not accompanied by loss of basal activity and it was concluded that this phenomenon is the expression of a desensitization reaction. A study of the dependency of desensitization on Mg (Figure 16) revealed that, at concentrations of Mg below 10 mM, LH stimulates (in the presence of cAMP) the desensitization process by diminishing the requirement for Mg. Thus, loss of LH–stimulated adenylyl cyclase activity was readily detectable at 5.0 mM of total MgCl$_2$ added to the reaction.

The half-maximum of desensitization by LH was found to occur at about 25 $\times$ 10^{-10} M, i.e., at about 4–6 times higher concentrations than needed for half-maximal stimulation. Residual LH-stimulated adenylyl cyclase exhibits the same dose dependency for LH stimulation as control adenylyl cyclase. As shown in Figure 17, desensitization is dependent not only on Mg but also on ATP. Half-maximal desensitization of the system to LH stimulation was obtained with 0.5–0.7 mM ATP in the incubation medium.

Preincubation of membranes in the presence of Mg, LH, and cAMP, with GTP or AMP-P(NH)P, did not result in desensitization unless ATP was also added, strongly suggesting that the biochemical process involved in establishment of a desensitized state is phosphorylation of one of the components of the

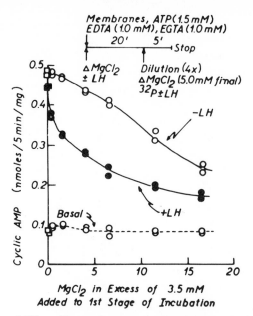

Figure 16. Effect of LH on Mg requirement of LH-stimulable desensitization of porcine Graafian follicle membranes. The basic protocol is described on the figure. Membrane particles were incubated in the absence (o) or presence (•) of LH (10 μg/ml) for 20 min at 30°C in 25 μl of medium containing all adenylyl cyclase reagents and the indicated concentrations of ATP, EDTA, EGTA, and $MgCl_2$ (the last shown as concentration in excess of 3.5 mM on the abscissa). The reaction media were then diluted with 75 μl of adenylyl cyclase assay reagents containing the same concentrations of ATP, EDTA, EGTA (totalling 3.5 mM), but varying concentrations of $MgCl_2$, to give a constant final concentration during the 5-min adenylyl cyclase assay of 5.0 mM (total added, i.e. 1.5 mM in excess of 3.5 mM): ——, 10 μg/ml of LH added to the second stage of incubation; – – –, no LH added to the second stage of incubation; □, no $MgCl_2$ during the first stage of incubation. Additional assay conditions and tissue preparation are described elsewhere (140).

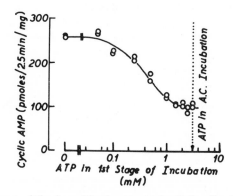

Figure 17. Dependency of the desensitization process in pig Graafian follicle membranes on ATP. The first stage of incubation was carried out in a medium containing chelating agents, $MgCl_2$, an ATP-regenerating system, buffer, 10 μg/ml of LH and membranes, and the indicated concentrations of ATP. After 20 min, the second 5-min stage of incubation was started by the addition of 10 μl of a medium containing labeled ATP, 5 μg of LH, and varying amounts of unlabeled ATP to give a final concentration of 1.5 mM (dotted line). Additional assay conditions and tissue preparation are described elsewhere.

adenylyl cyclase system. However, direct determination of phosphorylation of a membrane component under the influence of LH has not yet been demonstrated.

In other experiments, it was demonstrated that the LH-stimulated desensitization is specific for LH. Neither stimulation by fluoride ion nor by PGE_1 were found to be significantly altered under conditions in which 60% of the LH-stimulated activity was lost.

Several aspects of the desensitization seen in pig follicle membranes are interesting. 1) The reaction is not dependent on LH, because at high enough Mg it proceeded without LH addition. 2) Whereas cAMP may be an absolute requirement for desensitization (the reaction has not been carried out under conditions where less than 10^{-7} M cAMP was produced during the first 5 min of incubation), LH-stimulation of desensitization is clearly *not mediated* by cAMP and must, therefore, be considered as an effect of hormone receptor that may be unrelated to adenylyl cyclase stimulation, and 3) it may be that membrane functions other than adenylyl cyclase are affected as well.

Is desensitization of this and other adenylyl cyclase systems mediated by the same receptors that mediate stimulation of adenylyl cyclase? Experiments with the pig follicle system showing that slightly but significantly higher concentrations of LH are needed for desensitization than for stimulation would suggest that there might be separate receptors involved. On the other hand, Lefkowitz and colleagues (184) showed that a catecholamine-mediated desensitization of β-receptor stimulation of adenylyl cyclase is blocked by β-blockers, thus suggesting that the same type of receptors are involved in stimulation and desensitization.

Is the desensitization process seen in isolated membranes the same as that seen in intact Graafian follicles? Although desensitization in isolated membranes was detected also in membranes from rabbit Graafian follicles (141) in which the in vivo effect was originally discovered and in which in vivo desensitization is very rapid, there is no direct evidence available to prove that the two phenomena are the same. To demonstrate this, it will be necessary to demonstrate common characteristics between adenylyl cyclase desensitized in vivo and that desensitized in vitro. DeVellis and Brooker (185) have recently shown that catecholamine-induced desensitization of a glioma cell line can be blocked and reversed by blockers of protein synthesis such as cyclohexamide. On the other hand, desensitization of adenylyl cyclase in an astrocytoma cell line is not blocked or reversed by cycloheximide (J. P. Perkins, personal communication). It will be interesting to determine effects of cycloheximide on in vitro desensitization in intact follicles and then test desensitization in membranes derived from such treated follicles.

From the above discussion, it follows that gonadotropins regulate adenylyl cyclase activities in two completely different ways. One type of regulation is the stimulation of adenylyl cyclase activity, the type of regulation recognized by Sutherland and coworkers in the late 1950s. It led to discovery of cAMP and its role in biology. The other type of regulation is more subtle and leads to desensitization of the adenylyl cyclase system to the stimulatory effect. Desensitization is definitely slower than stimulation, may be mediated by phosphorylation of one of the components of adenylyl cyclase, possibly the receptor, and results in loss of receptor activity. Because desensitization is observed not only with adenylyl cyclase stimulating hormones but also with hormones that apparently do not affect this enzyme (186), stimulation of the mechanism by which desensitization occurs may be even more general than stimulation of adenylyl cyclase.

TRANSLATION OF BINDING SIGNAL
TO ACTIVATION OF ADENYLYL CYCLASE

There can be little doubt from the foregoing that receptor binding and adenylyl cyclase activation are early and related events in the production of at least some of the biological actions of gonadotropins. At least two major questions arise from these events. The first, how does cAMP mediate the specific biological effects of the gonadotropins, is beyond the scope of this chapter. The second, how does the event of hormone binding to receptor lead to activation of the cyclase system, will receive some additional comment, albeit speculative.

The need for understanding these mechanisms is illustrated by the data of Catt and Dufau (45). In this study, they showed that the total number of hCG receptor sites in the rat testis exceeded the number required for maximal production of cAMP by a factor of about 10 and for the maximal production of testosterone by a factor greater than 100. These data give rise to the question of spare or redundant receptors, which is an echo of past considerations with respect to drug action.

A number of models have been proposed to explain the translation of the binding event into the activation of the adenylyl cyclase enzyme. These models need to take into account the concept of spare receptors, the role of nucleotides, burst and lag phenomena, fluoride activation, basal activity, the role of divalent ions and membrane lipids, and the desensitization phenomenon. At the present state of knowledge, none of the available models can fully account for these variables. Two functional models have been discussed above. Several physical models remain to be discussed.

Robison et al. (187) proposed a two subunit model in which the receptor (or regulatory or discrimination) subunit projects on the outer surface of the plasma membrane and makes physical contact with the catalytic unit on the inner portion of the membrane. The model can, of course, be modified by the placement of a transducer or coupler subunit between the receptor and catalytic units.

One postulate concerning these subunit models is the notion that the catalytic unit is repressed until it is in some way perturbed, analogous to the protein kinases. The simplest form of activation, hormone binding leading to dissociation or perturbation of the catalytic unit, is insufficient to explain the effects of fluoride and nucleotides. Schramm and Naim (188) noted that fluoride activation of rat parotid adenylyl cyclase persisted after removal of fluoride. This lead them to postulate the existence of an inhibitor protein which bound both hormone and fluoride, either binding event resulting in depression of the cyclase activity. Constantopoulos and Najjar (189) postulated that the catalytic unit is repressed by phosphorylation and that fluoride and/or hormone leads to dephosphorylation and activation.

Pfeuffer and Helmreich (190) noted that metabolically stable GTP analogues activated adenylyl cyclase in pigeon erythrocyte membranes and the activation persisted after removal of the analogue. Furthermore, they were able to identify and partially purify a high affinity guanyl nucleotide-binding protein which was of different molecular size that the adenylyl cyclase enzyme. This lead them to postulate that the nucleotide binding protein interacts with the receptor and adenylyl cyclase inside the cell membrane. In this view, the nucleotide binding protein is associated with the cyclase and the nucleotide binding event releases and activates the cyclase unit. The activated cyclase is then available for binding to and further activation by hormone receptor complexes. The last reaction is dependent upon membrane lipids, because hormonal activation of adenylyl cyclase requires the presence of lipids (190), particularly phospholipids (191).

One difficulty with these allosteric subunit models relates to the notion of spare or redundant receptor sites mentioned above. The concept of spare receptors assumes that there is a 1:1 relationship between a limited number of receptor molecules and an equal number of cyclase units. Any receptor sites in excess of the number of cyclase units are unrelated and may simply be redundant or subserve other functions. However, data from the rat hepatocyte membrane system (124) would seem to indicate that all of the glucagon binding sites are functionally related to the adenylyl cyclase system, but in a hyperbolic rather than a linear fashion. Furthermore, in the fat cell system there is evidence that a single adenylyl cyclase system subserves several hormone receptor systems (107, 113).

Perkins (192) has proposed a model which takes into account the fluid lipid matrix of the plasma membrane. In this view, the receptor and cyclase molecules are separated, or at least do not need to be linked. Hormone binding to the receptor on the extracellular surface results in either 1) lateral motion of the receptor in the lipid matrix to a position favoring perturbation and activation of the cyclase, or 2) an alteration of the lipid matrix which by a field effect leads to activation of the cyclase. In this model, there is no need to postulate spare receptors, because the degree of cyclase activation would be determined by the number of receptors occupied and set in motion or the strength of the field effect.

Cuatrecasas (193) expanded on the fluidity model, at least with respect to the variation concerning lateral motion of the receptor. He visualized two separate, successive, reversible reactions involving hormone (H), receptor (R), and adenylyl cyclase (Ac):

$$H + R \rightleftharpoons H \cdot R \text{ and } H \cdot R + Ac \rightleftharpoons H \cdot R - Ac.$$

This schema predicts discrepancies in the kinetics of hormone binding and activation of the biological response (adenylyl cyclase activity). If the rate of reaction of $H \cdot R$ with Ac is sufficiently slow, then lags would be anticipated. This second reaction may also depend, at least in part, on the state of fluidity of the membrane and the rate of change of phase transitions. If the rate of change from low fluidity to high fluidity is slow, then lags would be anticipated. If it is rapid, burst phenomena may occur. Changes in the state of fluidity would be affected by the lipid composition of the membrane and the temperature. Temperature has been reported to affect lags in hormonal activation of adenylyl cyclase (193). It is also possible that these phase transitions may occur in only limited segments of the membrane.

There is no a priori reason to believe that the dissociation of the $H \cdot R - Ac$ complex needs to follow the same sequence as the association. Thus, it may be possible that the dissociation of H from the $H \cdot R - Ac$ complex is slower than the dissociation of H from $H \cdot R$. This kind of phenomenon might explain the biphasic dissociation of hCG from rat ovarian receptors referred to above. Furthermore, the rate of dissociation of hormone may be more rapid than the dissociation of $R - Ac$.

These various association and dissociation reactions offer possibilities for explaining the effects of nucleotides and divalent ions on hormone binding and cyclase activation. For example, in the glucagon hepatocyte membrane system, GTP might favor the formation of $H \cdot R - Ac$ complexes and thus enhance glucagon-stimulated cyclase activity, while decreasing the affinity of H for R. It is also possible that nucleotides and divalent cations (by binding to polar head groups of lipids) could alter the state of membrane fluidity. These authors are not aware, however, of data indicating an effect of nucleotides on membrane fluidity. Alternatively, nucleotides and ions could affect microfilaments and microtubules and indirectly affect membrane organization and fluidity.

The fluidity model also allows the possibility of a "coded" membrane response to a single hormone binding event and thus a further mechanism for hormone specificity. The observations that certain hormonal stimuli increase cGMP while decreasing cAMP (194) might be a consequence of lateral motion of the receptor hormone complexes leading to sequential differential effects on the two cyclase enzymes. Similarly, there might be different effects on various membrane enzymes and transport processes. This scheme would be in contradistinction to the notion that each different cyclase system has its own receptor and the Hollenberg-Cuatrecasas (195) model which postulates a single cyclase catalytic unit whose substrate specificity is determined by the nature of the hormone being bound. Other possibilities have also been suggested (196).

The desensitization phenomenon described above may also be related to this fluidity model. The large amounts of hormone and relatively long time periods required to induce desensitization might result in clustering or capping of receptors and thus render them inactive until they diffuse or otherwise return to their original relationship to the cyclase units. The requirements for ATP and Mg^{2+} may be related to the kinetic and fluidity phenomena already discussed.

There are a variety of evidences for lateral motion of membrane proteins. Capping of ConA receptors (197) and cholera toxin receptors (198) on lymphocytes has been reported. The capping phenomenon appears to require a divalent ligand and suitable geometry so that 1 mol of ligand can bind to two or more receptors (198). Kinetic data for LH-hCG binding to rat ovarian receptors suggest that the gonadotropin is univalent. Thus far, there are no data to demonstrate either active or passive hormone-receptor mobility. Lateral diffusion of membrane protein has, however, been demonstrated in several systems (199).

Gilman and Minna studied clonal cell lines cultured in vitro for cAMP responsiveness to catecholamines (200), theophylline, and prostaglandins (210) and subjected these cell lines to somatic cell hybridization. These studies indicate the existence of hereditable control mechanisms concerned with cAMP responsiveness and suggested different types of mechanisms for catecholamines and prostaglandins. Clearly, more data will be required before the mechanisms by which hormones activate adenylate cyclase is fully understood.

ACKNOWLEDGMENT

We are particularly grateful to Mrs. Evelyn Gardner for her help in the preparation of the manuscript.

REFERENCES

1. Bishop, W. H., Nureddin, A., and Ryan, R. J. (1976). Pituitary luteinizing and follicle-stimulating hormones. *In* J. Parsons (ed.), Peptide Hormones, pp. 95–102. Macmillan Co., London.
2. Papkoff, H., Ryan, R. J., and Ward, D. N. The gonadotrophic hormones, LH(ICSH) and FSH: major advances, current status and outstanding problems. *In* R. O. Greep (ed.), Ford Foundation Review of Reproductive Biology and Contraceptive Development. In press.
3. Labhsetwar, A. P. (1973). Pituitary gonadotrophic function (FSH and LH) in various reproductive states. Adv. Reprod. Physiol. 6:97.
4. Faiman, C., and Winter, J. S. D. (1974). Gonadotropins and sex hormone patterns in puberty: clinical data. *In* M. M. Grumbach, G. D. Grave, and F. E. Mayer (eds.), Control of the onset of puberty, pp. 32–55. Wiley, New York.
5. Gay, V. L., Rebar, R. W., and Midgley, A. R., Jr. (1969). Constant monitoring of plasma luteinizing hormone by radioimmunoassay in individual rats following injection of hypothalamic extract. Proc. Soc. Exp. Biol. Med. 130:1344.
6. Nankin, H. R., and Troen, P. (1971). Repetitive luteinizing hormone elevations in serum of normal men. J. Clin. Endocrinol. Metab. 33:558.
7. Boyer, R. M., Finkelstein, J., Rottwarg, H., and Hellman, L. (1972). Synchronization of augmented LH secretion with sleep during puberty. N. Engl. J. Med. 287:582.
8. Ross, G. T., Cargille, C. M., Lipsett, M. B., Rayford, P. L., Marshall, J. R., Strott, C. A., and Rodbard, D. (1970). Pituitary and gonadal hormones in women during spontaneous and induced ovulatory cycles. Recent Prog. Horm. Res. 26:1.
9. Vande Wiele, R. L., Bogumil, J., Dyenfurth, I., Ferin, M., Jewelewicz, R., Warren, M., Rizkallah, T., and Mikhail, G. (1970). Mechanisms regulating the menstrual cycle in women. Recent Prog. Horm. Res. 26:63.
10. Spies, H. G., and Quadri, S. K. (1967). Regression of corpora lutea and interruption of pregnancy in rabbits following treatment with rabbit serum to ovine LH. Endocrinology 80:1127.

11. Spies, H. G., Hilliard, J., and Sawyer, C. H. (1968). Maintenance of corpora lutea and pregnancy in hypophysectomized rabbits. Endocrinology 83:354.
12. Rennie, P. (1968). Luteal-hypophyseal interrelationship in the rabbit. Endocrinology 83:323.
13. Saldarini, R. J., Hilliard, J., Abraham, G. E., and Sawyer, C. H. (1974). Relative potencies of 17α- and 17β-estradiol in the rabbit. Biol. Reprod. 3:105.
14. Holt, J. A., Keyes, P. L., Brown, J. M., and Miller, J. B. (1975). Premature regression of corpora lutea in pseudopregnant rabbits following the removal of estradiol. Endocrinology 97:76.
15. Moor, R. M., Rowson, L. E. A., Hay, M. F., and Caldwell, B. V. (1969). The effect of exogenous gonadotropins on the conceptus and corpus luteum in pregnant sheep. J. Endocrinol. 44:495.
16. MacDonald, G. J., Tashjian, J. A. H., and Greep, R. O. (1970). Influence of exogenous gonadotrophins, antibody formation and hysterectomy on the duration of luteal function in hypophysectomized rats. Biol. Reprod. 2:202.
17. Banik, U. K. (1975). Pregnancy terminating effect of human chorionic gonadotrophin in rats. J. Reprod. Fertil. 42:67.
18. Albert, A., and Berkson, J. (1951). Clinical bio-assay for chorionic gonadotropin. J. Clin. Endocrinol. Metab. 11:805.
19. Kobayashi, Y., Kupelian, J., and Maudsley, D. V. (1971). Ornithine decarboxylase stimulation in rat ovary by luteinizing hormone. Science 172:379.
20. Parlow, A. F. (1961). Bioassay of pituitary luteinizing hormone by depletion. of ovarian ascorbic acid. In A. Albert (ed.), Human Pituitary Gonadotrophins, pp. 300–310. Charles C Thomas Co., Springfield, Illinois.
21. Bell, E. T., Mukerji, S., and Lorraine, J. A. (1964). A new bioassay method for luteinizing hormone depending on the depletion of rat ovarian cholesterol. J. Endocrinol. 28:321.
22. Witschi, E. (1955). Vertebrate gonadotrophins. Mem. Soc. Endocrinol. 4:149.
23. Van Hall, E. V., Vaitikaitis, J. L., Ross, G. T., Hickman, J. W., and Ashwell, G. (1971). Effects of·progressive desialylation on the rate of disappearance of immunoreactive hCG from plasma in rats. Endocrinology 89:11.
24. Hansson, V., Reusch, E., Trygstad, O., and Torgersen, O. (1973). FSH stimulation of testicular androgen binding protein. Nature (New Biol.) 246:56.
25. Moon, Y. S., Dorrington, J. H., and Armstrong, D. T. (1975). Regulation of ovarian estradiol-17β synthesis by FSH and testosterone. Soc. Study Reprod., Abstract No. 19.
26. Williams, P. C. (1956). The history and fate of redundant follicles. In G. E. W. Wolstenholme and E. P. C. Millar (eds.), Ciba Found. Colloq. on Ageing, Vol. 2, pp. 59–68. Little, Brown and Co., Boston.
27. Clark, A. J. (1937). General pharmacology. In W. Heubner and T. Schüller (eds.), Heffter's Handbuch der experimentellen Pharmakologie: Erganzungswerk, Bd. 4, pp. 61–142. Springer, Berlin.
28. Stephenson, R. P. (1956). A modification of receptor theory. Br. J. Pharmacol. 11:379.
29. Paton, W. D. M. (1961). A theory of drug action based on the rate of drug receptor combination. Proc. R. Soc. Biol. 154:21.
30. Rodbard, D. (1973). Theory of hormone-receptor interaction. III. The endocrine target cell as a quantal response unit; a general control mechanism. In B. W. O'Malley and A. R. Means (eds.), Receptors for Reproductive Hormones, pp. 342–364. Plenum Press, New York.
31. deKrester, D. M., Catt, K. J., and Paulsen, C. A. (1971). Studies in vitro testicular binding of iodinated luteinizing hormone in rats. Endocrinology 88:332.
32. Moudgal, N. R., Moyle, W. R., and Greep, R. O. (1971). Specific binding of luteinizing hormone to Leydig tumor cells. J. Biol. Chem. 246:4983.
33. Lee, C. Y., and Ryan, R. J. (1971). The uptake of human luteinizing hormone (hLH) by slices of luteinized rat ovaries. Endocrinology 89:1515.
34. Lee, C. Y., and Ryan, R. J. (1972). Luteinizing hormone receptor: specific binding of human luteinizing hormone to homogenates of luteinized rat ovaries. Proc. Nat. Acad. Sci. U.S.A. 69:3520.

35. Lee, C. Y., and Ryan, R. J. (1973). Luteinizing hormone receptors in luteinized rat ovaries. *In* B. W. O'Malley and A. R. Means (eds.), Receptors for Reproductive Hormones, pp. 419–430. Plenum Press, New York.

36. Lee, C. Y., and Ryan, R. J. (1973). Interaction of ovarian receptors with human luteinizing hormone and human chorionic gonadotropin. Biochemistry 12:4609.

37. Bellisario, R., and Bahl, O. P. (1975). Human chorionic gonadotropin. V. Tissue specificity of binding and partial characterization of soluble human chorionic gonadotropin-receptor complex. J. Biol. Chem. 250:3837.

38. Gospodarowicz, D. (1973). Properties of the luteinizing hormone receptor of isolated bovine corpus luteum plasma membrane. J. Biol. Chem. 248:5042.

39. Gospodarowicz, D. (1973). Preparation and characterization of plasma membranes from bovine corpus luteum. J. Biol. Chem. 248:5050.

40. Rao, C. V. (1974). Properties of gonadotropin receptor in the cell membrane of bovine corpus luteum. J. Biol. Chem. 249:2864.

41. Lee, C. Y., Coulam, C. B., Jiang, N. S., and Ryan, R. J. (1973). Receptors for human luteinizing hormone in human corpora luteal tissue. J. Clin. Endocrinol. Metab. 36:148.

42. Kammerman, S., Canfield, R. E., Kolena, J., and Channing, C. P. (1972). The binding of iodinated hCG to porcine granulosa cells. Endocrinology 91:65.

43. Lee, C. Y. (1974). Human chorionic gonadotropin (hCG) binding and stimulation of adenyl cyclase (AC) of porcine granulosa cells during follicle development (Abstr.). In Vitro 10:343.

44. Catt, K. J., Tsuruhara, T., and Dufau, M. L. (1972). Gonadotrophin binding sites of the rat testis. Biochim. Biophys. Acta 279:194.

45. Catt, K. J., and Dufau, M. L. (1973). Interactions of LH and hCG with testicular gonadotrophin receptors. *In* B. W. O'Malley and A. R. Means (eds.), Receptors for Reproductive Hormones, pp. 379–418. Plenum Press, New York.

46. Catt, K. J., and Dufau, M. L. (1973). Spare gonadotrophin receptors in rat testis. Nature (New Biol.) 244:219.

47. Leidenberger, F., and Reichert, L. E., Jr. (1973). Species differences in LH as inferred from slope variations in a radioligand receptor assay. Endocrinology 92:646.

48. Means, A. R. (1973). Specific interaction of 3H-FSH with rat testis binding sites. *In* B. W. O'Malley and A. R. Means (eds.), Receptors for Reproductive Hormones, pp. 431–448. Plenum Press, New York.

49. Bhalla, V. K., and Reichert, L. E., Jr. (1974). Properties of follicle-stimulating hormone-receptor interactions. J. Biol. Chem. 249:43.

50. Bhalla, V. K., and Reichert, L. R., Jr. (1974). FSH receptors in rat testes: chemical properties and solubilization studies. *In* M. L. Dufau and A. R. Means (eds.), Hormone Binding and Target Cell Activation in the Testis, pp. 201–220. Plenum Press, New York.

51. Cheng, K.-W. (1975). Properties of follicle-stimulating hormone receptor in cell membranes of bovine testis. Biochem. J. 149:123.

52. Midgley, A. R., Jr. (1973). Autoradiographic analysis of gonadotropin binding to rat ovarian tissue reactions. *In* B. W. O'Malley and A. R. Means (eds.), Receptors for Reproductive Hormones, pp. 365–378. Plenum Press, New York.

53. Midgley, A. R., Jr. (1972). Gonadotropin binding to frozen sections of ovarian tissue. *In* B. B. Saxena, C. G. Beling, and H. H. Gandy (eds.), Gonadotropins, pp. 248–260. Wiley-Interscience, New York.

54. Rajaniemi, H. J., Hirshfield, A. N., and Midgley, A. R., Jr. (1974). Gonadotropin receptors in rat ovarian tissue. I. Localization of LH binding sites by fractionation of subcellular organelles. Endocrinology 95:579.

55. Han, S. S., Rajaniemi, H. J., Cho, M. I., Hirshfield, A. N., and Midgley, A. R., Jr. (1974). Gonadotropin receptors in rat ovarian tissue. II. Subcellular localization of LH binding sites by electron microscopic radioautography. Endocrinology 95:589.

56. Jungmann, R. A., and Schweppe, J. (1972). Mechanism of action of gonadotropins. II. Control of ovarian nuclear ribonucleic acid polymerase activity and chromatin template capacity. J. Biol. Chem. 247:5535.

57. Ryan, R. J., and Lee, C. Y. (1976). The role of membrane bound receptors. Biol. Reprod. 14:16.

146   Ryan, Birnbaumer, Lee, and Hunzicker-Dunn

58. Ketelslegers, J. M., Knott, G. D., and Catt, K. (1975). Kinetics of gonadotropin binding by receptors of the rat testis: analysis by a non-linear curve-fitting method. Biochemistry 14:3075.
59. Yang, K. P. P., Samaan, N. A., and Ward, D. N. (1975). Partial characterization of an inhibitor for luteinizing hormone receptor site binding. Soc. Study Reprod., Abstract No. 5.
60. Rodbell, M., Birnbaumer, L., Pohl, S. L., and Kraus, H. M. J. (1971). The glucagon-sensitive adenylate cyclase system in plasma membranes of rat liver. J. Biol. Chem. 246:1872.
61. Kammerman, S., and Ross, J. (1975). Increase in numbers of gonadotropin receptors on granulosa cells during follicle maturation. J. Clin. Endocrinol. Metab. 41:546.
62. Greenwood, F. C., Hunter, W. M., and Glover, J. S. (1963). The preparation of ^{131}I-labelled human growth hormone of high specific radioactivity. Biochem. J. 89:114.
63. Phillips, D. R., and Morrison, M. (1971). Exposed protein on the intact human erythrocyte. Biochemistry 10:1766.
64. Hubbard, A. L., and Cohn, Z. A. (1972). The enzymatic iodination of the red cell membrane. J. Cell Biol. 55:390.
65. Lee, C. Y., Tateishi, K., Ryan, R. J., and Jiang, N. S. (1975). Binding of human chorionic gonadotropin by rat ovarian slices: dependence on the functional state of the ovary. Proc. Soc. Exp. Biol. Med. 148:505.
66. Channing, C. P., and Ledwitz-Rigby, F. (1975). Methods for assessing hormone-mediated differentiation of ovarian cells in culture and in short-term incubation. In J. G. Hardman and B. W. O'Malley (eds.), Methods in Enzymology, Vol. 39, Part D, pp. 183–230. Academic Press, New York.
67. Lee, C. Y., and Ryan, R. J. (1974). Purification of the LH-hCG receptor from luteinized rat ovaries. In N. R. Moudgal (ed.), Gonadotropins and Gonadal Function, pp. 444–459. Academic Press, New York.
68. Dufau, M. L., and Catt, K. J. (1973). Extraction of soluble gonadotropic receptors from rat testis. Nature (New Biol.) 242:246.
69. Pohl, S. L., and Crofford, O. B. (1975). Techniques for the study of polypeptide hormone inactivation of receptor sites. In B. W. O'Malley and J. G. Hardman (eds.), Methods in Enzymology, Vol. 37, pp. 198–211. Academic Press, New York.
70. Lunenfeld, B., and Eskol, A. (1967). Immunology of human chorionic gonadotropin (hCG). In R. S. Harris, I. G. Wool, and J. A. Loraine (eds.), Vitamins and Hormones, Vol. 25, pp. 137–190. Academic Press, New York.
71. Espeland, D. H., Naftolin, F., and Paulsen, C. A. (1968). Metabolism of labeled ^{125}I-hCG by the rat ovary. In E. Rosemberg (ed.), Gonadotropins, pp. 177–184. Geron-S, Inc., Los Angeles.
72. Chang, S. C., Anderson, W., Kang, Y. H., and Ryan, R. J. The porcine ovarian follicle. III. Histology and fine structure at different stages of development. In press.
73. Scatchard, G. (1949). The attractions of proteins for small molecules and ions. Ann. N. Y. Acad. Sci. 51:660.
74. Danzo, B. J. (1973). Characterization of a receptor for human chorionic gonado-tropin in luteinized rat ovaries. Biochem. Biophys. Acta 304:560.
75. Haour, F., and Saxena, B. B. (1974). Characterization and solubilization of gonado-tropin receptor of bovine corpus luteum. J. Biol. Chem. 249:2195.
76. Louvet, J. P., and Vaitukaitis, J. L. (1975). Induction of follicle stimulating hormone (FSH) receptors in rat ovaries by estrogen priming (abstract). Abstracts of the 57th Annual Meeting of the Endocrine Society, p. 135.
77. Channing, C. P., and Kammerman, S. (1973). Characteristics of gonadotropin recep-tors of porcine granulosa cells during follicle maturation. Endocrinology 92:531.
78. Chang, S. C. S., Jones, J. D., Ellefson, R. D., and Ryan, R. J. The porcine ovarian follicle. II. Selected chemical analysis of follicular fluid at different developmental stages of the Graafian follicle. In press.
79. Lee, C. Y., and Ryan, R. J. (1974). Estrogen stimulation of human chorionic gonadotropin binding by luteinized rat ovarian slices. Endocrinology 95:1691.
80. Richards, J. A., and Midgley, A. R. (1976). Protein hormone action: a key to understanding ovarian follicular and luteal development. Biol. Reprod. 14:82.

81. Dufau, M. L., Ryan, D., and Catt, K. J. (1974). Disulfide groups of gonadotropin receptors are essential for specific binding of human chorionic gonadotropin. Biochim. Biophys. Acta 343:417.

82. Dufau, M. L., Charreau, F. H., Ryan, D., and Catt, K. J. (1974). Soluble gonadotropin receptors in the rat ovary. FEBS Letts. 39:149.

83. Catt, K. J., Dufau, M. L., and Tsuruhara, T. (1973). Absence of intrinsic biological activity in LH and hCG subunits. J. Clin. Endocrinol. Metab. 36:73.

84. Bishop, W. H., and Ryan, R. J. (1975). An investigation of relative subunit motility in human luteinizing hormone molecules using depolarization of fluorescence. Biochem. Biophys. Res. Commun. 65:1184.

85. Marsh, J. M., and Savard, K. (1964). The effect of 3',5'-AMP on progesterone synthesis in the corpus luteum. Fed. Proc. 23:462.

86. Marsh, J. M., and Savard, K. (1964). The activation of luteal phosphorylase by luteinizing hormone. J. Biol. Chem. 239:1.

87. Marsh, J. M., and Savard, K. (1966). The stimulation of progesterone synthesis in bovine corpora lutea by adenosine 3',5'-monophosphate. Steroids 8:133.

88. Sandler, R., and Hall, P. F. (1966). Stimulation in vitro by adenosine-3',5'-cyclic monophosphate of steriodogenesis in rat testis. Endocrinology 79:647.

89. Tsafriri, A., Lindner, H. R., Zor, U., and Lamprecht, S. A. (1972). In vitro induction of meiotic division in follicle-enclosed rat oocytes by LH, cyclic AMP, and prostaglandins E_2. J. Reprod. Fertil. 31:39.

90. Kolena, J., and Channing, C. P. (1972). Stimulatory effects of LH, FSH and prostaglandins upon cyclic 3',5'-AMP levels in porcine granulosa cells. Endocrinology 90:1543.

91. Marsh, J. M., Butcher, R. W., Savard, K., and Sutherland, E. W. (1966). The stimulatory effect of luteinizing hormone on adenosine 3',5'-monophosphate accumulation in corpus luteum slices. J. Biol. Chem. 241:5436.

92. Means, A. R., MacDougall, E., Soderling, T. R., and Corbin, J. D. (1974). Testicular adenosine 3',5'-monophosphate-dependent protein kinase: regulation by follicle stimulating hormone. J. Biol. Chem. 249:1231.

93. Dorrington, J. H., Roller, N. F., and Fritz, I. B. (1974). The effects of FSH on cell preparations from the rat testis. In M. L. Dufau and A. R. Means (eds.), Hormone Binding and Target Cell Activation in the Testis, pp. 237–241. Plenum Press, New York.

94. Marsh, J. M. (1970). The stimulatory effect of luteinizing hormone on adenyl cyclase in bovine corpora lutea. J. Biol. Chem. 245:1596.

95. Dorrington, J. F., and Baggett, B. (1969). Adenyl cyclase activity in the rabbit ovary. Endocrinology 84:989.

96. Marsh, J. M. (1975). The role of cyclic AMP in gonadal function. In P. Greengard and G. A. Robison (eds.), Advances in Cyclic Nucleotide Research, Vol. 6, pp. 100–137. Raven Press, New York.

97. Davoren, P. R., and Sutherland, E. W. (1963). The cellular location of adenyl cyclase in pigeon erythrocytes. J. Biol. Chem. 238:3016.

98. Pohl, S. L., Birnbaumer, L., and Rodbell, M. (1969). Glucagon sensitive adenyl cyclase in plasma membrane of hepatic parenchymal cells. Science 164:566.

99. Hechter, O., Bar, H.-P., Matsuba, M., and Soifer, D. (1969). ACTH sensitive adenyl cyclase in bovine adrenal cortex membrane fractions. Life Sci. 8:935.

100. Sutherland, E. W., and Rall, T. W. (1962). Adenyl cyclase. I. Distribution, preparation and properties. J. Biol. Chem. 237:1220.

101. Rall, T. W., and Sutherland, E. W. (1962). Adenyl cyclase. II. The enzymatically catalyzed formation of adenosine 3',5'-monophosphate and inorganic pyrophosphate from adenosine triphosphate. J. Biol. Chem. 237:1228.

102. Birnbaumer, L., Pohl, S. L., Rodbell, M. (1969). Adenyl cyclase in fat cells. I. Properties and the effects of adrenocorticotropin and fluoride. J. Biol. Chem. 244:3468.

103. Drummond, G. I., and Duncan, L. (1970). Adenyl cyclase in cardiac tissue. J. Biol. Chem. 245:976.

104. Rodbell, M., Birnbaumer, L., Pohl, S. L., and Krans, H. M. J. (1971). The glucagon-

sensitive adenyl cyclase system in plasma membranes of rat liver. V. An obligatory role of guanyl nucleotide in glucagon action. J. Biol. Chem. 246:1877.

105. Birnbaumer, L., Pohl, S. L., and Rodbell, M. (1971). The glucagon-sensitive adenyl cyclase system in plasma membranes of rat liver. II. Comparison between glucagon- and fluoride-stimulated activities. J. Biol. Chem. 246:1857.

106. Menon, K. M. J., and Smith, M. (1971). Characterization of adenyl cyclase from the testis of chinook salmon. Biochemistry 10:1186.

107. Birnbaumer, L., and Rodbell, M. (1969). Adenylyl cyclase in fat cells. II. Hormone receptors. J. Biol. Chem. 244:3477.

108. Bar, H.-P., and Hechter, O. (1969). Adenyl cyclase and hormone action. III. Calcium requirement for ACTH stimulation of adenyl cyclase. Biochem. Biophys. Res. Commun. 35:68.

109. Braun, T., and Dods, R. T. (1975). Development of a Mn^{2+}-sensitive "soluble" adenylate cyclase in rat testis. Proc. Natl. Acad. Sci. U.S.A. 72:1097.

110. Flawia, M. M., and Torres, H. N. (1973). Adenylate cyclase activity in *Neurospora crassa*. III. Modulation by glucagon and insulin. J. Biol. Chem. 248:4517.

111. Birnbaumer, L., Pohl, S. L., and Rodbell, M. (1972). The glucagon-sensitive adenylate cyclase system in plasma membranes of rat liver. VII. Hormonal stimulation: reversibility and dependence on concentration of free hormone. J. Biol. Chem. 247:2038.

112. Kauman, A. J., and Birnbaumer, L. (1974). Studies on receptor-mediated activation of adenylyl cyclase. IV. Characteristics of the adrenergic receptor coupled to myocardial adenylyl cyclase: stereospecificity for ligands and determination of apparent affinity constants for β-blockers. J. Biol. Chem. 249:7874.

113. Rodbell, M., Birnbaumer, L., and Pohl, S. L. (1970). Adenyl cyclase in fat cells. III. Stimulation by secretin and the effects of trypsin on the receptor lipolytic hormones. J. Biol. Chem. 245:718.

114. Birnbaumer, L., and Yang, P.-C. (1974). Studies on receptor-mediated activation of adenylyl cyclase. III. Regulation by purine nucleotides of the activation of adenylyl cyclases from target organs for prostaglandins, luteinizing hormone, neurohypophyseal hormones and catecholamines. Tissue- and hormone-dependent variations. J. Biol. Chem. 249:7867.

115. Rodbell, M., Birnbaumer, L., Pohl, S. L., and Krans, H. M. J. (1971). Regulation of glucagon action at its receptor. *In* M. Margoulies and H. Greenwood (eds.), Polypeptide and Protein Hormones, No. 241, Vol. I, pp. 199–211. Excerpta Medica, Amsterdam.

116. Salomon, Y., Lin, M. C., Londos, C., Rendell, M., and Rodbell, M. (1975). The hepatic adenylate cyclase system. I. Evidence for transition states and structural requirements for guanine nucleotide activation. J. Biol. Chem. 250:4239.

117. Wolff, J., and Cook, G. H. (1973). Activation of thyroid membrane adenylate cyclase by purine nucleotides. J. Biol. Chem. 248:350.

118. Birnbaumer, L., Yang, P.-C., Hunzicker-Dunn, M., Bockaert, J., and Duran, J. M. (1976). Adenylyl cyclase activities in ovarian tissues. I. Homogenization and conditions of assay in Graafian follicles and corpora lutea of rabbits, rats and pigs: regulation by ATP and some comparative properties. Endocrinology 99:163.

119. Birnbaumer, L., Nakahara, T., and Yang, P.-C. (1974). Studies on receptor-mediated activation of adenylyl cyclases. II. Nucleotide and nucleoside regulation of the activities of the beef renal medullary adenylyl cyclase and their stimulation by neurohypophyseal hormones. J. Biol. Chem. 249:7857.

120. Harwood, J. P., and Rodbell, M. (1973). Inhibition of fluoride ion of hormonal activation of fat cell adenylate cyclase. J. Biol. Chem. 248:4901.

121. Goldfine, I. D., Roth, J., and Birnbaumer, L. (1972). Glucagon receptors in β cells: binding of ^{125}I-glucagon and activation of adenylate cyclase. J. Biol. Chem. 247:1211.

122. Krishna, G., Harwood, J. P., Barber, A. J., and Jamieson, A. J. (1972). Requirement for guanosine triphosphate in the prostaglandin activation of adenylate cyclase of platelet membranes. J. Biol. Chem. 247:2253.

123. Leray, F., Chambaut, A.-M., and Hanoune, J. (1972). Role of GTP in epinephrine and glucagon activation of adenyl cyclase of liver plasma membrane. Biochem. Biophys. Res. Commun. 48:1385.

124. Rodbell, M., Lin, M. C., and Salomon, Y. (1974). Evidence for interdependent action of glucagon and nucleotides on the hepatic adenylate cyclase system. J. Biol. Chem. 249:59.
125. Londos, C., Salomon, Y., Lin, M. C., Harwood, J. P., Schramm, M., Wolff, J., and Rodbell, M. (1974). 5'-Guanylylimidodiphosphate, a potent activator of adenylate cyclase systems in eukaryotic cells. Proc. Natl. Acad. Sci. U.S.A. 71:3087.
126. Lefkowitz, R. J. (1974). Stimulation of catecholamine-sensitive adenylate cyclase by 5'-guanylyl-imidodiphosphate. J. Biol. Chem. 249:6119.
127. Cuatrecasas, P., Jacobs, S., and Bennet, V. (1975). Activation of adenylate cyclase by phosphoramidate and phosphorate analogs of GTP: possible role of covalent enzyme-substrate intermediates in the mechanism of hormone action. Proc. Natl. Acad. Sci. U.S.A. 72:1739.
128. Lin, M. C., Salomon, Y., Rendell, M., and Rodbell, M. (1975). The hepatic adenylate cyclase system. II. Substrate binding and utilization and the effects of magnesium ion and pH. J. Biol. Chem. 250:426.
129. Rendell, M., Salomon, Y., Lin, M. C., Rodbell, M., and Berman, M. (1975). The hepatic adenylate cyclase system. III. A mathematical model for the steady state kinetics of catalysis and nucleotide regulation. J. Biol. Chem. 250:4235.
130. Drummond, G. I., Severson, D. L., and Duncan, L. (1971). Adenyl cyclase: kinetic properties and nature of fluoride and hormone stimulation. J. Biol. Chem. 246:4166.
131. Perkins, P., and Moore, M. M. (1971). Adenyl cyclase in rat cerebral cortex: activation by sodium fluoride and detergents. J. Biol. Chem. 246:62.
132. Garbers, D. L., and Johnson, R. A. (1975). Metal and metal-ATP interactions with brain and cardiac adenylate cyclases. J. Biol. Chem. 250:8449.
133. Cleland, W. W., Gross, M., and Folk, J. E. (1973). Inhibition patterns obtained where an inhibitor is present in constant proportion to variable substrate. J. Biol. Chem. 248:6541.
134. DeHaën, C. (1974). Adenylate cyclase: a new kinetic analysis of the effects of hormones and fluoride ion. J. Biol. Chem. 249:2756.
135. Murad, F., Strauch, B. S., and Vaughan, M. (1969). The effect of gonadotropins on testicular adenyl cyclase. Biochim. Biophys. Acta 177:591.
136. Birnbaumer, L., Duran, J., Nakahara, T., and Kauman, A. Adenylyl cyclases: stimulation by hormones and regulation by nucleotides. In J. R. Symphies (ed.), Receptors in Pharmacology. Marcel Deckker, New York. In press.
137. Kolena, J., and Channing, C. P. (1971). Stimulatory effects of gonadotropins on the formation of cyclic adenosine 3',5'-monophosphate by porcine granulosa cells. Biochim. Biophys. Acta 252:601.
138. Marsh, J. M., Mills, T. M., and LeMaire, W. J. (1972). Cyclic AMP synthesis in rabbit Graafian follicles and the effect of luteinizing hormone. Biochim. Biophys. Acta 273:389.
139. Marsh, J. M., Mills, T. M., and LeMaire, W. J. (1973). Preovulatory changes in the synthesis of cyclic AMP by rabbit Graafian follicles. Biochim. Biophys. Acta 304:197.
140. Braun, T., and Sepsenwol, S. (1974). Stimulation of ^{14}C-cyclic AMP accumulation by FSH and LH in testis from mature and immature rats. Endocrinology 94:1028.
141. Bockaert, J., Hunzicker-Dunn, M., and Birnbaumer, L. (1976). Hormone-stimulated desensitization of hormone-dependent adenylyl cyclase. J. Biol. Chem. 251:2653.
142. Neer, E. J. Adenylate cyclase: approaches and problems. In L. Birnbaumer and B. W. O'Malley (eds.), Hormone Receptors, Vol. 2. Academic Press, New York. In press.
143. Hunzicker-Dunn, M., and Birnbaumer, L. (1976). Adenylyl cyclase activities in ovarian tissues. II. Regulation of responsiveness to LH, FSH and PGE$_1$ in the rabbit. Endocrinology 99:185.
144. Hunzicker-Dunn, M., and Birnbaumer, L. (1976). Adenylyl cyclase activities in ovarian tissues. III. Regulation of responsiveness to LH, FSH and PGE$_1$ in the prepubertal, cycling, pregnant and pseudopregnant rat. Endocrinology 99:198.
145. Zor, U., Bauminger, S., Lamprecht, S. A., Koch, Y., Clobsieng, P., and Lindner, H. R. (1973). Stimulation of cyclic AMP production in the rat ovary by luteinizing hormone: independent of prostaglandin mediation. Prostaglandins 4:449.
146. Kuehl, F. A., Jr., Humes, J. L., Tarnoff, J., Cirillo, V. J., and Ham, E. A. (1970).

Prostaglandin receptor site: evidence for an essential role in the action of luteinizing hormone. Science 169:883.

147. LeMaire, W. J., Yang, N. S. T., Behrman, H. H., and Marsh, J. M. (1973). Preovulatory changes in the concentration of prostaglandins in rabbit Graafian follicles. Prostaglandins 3:367.

148. Armstrong, D. T., Moon, Y. S., and Zamecnik, J. (1974). Evidence for a role of ovarian prostaglandins in ovulation. In N. R. Moudgal (ed.), Gonadotropins and Gonadal Function, pp. 345–356. Academic Press, New York.

149. Lamprecht, S. A., Zor, U., Tsafriri, A., and Lindner, H. R. (1974). Action of prostaglandin E_2 and luteinizing hormone on ovarian adenylate cyclase, protein kinase and ornithine decarboxylase activity during postnatal development and maturity in the rat. J. Endocrinol. 57:217.

150. Menon, K. M. K., Kawano, A., and Gunaga, K. P. (1975). Luteinizing hormone-induced accumulation of cyclic AMP and progesterone in rat interstitial cell suspension. In G. I. Drummond, P. Greengard, and G. A. Robison (eds.), Advances in Cyclic Nucleotide Research, Vol. 5, p. 803. Raven Press, New York.

151. Ratner, A., Weiss, G. K., and Duszynski, C. R. (1975). Adrenergic receptors in the regulation of ovarian cyclic AMP. In P. Greengard and G. A. Robison (eds.), Advances in Cyclic Nucleotide Research, Vol. 5, pp. 803–804. Plenum Press, New York.

152. Lee, C. Y. (1975). Gonadotropin and prostaglandin stimulation of adenylyl cyclase (AC) of porcine granulosa cells during follicle development. In the Program of the 57th Annual Endocrine Society Meeting, June 18–20, New York City.

153. Miller, J. B., and Keyes, P. L. (1974). Initiation of luteinization in rabbit Graafian follicles by dibutryl cyclic-AMP in vitro. Endocrinology 95:253.

154. Tsafriri, A., Lieberman, M. E., Barnes, A., Bauminger, S., and Lindner, H. R. (1973). Induction by luteinizing hormone of ovum maturation and steroidogenesis in isolated Graafian follicles of the rat: role of RNA and protein synthesis. Endocrinology 93:1378.

155. Mendelson, C., Dufau, M., and Catt, K. (1975). Gonadotropin binding and stimulation of cyclic adenosine $3',5'$-monophosphate and testosterone production in isolated Leydig cells. J. Biol. Chem. 250:8818.

156. Birnbaumer, L. (1973). Hormone-sensitive adenylyl cyclases: useful models for studying hormone receptor functions in cell-free systems. Biochim. Biophys. Acta 300:129.

157. Braun, T. (1974). Evidence for multiple, cell specific, distinctive adenylate cyclase systems in rat testis. In M. L. Dufau and A. R. Means (eds.), Hormone Binding and Target Cell Activation in the Testis, pp. 243–265. Plenum Press, New York.

158. Dorrington, J. F., and Fritz, I. B. (1974). Effects of gonadotropins on cyclic AMP production by isolated seminiferous tubule and interstitial cell. Endocrinology 94:395.

159. Fontaine, U.-A., Salmon, C., Fontaine-Bertrand, E., and Delurue-LeBelle, N. (1973). Age-related change in FSH- and in LH-sensitive ovarian adenyl cyclase. Horm. Metab. Res. 5:376.

160. Smith, M. S., Freeman, M. E., and Neill, J. D. (1975). The control of progesterone secretion during the estrous cycle and early pseudopregnancy in the rat: prolactin, gonadotropin and steroid levels associated with rescue of the corpus luteum of pseudopregnancy. Endocrinology 96:219.

161. Ying, S.-Y., and Greep, R. O. (1972). Effect of a single injection of estradiol benzoate (EB) on ovulation and reproductive function in 4-day cyclic rats. Proc. Soc. Exp. Biol. Med. 139:741.

162. Means, A. R., and Huckins, C. (1974). Coupled events in the early biochemical actions of FSH on the Sertoli cells of the testis. In M. L. Dufau and A. R. Means (eds.), Hormone Binding and Target Cell Activation in the Testis, pp. 145–163. Plenum Press, New York.

163. Dorrington, J. H., Roller, N. F., and Fritz, I. B. (1975). Effects of follicle stimulating hormone on cultures of Sertoli cell preparations. Mol. Cell. Endocrinol. 3:57.

164. Scott, R. S., and Rennie, P. I. C. (1970). Factors controlling the life span of the corpora lutea in the pseudopregnant rabbit. J. Reprod. Fertil. 23:415.

165. Anderson, R. N., Schwartz, F. L., and Ulberg, L. C. (1974). Adenylate cyclase activity of porcine corpora lutea. Biol. Reprod. 10:321.

166. Hunzicker-Dunn, M., and Birnbaumer, L. Desensitization of adenylyl cyclase in ovarian tissues by gonadotrophins. *In* L. Birnbaumer and B. W. O'Malley (eds.), Hormone Receptors, Vol. 2. Academic Press, New York. In press.
167. Challis, J. R. G., Davies, I. J., and Ryan, K. J. (1973). The concentration of progesterone, estrone and estradiol-17β in the plasma of pregnant rabbits. Endocrinology 93:971.
168. Morishige, W. K., Pepe, G. J., and Rothchild, I. (1972). Serum luteinizing hormone, prolactin and progesterone levels during pregnancy in the rat. Endocrinology 92:1527.
169. Pepe, G. J., and Rothchild, I. (1974). A comparative study of serum progesterone levels in pregnancy and in various types of pseudopregnancy in the rat. Endocrinology 95:275.
170. Morishige, W. K., and Rothchild, I. (1974). Temporal aspects of the regulation of corpus luteum function by luteinizing hormone, prolactin and placental luteotrophin during the first half of pregnancy in the rat. Endocrinology 95:260.
171. Channing, C. P., and Seymour, J. F. (1970). Effects of dibutryl cyclic-3′,5′-AMP and other agents upon luteinization of porcine granulosa cells in culture. Endocrinology 87:165.
172. Erickson, G. F., and Ryan, K. J. (1975). The effect of LH/FSH, dibutryl cyclic AMP and prostaglandins on the production of estrogens by rabbit granulosa cells. Endocrinology 97:108.
173. Miller, J. P., and Keyes, P. L. (1975). Progesterone synthesis in developing rabbit corpora lutea in the absence of follicular estrogens. Endocrinology 97:83.
174. Hilliard, J., and Eaton, W. E., Jr. (1971). Estradiol-17β, progesterone and 20α-hydroxypregn-4-ene-3-one in rabbit ovarian venous plasma. II. From mating through implantation. Endocrinology 89:522.
175. Mills, T. M., and Savard, K. (1973). Steroidogenesis in ovarian follicles isolated from rabbits before and after mating. Endocrinology 92:788.
176. Mills, T. M., Telegdy, G., and Savard, K. (1972). The synthesis and secretion of progesterone and 20α-hydroxy-Δ⁴-pregn-3-one by the rabbit ovary at various intervals after a single injection of hCG. Steroids 19:621.
177. Hunzicker-Dunn, M., and Birnbaumer, L. (1976). Adenylyl cyclase activities in ovarian tissues. IV. HCG-induced desensitization of responsiveness to LH in corpora lutea of rats and rabbits. Endocrinology 99:211.
178. Remold-O'Donnel, E. (1974). Stimulation and desensitization of macrophage adenylate cyclase by prostaglandins and catecholamines. J. Biol. Chem. 249:3615.
179. Franklin, T. J., and Foster, S. J. (1973). Hormone-induced desensitization of hormonal control of cyclic AMP levels in human diploid fibroblasts. Nature (New Biol.) 246:146.
180. Flint, A. P. F., Grinwich, D. L., Kennedy, T. F., and Armstrong, D. T. (1974). Metabolism of the corpus luteum during luteolysis in the pseudopregnant rabbit. Endocrinology 94:509.
181. Hill, R. T. (1934). Variation in the activity of the rabbit hypophysis during the reproductive cycle. J. Physiol. 83:129.
182. Linkie, D. M., and Niswender, G. D. (1972). Serum levels of prolactin luteinizing hormone and follicle stimulating hormone during pregnancy in the rat. Endocrinology 90:632.
183. Krey, L. C., and Everett, J. W. (1973). Multiple ovarian responses to sinele estrogen injections early in rat estrous cycles: impaired growth, luteotrophic stimulation and advanced ovulation. Endocrinology 93:377.
184. Mucherjee, C., Caron, M. G., and Lefkowitz, R. J. (1975). Catecholamine-induced subsensitivity of adenylate cyclase associated with loss of β-adrenergic receptor binding sites. Proc. Natl. Acad. Sci. U.S.A. 72:1945.
185. deVellis, J., and Brooker, G. (1974). Reversal of catecholamine refractoriness by inhibitors of RNA and protein synthesis. Science 186:1221.
186. Roth, J., Kahn, C. R., Lesniak, M. A., Gordon, P., DeMeyts, O., Megyisi, K., Neville, D. M., Jr., Gavin, J. R., III, Soll, A. H., Freychet, P., Goldfine, I. D., Bar, R. S., and Archer, J. A. (1975). Receptors for insulin NSILA-s, and growth hormone: applica-

tions to disease states of man. *In* R. O. Greep (ed.), Recent Progress in Hormone Research, Vol. 31, pp. 95–139. Academic Press, New York.

187. Robison, G. A., Butcher, R. W., and Sutherland, E. W. (1967). Adenyl cyclase as an adrenergic receptor. Ann. N. Y. Acad. Sci. 139:703.

188. Schramm, M., and Naim, E. (1970). Adenyl cyclase of rat parotid gland: activation by fluoride and norepinephrine. J. Biol. Chem. 245:3225.

189. Constantopoulos, A., and Najjar, V. A. (1973). The activation of adenylate cyclase. II. The postulated presence of (A) adenylate cyclase in a phospho (inhibited) form (B) and a dephospho (activated) form with a cyclic adenylate stimulated membrane protein kinase. Biochem. Biophys. Res. Commun. 53:794.

190. Pfeuffer, T., and Helmreich, E. J. M. (1975). Activation of pigeon erythrocyte membrane adenylate cyclase by guanylnucleotide analogues and separation of a nucleotide binding protein. J. Biol. Chem. 250:867.

191. Levey, G. S., Fletcher, M. A., Klein, I., Ruiz, E., and Schenk, A. (1975). Characterization of ^{125}I-glucagon binding in a solubilized preparation of cat myocardial adenylate cyclase: further evidence for a dissociable receptor site. J. Biol. Chem. 249:2665.

192. Perkins, J. P. (1973). Adenyl cyclase. *In* O. Greengard and G. A. Robison (eds.), Advances in Cyclic Nucleotide Research, Vol. 3, pp. 1–64. Raven Press, New York.

193. Cuatrecasas, P. (1974). Membrane Receptors. Ann. Rev. Biochem. 43:169.

194. Goldberg, N. D., O'Dea, R. F., and Haddox, M. K. (1973). Cyclic GMP. *In* P. Greengard and G. A. Robison (eds.), Advances in Cyclic Nucleotide Research, Vol. 3, pp. 155–223. Raven Press, New York.

195. Hollenberg, M. D., and Cuatrecasas, P. (1975). Insulin: interaction with membrane receptors and relationship to cyclic purine nucleotides and cell growth. Fed. Proc. 34:1556.

196. Illiano, G., Tell, G. P. E., Siegel, M. I., and Cuatrecasas, P. (1973). Guanosine 3′:5′-cyclic monophosphate and the action of insulin and acetylcholine. Proc. Natl. Acad. Sci. U.S.A. 70:2443.

197. Ingbar, M., and Sachs, L. (1973). Mobility of carbohydrate containing sites on the surface membrane in relation to the control of cell growth. FEBS Lett. 32:124.

198. Craig, S. W., and Cuatrecasas, P. (1975). Mobility of cholera toxin receptors on rat lymphocyte membranes. Proc. Natl. Acad. Sci. U.S.A. 72:3844.

199. Edidin, M., Zagyansky, Y., and Lardner, T. J. (1976). Measurement of membrane protein lateral diffusion in single cells. Science 191:466.

200. Gilman, A. G., and Minna, J. D. (1973). Expression of genes for metabolism of cyclic adenosine 3′:5′:monophosphate in somatic cells. I. Responses to catecholamines in parental and hybrid cells. J. Biol. Chem. 248:6610.

201. Minna, J. D., and Gilman, A. G. (1973). Expression of genes for metabolism of cyclic adenosine 3′:5′-monophosphate in somatic cells. J. Biol. Chem. 248:6618.

International Review of Physiology
Reproductive Physiology II, Volume 13
Edited by Roy O. Greep
Copyright 1977   University Park Press   Baltimore

5
Recent Advances
in the Control of
Male Reproductive Functions

M. R. N. PRASAD AND M. RAJALAKSHMI

Department of Zoology, University of Delhi, Delhi-110007, India

Considerable attention has been paid in recent years to finding new approaches to control of fertility in the male, based on elucidation of the ultrastructure of the testis, kinetics and hormonal regulation of spermatogenesis, blood-testis barrier, physiology of the epididymis, sperm maturation, biochemistry of the accessory gland secretions and semen, and the mechanism of action of androgens. There are two steps in the reproductive processes in the male which possibly can be interfered with and modified by alteration of the endogenous hormonal and nonhormonal factors that regulate their function. These are (*a*) inhibition of spermatogenesis and sperm transport and (*b*) inhibition of sperm maturation and their viability in the epididymis.

Two basic problems inherent in the application of either of these methods to control fertility in the human male are 1) the long time interval between the initiation of any treatment and the onset of infertility, which is sometimes not evident until 8–10 weeks after vasectomy, and 2) difficulty in selective interference with sperm production or sperm maturation without affecting libido or functions of the accessory glands. Recent advances in the regulation of testicular and epididymal function and their relevance to control of fertility in the male are discussed in this chapter.

CONTROL OF SPERMATOGENESIS

The hormonal requirements for maintenance of spermatogenesis have been reviewed extensively (1). Spermatogenesis can be maintained in hypophysectomized adult males by testosterone in the absence of gonadotropins, but its initiation at puberty requires follicle-stimulating hormone (FSH) (2). Reinitiation of spermatogenesis in the regressed testis following hypophysectomy also requires the presence of FSH (3). FSH alone has apparently no influence on the stimulation of spermatogenesis, but it synergizes with testosterone in the quantitative expression of spermiogenesis (1, 4). Testosterone or its 5α reduced metabolites are involved in initiation and completion of the meiotic division and other crucial stages of the spermatogenic process (2, 5).

The role of FSH in spermatogenesis is not yet clear. There are two different views on the relationship between spermatogenesis and FSH secretion: 1) specific stages in spermatid formation during spermiogenesis (6) or at the level of maturation of spermatids (1, 7) are involved in testicular feedback; 2) there is no relation between the stage of spermatogenesis and FSH levels, but there appears to be an inverse correlation between the severity of reduction of the germinal cells from spermatogonia to late spermatids and the levels of FSH. The poorer the germ cell population, the higher the levels of FSH and the response to luteinizing hormone-releasing hormone (LHRH) (8–15). It is likely that maturation of spermatids induces or permits the formation of factors controlling FSH secretion; alternatively, severe depression of the basal germ cell population may result in increased FSH levels (8). These observations support the concept of McCullagh (16) that the substance regulating FSH secretion may be of testicular origin.

"Inhibin"

There has been a resurgence of interest in the identification of the substance which was termed "inhibin" (16, 17). A protein with a molecular weight of 100,000 free from estrogen and testosterone, which causes suppression of FSH, has been purified from the water-soluble extract of bovine testis (14). A similar aqueous extract following extraction with diethylether and ultracentrifugation through membranes causes a selective decrease of FSH (42 and 85% of baseline levels) which remains suppressed for 24 hr; plasma levels of testosterone or estradiol are not altered (18). A crude ovine testicular extract, which is heat stable, lyophilizable, nondialyzable, and free from steroid contamination, suppresses FSH levels in 3–6 hr in rats, but not below tonic level; luteinizing hormone (LH) levels are not affected (19). The suppressive effect of the crude ovine testicular extract is dose-dependent, reaching a maximum at a dose of 100 mg of testicular extract. The suppression caused by a single injection of the testicular extract wanes with time and totally disappears by 36 hr.

Extracts from the rete testis fluid of rams contain two active materials, one with a molecular weight of 25,000 and another with a molecular weight of 80,000, which cause a decrease in plasma FSH in 35-day castrate rats (B. P. Setchell, personal communication). Human and bull seminal plasma also contain substance(s) proteinaceous in nature, which decrease FSH in rats (8, 20).

The mechanisms of action of "inhibin" are not clear. Injected intraperitoneally, "inhibin" extracted from seminal plasma decreases the FSH response to LHRH while, injected intravenously, it produces a decrease in response of both FSH and LH to LHRH; the reasons for such difference in response are not clear (8). An isolated increase in serum FSH in response to exogenous LHRH occurs in cryptorchid patients after correction of cryptorchidism, which suggests that "inhibin" mediates the secretion of FSH by an action on the pituitary rather than by modifying exogenous LHRH production by the hypothalamus (21). However, extracts from the rete testis fluid which inhibit FSH in vivo have no effect on endogenous release of FSH by pituitaries incubated in vitro (B. P. Setchell, personal communication).

Several model systems have been proposed to test "inhibin" activity experimentally. Two direct approaches are 1) continuous infusion of ovine testicular extract over a 24-hr period in rams castrated at birth, resulting in significant suppression of FSH by 12 hr (18); and 2) a rat model system in which ovine rete testicular fluid is injected subcutaneously into castrated adult male and immature female rats, resulting in suppression of FSH levels 48 hr later (22). Nandini et al. (19), have developed a highly sensitive, quick assay system in rats for testing the "inhibin" activity of ovine testicular extracts. Their model is based on the observation that plasma FSH reaches a peak in immature male rats on day 35 and declines thereafter (23). Administration of the ovine testicular extract to 35-day-old male rats, orchidectomized just prior to use, results in prevention of elevation of plasma FSH levels as seen 10½ hr after treatment. A single injection suppresses FSH levels within 3–6 hr, whereas LH levels are unaffected (19). The availability of this rapid assay for "inhibin" should be of great advantage in testing the potency of a number of preparations.

Role of FSH in Spermatogenesis: Site of Action

Sertoli Cells FSH by itself does not maintain spermatogenesis but acts synergistically with small doses of testosterone (1, 4). Successful treatment of cases of infertility in men with gonadotropins or clomiphene emphasizes the role of FSH in restoration of spermatogenesis (24–28). The role of FSH in stimulating the synthesis of a number of substances in the Sertoli cells and in regulating differentiation of germ cells has been highlighted in recent studies. The luminal processes of the Sertoli cells extend between the differentiating germ cells and partially surround them; the Sertoli-Sertoli cell junctions constitute the blood-testis barrier (29, 30). On the basis of junctional specializations of the boundaries between Sertoli cells and the germ cells, Fawcett (30) speculated that "any influence of the Sertoli cells on the differentiation of the germ cells must be through the release of small molecules to which the germ cell membranes are permeable or through alterations in the extra-testicular fluid environment in which the germ cells develop." The Sertoli cells may be affected by FSH, as shown by changes in the morphology of the cell (31) or binding of ferritin or ^{3}H/ ^{125}I-labeled FSH (32–36). FSH stimulates the synthesis of new messenger RNA and protein in the testis of immature rats (37–39). FSH binds to plasma membranes of the cells of the seminiferous epithelium and of Sertoli cells, resulting in elevation of cyclic adenosine $3':5'$-monophosphate (cAMP) and activation of cAMP dependent protein kinase (40–43). Temporal sequence of stimulation of testis by FSH including synthesis of rapidly labeled nuclear RNA in 15 min and of protein synthesis in 1–2 hr has been postulated and has been related to its effects on spermatogonia (42). An increase in mitosis of spermatogonia occurs during the period when protein synthesis is maximal; administration of FSH decreases markedly by 9 hr the degeneration of spermatogonia, which normally occurs in the seminiferous epithelium of immature rats. Means (42) postulated that prevention of degeneration of spermatogonia not only increases the number of viable cells which divide, but affects the number of spermatogonia which continue to differentiate. These data provide evidence for a rapid and direct effect of FSH on the proliferating spermatogonial population. The cell type on which FSH acts is not clear. The Sertoli cells may be the target cell for the action of FSH to initiate the biochemical changes culminating in the effects on spermatogonial mitosis. This is indicated by the electron microscopic studies of Fawcett (44), who postulated that the "Sertoli cells may release into the intercellular spaces of the adluminal compartment diffusible substances affecting germ cell differentiation." The nature of such substances transferred from the Sertoli cells to the germ cells is not clear.

Androgen-binding Protein   FSH has been implicated in the mechanisms that regulate the availability of androgens to the seminiferous epithelium. The action of FSH on maintenance of spermatogenesis is facilitated by androgens (1, 4) and is inhibited by treatment with antiandrogens, indicating a role for FSH which may be mediated by the action of androgens. The testes of several mammals have been shown to produce an androgen-binding protein (ABP) which is secreted into the testicular fluid (45–50). The ABP is possibly secreted by the Sertoli cells, and its secretion is stimulated by FSH (47, 51–53). ABP has been demonstrated in the 105,000 X g supernatants of the testis; it is a glycoprotein (MW 65–68,000) containing 20–30 carbohydrates and 70–80% amino acids. The testicular ABP is similar immunologically to testosterone-binding globulin in rabbit serum and may be the same protein; the two may be produced at different sites in the body by different hormonal mechanisms (54, 55). The stimulation of ABP by FSH may serve as a regulatory mechanism for the concentration of androgen from the extratubular testicular lymph into the seminiferous tubules to make available more androgen to the target cells (50). Androgens also stimulate the production of ABP by the Sertoli cells (50). The seminiferous tubules contain specific androgen receptors, different from ABP, which bind testosterone or dihydrotestosterone. Such cytoplasmic binding of androgens may constitute a mechanism for translocation of the hormone to the nuclei of the germ cells and Sertoli cells (49, 50, 56–58). It is not clear whether the transport and binding of androgen by the germ cells in the seminiferous epithelium is direct or mediated by the Sertoli cells. Tritiated testosterone can be localized autoradiographically in the Sertoli cells, spermatogonia, spermatocytes, and also in the Leydig cells (59). ABP is a carrier protein for the transport of androgens into the epididymis by way of the rete testis and efferent ducts (50, 60).

Interference with Spermatogenesis

Characteristics of Inhibin   Attempts to control fertility in the male by lowering levels of FSH by administration of "inhibin" are based on the assumption that deprivation of FSH interferes with spermatogenesis. A number of basic questions need to be answered, especially with regard to the nonhuman primates, before the validity of this assumption is proven and "inhibin," when available in purified form, is used for control of fertility. These questions relate to 1) the extent to which pituitary FSH can be decreased by inhibition of synthesis and/or release, following action of "inhibin," and 2) the extent of the reduction and duration of inhibition of FSH which is necessary to interfere with the cycle of the seminiferous epithelium. It is postulated that interference with the synthesis of ABP as a sequel to suppression of FSH and the transport of androgens into the seminiferous tubules and epididymis may inhibit spermatogenesis and epididymal function (60). This postulate needs critical examination. "Inhibin" presumably acts selectively by reduction of FSH, and not LH; plasma testosterone levels are not altered (18). Because androgens also stimulate the production of ABP by the Sertoli cells (60), it is likely that the transport of androgens into the seminiferous tubules and epididymis is not altered. Whether maintenance of spermatogenesis and/or the integrity of epididymal function are inhibited under these conditions following interference with FSH needs to investigated. Eshkol and Lunenfeld (4, 61) showed that spermatogenesis is maintained in adult rats deprived of gonadotropins by administration of human chorionic gonadotropin (hCG) or testosterone. Studies are in progress in non-human primates designed to answer the above questions by interfering with the action of FSH on the testis by passive-active immunization using highly purified FSH antisera (M. Prasad, N. R. Moudgal, and A. Kumar, unpublished observations).

Spermatogonial Chalone   The mechanisms regulating spermatogonial mitosis are not clear. In the rat irradiated with x-rays (300 rads), the renewing type A cells disappear in 12–14 days, followed by rapid division of reserve stem cells (AO cells) which divide and repopulate the seminiferous epithelium. On the basis of these studies, the existence of a mitotic inhibitory substance in the rat seminiferous epithelium has been suggested (62, 63). Testicular extracts produce a significant decrease in the uptake of [^{3}H] thymidine by the irradiated testis with no difference in the labeling indices of the intermediate or type B spermatogonia or preloptotene spermatocytes; these results have been considered to be evidence of specific inhibition of mitosis of type A spermatogonia by the action of a substance that is termed "spermatogonial chalone" (64). Similar results have been obtained following administration of testicular extract in immature rats (65). These interesting observations in the rat need to be confirmed in other species. Furthermore, the nature of the substance(s) involved in the inhibition needs to be established.

Inhibition of Spermatogenesis by Steroids   A number of steroidal and non-steroidal compounds have been tested for their effects on inhibition of spermatogenesis. Discussion of these compounds is omitted, because they have been reviewed extensively (66, 67). A few benzimidazole derivatives have been tested recently in rats. 2,3-Dihydro-2(1-naphtlyl)-4-(^{1}HO-quinazolinone (U-29409) causes rapid exfoliation of immature germ cells, which are eliminated through the epididymis; this effect may be due to disruption of intercellular bridges connecting the germ cells and the Sertoli cells (68). With the use of a "transillumination test" to study the early effects of such drugs on the morphology of the seminiferous tubules (69), it has been shown that a number of benzimidazole analogues (urea, 1-dimethyl-3-(-methyl-2-benzimidazolyl) hydrochloride, hydrate, U-32422 E and 2-benzimidazolecarbamic acid, methyl ester, and U-32104) disrupt contact of stages VII and IX with the Sertoli cells within 4 hr of administration of the compounds; they also cause death of the young spermatids (70). Of the many compounds that are antispermatogenic, none has the possibility of use in control of fertility in the human male for a number of obvious reasons (71): 1) irreversible action, damaging the cell types in the seminiferous epithelium (e.g., busulfan and other related derivatives of sulfonic acid); 2) lack of specific action on the testis and generalized metabolic effects (e.g., colchicine); 3) mutagenic hazards (e.g., methyl methane sulfonate and other sulphonic acid derivatives); 4) toxic effects (e.g., dinitropyrroles); 5) undesirable side effects (antabuse type of side effect, e.g., diamines), and 6) ineffectiveness of compounds in primates or man (e.g. deladroxone, ambilhar, etc.), although effective in rodents.

Recent reports of clinical trials with combinations of steroids for inhibition of spermatogenesis are discussed below.

A number of progestational compounds have been used (Norlutin, 17-ethynyl-19-nortestosterone (15 mg daily) or Enovid, 17-ethynyl-17-hydroxy-5,10-estren-3-one (15 mg twice daily) to produce irreversible azoospermia in men with concurrent loss of libido (72, 73).

Chlormadinone acetate (10 mg/day for 6 weeks) causes reduction in weight of the testis and accessory glands without any effects on sexual activity; these effects are reversible (74).

Depoprovera (17α-acetoxy-6α-methyl-pregn-4-ene-3,20-dione), a long-acting progestogen, suppresses spermatogenesis in rabbit and ram (75). MacLeod (76) observed spermatogenic arrest without any feminizing effect or loss of libido in men following a single injection of depoprovera. These results have not been confirmed by later work (67). Depoprovera causes reduction in weight of testes and sex accessories in rats without any effects on spermatogenesis, which may be due to its inhibitory effects on the pituitary or to its antiandrogenic activity (77). There is renewed interest in the use of depoprovera, in combination with androgens, for fertility control in men.

Norgestrel (1-13-β-ethyl-17-ethynyl-19-nortestosterone) suppresses spermatogenesis and accessory gland function in rats (78), although high doses of norgestrel (3.5 mg/day for 94 days) produce inconsistent results in monkeys (79). Norgestrel (250 μg/day) and testosterone enanthate (200 mg/month) are currently undergoing clinical trials. Norethindrone (25 mg/day orally for 3 weeks) depresses spermatogenesis and reduces plasma testosterone to levels which are compatible with normality of libido and potency (80).

A new approach to control of fertility in the male is the use of progestational compounds in combination with testosterone for suppression of spermatogenesis and maintenance of libido and accessory gland function, respectively. The effectiveness of such combinations has been demonstrated in rats; 10 μg of testosterone/day maintain accessory gland function in rats treated with depopro-vera whereas 100 μg/day of testosterone maintain accessory glands in rats treated with ethynodiol acetate (81). Similar results have been reported in rats using provera and testosterone (82).

A number of clinical trials using combinations of progestogens and testos-terone have been initiated in several countries. Progestational compounds are administered orally, whereas testosterone or the long-acting derivatives of testosterone are released from subcutaneously implanted silastic capsules or are injected. Norethandrolone (17-α-ethyl-17-hydroxy-19-nor-4-androsten-3-one) along with testosterone did not suppress spermatogenesis, although the combina-tion reduced pituitary gonadotropins (83). Oral administration of megesterol acetate (30 mg/day) or norethindrone (15 mg/day) along with four to six silastic capsules filled with testosterone caused azoospermia in 8–12 weeks; sperm counts returned to normal levels following withdrawal of treatment (84). Implants of four to five silastic capsules containing megesterol acetate (17 α-acetoxy-6-methyl-6-dehydroprogesterone) along with testosterone implants caused suppression of spermatogenesis. Levels of plasma LH and endogenously produced testosterone were markedly reduced. Although the amount of testos-terone released from silastic capsule implants was not equal to the normal level of plasma testosterone, it was sufficient to maintain libido (85). Coutinho and Melo (86) implanted subdermally in men two to five capsules of testosterone, followed by 100 mg/week of norgestrienone (17 α-ethynyl-17α-hydroxy-estra-4,9,11-trien-3-one) or 13-ethylnorgestrinone administered orally. A marked reduction in sperm count occurred in 8–12 weeks with no loss of libido or potency, but it was accompanied by weight gain, reduction in testicular mass, and gynecomastia in some cases. To overcome these limitations, Coutinho (87) suggested the use of the above regimen by the male for 1 year, with the same compound being used by the female partner on a weekly regime at one-tenth of the dose used by men for inhibition of fertility. This "unisex pill" is now undergoing clinical trials.

Nelevar (norethandrolone; 17α-ethyl-19-nortestosterone, 10 mg orally twice a day for 10 weeks or longer) and 200 mg of delatestryl (testosterone enanthate, 200 mg intramuscularly three times at weekly intervals) caused azoospermia which was reversible. Sexual potency was maintained throughout (88).

The effects of testosterone on the testis are dose dependent in the rat (89). Reddy and Prasad (90) demonstrated that implantation of two silastic capsules filled with testosterone (releasing 132 μg/day) inhibited spermatogenesis by suppression of gonadotropins, although the accessory gland function and libido were maintained. Similar results have been reported in rats (91). Doses of testosterone needed to suppress spermatogenesis may be high in the human. High doses of testosterone enanthate or testosterone induce reversible suppres-sion of spermatogenesis by inhibition of pituitary gonadotropins (76, 92, 93). Testosterone or testosterone propionate decrease the levels of plasma LH and FSH and excretion of gonadotropins (94, 95).

Danazol, a derivative of ethinyl testosterone (17α-pregnene-20-one-isoxazol-17-ol) is an orally active gonadotropin inhibitor devoid of estrogenic or progestational activity (96). Danazol (600 mg/day) administered orally for 4 months induced a slight reduction in sperm counts with a reduction in serum levels of LH and testosterone and consequent loss of libido. Danazol in combination with 10 mg of testosterone propionate (10 mg intramuscularly three times a week) caused reduction of spermatogenesis to oligospermic levels while libido was maintained. Administration of 200 mg of testosterone enanthate once a month with danazol (daily) caused even greater suppression of spermatogenesis than with danazol alone or with testosterone propionate (97). Administration of different combinations of danazol and testosterone caused reversible suppression of spermatogenesis, the most effective combination being 600 mg/day of danazol and 200 mg of testosterone enanthate, administered intramuscularly once a month. Infertility occurs after 12 weeks of treatment, with a sperm count in the oligospermic range of 5–10 million/ml and may be due to increase in the percentage of immature germ cells (98). Administration of 800 mg/day of danazol, combined with 200 mg of testosterone enanthate for 8 months, caused reduction in sperm counts which was associated with significant decrease in levels of testosterone in plasma and seminal plasma, whereas dihydrotestosterone levels remained unaffected (99). The mechanisms by which danazol decreases sperm counts is not clear; it may inhibit testicular steroidogenesis independent of its effects on gonadotropins (96) or by inhibiting spermatogenesis at the spermatocyte-spermatid level (98).

Testosterone enanthate (250 mg/week), administered for 21 weeks to normal males, caused no change in libido, sexual potency, or frequency of sexual intercourse; however, there was a reversible gain in body weight. Serum testosterone levels increased, whereas serum FSH and LH were suppressed (100). Decrease in sperm count was paralleled by a decrease in seminal fructose, although seminal plasma fluid volume was not changed. Recovery of spermatogenesis was variable from 13–16 weeks in some to 25–28 weeks in others. These results indicate need for further studies to clarify the mode of action of testosterone enanthate following different doses and methods of administration.

These results show that androgens per se may not be of use in control of fertility in the human male, but are needed in small amounts in combination with progestogens for maintenance of libido, potency, and accessory gland function. Long acting androgens are being tested to provide needed levels of androgens following oral or intramuscular administration. Oral administration of testosterone undecanoate caused a marked increase in plasma testosterone and androstenedione to levels attained following parenteral administration of testosterone; the effects of testosterone undecanoate may be due to absorption via the lymph, rather than via the portal, vessels (101). Testosterone undecanoate promises to be an effective, orally active androgen.

Although control of fertility in the human male by suppression of spermatogenesis by progestogens and maintenance of libido and accessory gland function with androgens is possible, a number of questions need to be answered before this promising approach is cleared for mass use in a family planning program. These questions relate to 1) the dose of progestogen needed to suppress spermatogenesis uniformly in all men; 2) long-term side effects of progestogens, viz. weight gain, gynecomastia, changes in liver function, and effects of androgens on liver function, etc.; 3) psychological complications following shrinkage in the size of the testes, and 4) number of capsules of testosterone to be implanted to ensure a level of circulating plasma testosterone high enough to maintain libido, potency, and functions of the accessory glands. Advances in drug delivery systems which provide a regulated, continuous release of the steroids to achieve the dual objectives of suppression of spermatogenesis and maintenance of libido are necessary to obviate the need for daily oral administration of steroids in men.

Role of Heat in Male Contraception

There has been recurrent interest since 1922 on the effect of heat on the seminiferous epithelium. Application of heat in the scrotum results in degenerative changes in the testis and decrease in sperm counts in men (102). Specific deleterious effects of heat are demonstrable on the kinetics of the seminiferous epithelium and changes in enzymes sensitive to heat in the testis (103). The effects of several sources of heat—hot water (60°C), infrared spot heating, microwave diathermy and ultrasound—on fertility have been tested in rats (104). The irradiated rats were infertile during a 10-month period, although plasma testosterone levels were not altered. The effects varied depending on the power and duration in the case of microwave and the dosage, duration, and frequency of application in the case of ultrasound. The possibility of reversible or irreversible sterilization of human males using modern methods of heat application has been suggested (104). A marginal increase in temperature to 37.6–37.9°C results in failure of the testicular thermoregulatory mechanism and increase in intrascrotal temperature (105).

RETE TESTIS

Recent studies on the ultrastructure and physiology of the rete testis have focused attention on the rete testis as an extraseminiferous tubular site for the action of compounds interfering with the viability of testicular spermatozoa. The epithelium of the rete testis is resorptive and secretory (106–109). The rete testis fluid (RTF) collected following either efferent duct ligation (110, 111) or directly by cannulating the extratesticular rete through the efferent ducts (112–114) is a mixture of the testicular fluid that is rich in potassium and bicarbonate and also sodium chloride equal in amount to that in plasma (115, 116). RTF contains a homogeneous suspension of relatively immotile sperm which are immature and possess a clear cytoplasmic droplet. The flow rate of

RTF is faster in smaller animals than in larger ones, but is constant from hour to hour and day to day. Although the RTF is practically free of glucose and fructose, it has a high concentration of inositol in all species except the wallaby (117). RTF contains a number of specific proteins, but all plasma proteins are not present. However, unlike in the seminiferous tubular epithelium, the proteins can readily pass through the epithelium of the rete testis. Testosterone- and dihydrotestosterone (DHT)-binding proteins have been detected in RTF (117). The testosterone concentration of RTF varies in the anesthetized monkey, rat, and the conscious ram, and is less than that in testicular and spermatic venous blood (111, 118). Among the other androgenic hormones, androstenedione, 5α-dihydrotestosterone, and dehydroepiandrosterone have been detected in the bull, rat, and ram (116). Testosterone and dehydroepiandrosterone enter the RTF unchanged, whereas androstenedione and progesterone are largely metabolized during transfer into RTF. However, cholesterol is totally excluded from the tubular and rete testis fluid even after being in circulation for up to 25 hr (119). Antifertility compounds such as methanesulfonic esters and halogenated alcohols like α-chlorohydrin and bromohydrin enter the RTF and equilibrate within 4–5 hr with the blood (116, 120, 121). These data show that the blood-testis barrier varies in permeability from total exclusion of substances like cholesterol from tubules to the free transport of molecules like androgens; the mechanisms by which these are regulated need to be clarified. The transport of androgens across the germinal epithelium into the seminiferous tubule appears to be dependent on molecular structure, metabolism of the steroid, rate of blood flow in the testis, and production of androgen-binding protein by Sertoli cells (118, 119, 122–126). Furthermore, it is not clear whether substances found in the RTF originate from the blood or the interstitial cells or are synthesized by the tubules themselves (117). The RTF, along with all its constituents, transports the spermatozoa to the epididymis, where selective absorption of a number of constituents occurs. In particular, the caput epididymis is dependent on the transport of androgens through the RTF for the maintenance of its functions (127); interference with androgen transport into the epididymis is a possible method of altering epididymal function and inhibition of sperm maturation.

In addition to these factors, the RTF may regulate the metabolism of spermatozoa by the supply of exogenous substrates and may also inhibit sperm motility, because the testicular and rete sperm possess all the necessary machinery for movement. The high potassium content of RTF and the epididymis may play a role in the retention of spermatozoa in these efferent ducts in a relatively immotile state. The energy reserves of the sperm are thus conserved until ejaculation, when the secretions of the accessory glands dilute the inhibitors present (115, 128). The rete testis thus offers an ideal extragonadal site for regulation of fertility by interference with its luminal physiology and consequent adverse effects on spermatozoa. However, no compound has been found to be effective selectively through the rete testis and its secretions.

PHYSIOLOGY OF EPIDIDYMIS

The epididymis is an organ in which the testicular spermatozoa undergo morphological and functional differentiation leading to the acquisition of characteristic patterns of motility and ability to fertilize ova (129–134).

Factors Regulating Epididymal Function

Androgens A number of hormonal and nonhormonal factors are involved in the regulation of epididymal function. Like the accessory glands, the structural and functional integrity of the epididymis is androgen-dependent (131, 132, 135–142). The epididymides of rat, hamster, and rhesus monkey (Tables 1 and 2) require relatively higher levels of androgens than the accessory glands for the maintenance of their morphological and functional characteristics (139–142). On the other hand, the epididymis has been shown to be more sensitive to testosterone than the accessory glands (143). The histology of the epididymis is maintained with lower doses of androgens than those required to attain dry weight (144). The structural integrity of stereocilia is maintained by lower levels of androgens than the secretory functions of the epididymal epithelial cells (134). The absorptive function of the epididymis is more sensitive to the levels of androgens than its secretory function (145).

Species differences exist in a number of mammals in the response of the caput and cauda epididymides to exogenously administered testosterone and DHT in maintaining growth, histoarchitecture, and secretory functions (Table 3).

The explanation for the differential responses of the caput and cauda epididymides of mammals to exogenous testosterone and DHT may reside in the amounts of androgens available to the different regions. The principal source of androgen supply to the cauda epididymidis is from the peripheral circulation, whereas the caput epididymidis receives, in addition, androgens from the testicular fluid. Within 60 min of administration of [^{3}H] testosterone through the femoral vein, the radioactivity rapidly appears in the epididymal fluid in excess of that found in tissue and plasma (146). Tritiated testosterone and [^{3}H] DHT are secreted into the cauda epididymidis and may be demonstrated by autoradiography (M. R. N. Prasad, M. Sar, and W. E. Stumpf, unpublished observations). The RTF entering the caput epididymidis contains testosterone in amounts slightly less than those in the spermatic venous blood (111, 116) and provides an additional source of androgens for the caput epididymidis. The androgens in the RTF may exist either as free testosterone or bound to the ABP produced by the Sertoli cells. The ABP acts as a carrier protein for the transport of androgens from the testis to the caput epididymidis, where the testosterone or DHT is taken up by the epithelial cells either by pinocytosis or by absorption (50, 60, 147–150). The mechanism of entry of testosterone into the epididymal epithelial cells and the fate of ABP in the epididymis is not clear. However, it has been shown that the ABP is absorbed principally in the caput epididymidis; very little ABP reaches the cauda epididymidis (50, 148, 151). Consequently, the epithelial

Table 1. Comparison of differential response of epididymis and accessory glands of rat and hamster to androgens[a]

Parameter	Hormone	DLP		Cowper's glands		Caput epididymidis		Cauda epididymidis	
		Rat	Hamster	Rat	Hamster	Rat	Hamster	Rat	Hamster
Weight	Testosterone[b]	50	10	50	10	200	200	200	400
	DHT[b,c]	25	25	25	25	200	200	200	200
Secretion[d] (content)	Testosterone[b]	50	NI	50	10	200	100	200	400
	DHT[b]	25	NI	200	5	100	200	200	200

From Karkun et al. (142).
[a]Minimal dose of androgens needed to maintain glands at control level.
[b]Dose in μg.
[c]The abbreviations are as follows: DHT, 5α-dihydrotestosterone; NI, not investigated.
[d]Sialic acid in epididymis and Cowper's glands; fructose in DLP in rat; and citric acid in DLP in hamster.

Table 2. Changes in weight and secretary activity of epididymis and accessory glands of castrated, subadult rhesus monkeys treated with testosterone-5α-dihydrotestosterone implants[a] (Means ± S.E.M.)

Treatment	Caput		Corpus epididymides		Cauda epididymides		Seminal vesicles		Bulbo-urethral glands	
	Weight[b] (mg)	Sialic acid[c]	Weight	Sialic acid	Weight	Sialic acid	Weight (g)	Fructose[d]	Weight	Sialic acid
Castrated control	664.0[e] ± 182.8	0.76[f] ± 0.14	892.0[f] ± 248.2	0.54 ± 0.18	753.7 ± 154.1	0.68[e] ± 0.17	3.9[f] ± 1.6	0.24[e] ± 0.11	124.0[f] ± 16.1	0.24[e] ± 0.06
Castrated + 4 implants of T	773.7[f] ± 250.6	0.90[f] ± 0.20	936.7[f] ± 117.5	1.55 ± 0.89	652.7[e] ± 132.1	0.67[e] ± 0.19	6.0[f] ± 1.2	1.50 ± 0.68	226.7 ± 60.6	0.52 ± 0.16
Castrated + 8 implants of T	596.0[e] ± 89.2	1.32 ± 0.10	802.0[e] ± 69.9	1.74 ±0.13	640.7[e] ± 41.8	0.77[e] ± 0.16	12.7 ± 1.7	25.33[e] ± 8.00	265.3 ± 51.8	0.71 ± 0.13
Castrated + 4 implants of DHT	721.0[e] ± 120.6	0.35[e] ± 0.09	1277.1[f] ± 12.2	0.79[f] ± 0.05	1126.4[f] ± 47.5	0.55[e] ± 0.12	14.5 ± 2.3	26.56[e] ± 10.98	304.0 ± 64.2	1.22 ± 0.25
Castrated + 8 implants of DHT	646.3[e] ± 172.7	0.16[e] ± 0.22	1421.1 ±	0.36[f] ±	1500.1 ±	0.41[e] ±	25.3[f] ±	74.20[e] ±	335.4 ±	1.04 ±
Intact control	1454.0 ± 172.7	1.39 ± 0.22	2197.7 ± 441.0	2.04 ± 0.61	1622.7 ± 240.6	1.84 ± 0.32	13.9 ± 4.3	2.09 ± 0.04	392.7 ± 159.0	0.75 ± 0.11

[a]Four or eight silastic capsules were implanted subcutaneously immediately after castration, Release rate of testosterone, 185 μg/day from four capsules and 325 μg/day from eight capsules; release rate of DHT, 190 μg/day from four capsules and 390 μg/day from eight capsules.
[b]Weight of paired organ.
[c]Content of sialic acid expressed as μmol/paired organ.
[d]Content of fructose expressed as mg/paired organ.
[e]Level of significance compared with intact control animals: $p < 0.05$.
[f]Level of significance compared with intact control animals: $p > 0.05$.

CAPUT EPIDIDYMIDIS

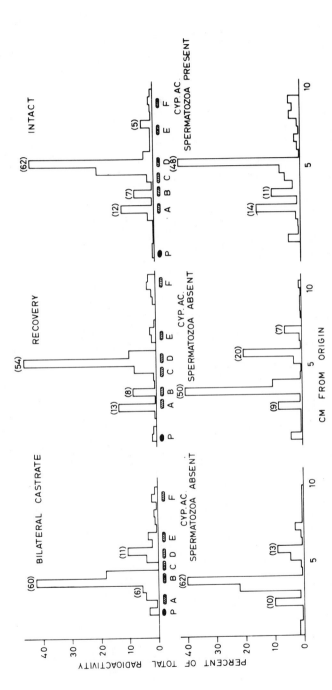

Figure 1. Androgen metabolites formed in slices of the caput epididymidis of the rhesus monkey after incubation with [³H] testosterone in vitro for 60 min. Steroids were separated by using ether-chloroform (1:9 v/v), and radioactivity was measured in 0.5-cm horizontal bands of the gel. Figures in parentheses indicate the radioactivity present in major metabolites, expressed as a percentage of the total radioactivity on the thin layer plate. *CYP.AC.* indicates treatment with cyproterone acetate. Subadult, male rhesus monkeys (6–10 kg) were used. Eight silastic capsules, each filled with 30 mg of cyproterone acetate, were implanted subcutaneously. Presence or absence of spermatozoa in semen was noted after electroejaculation. Monkeys were autopsied when they had reached the oligospermic stage (*CYP.AC.*, spermatozoa present) or azoospermic stage (*CYP.AC.* absent). In monkeys with azoospermia the capsules were removed, and the spermatogenesis was allowed to recover to normal (recovery). Intact animals and bi-lateral castrates (45 days) were used as controls. *P*, polar metabolite; *A*, androstanediol; *B*, testosterone; *C*, androsterone; *D*, dihydrotestosterone; *E*, androstenedione; and *F*, androstanedione.

Table 3. Comparative effects of testosterone and dihydrotestosterone on maintenance of epididymis and accessory glands in castrated rat, hamster, and monkey

	Weight		Histology		Secretion		Cowper's glands	
	Caput	Cauda	Caput	Cauda	Caput	Cauda	Weight	Secretion
Rat	T^a = DHT	T = DHT	DHT > T	DHT > T	DHT > T	T > DHT	DHT > T	T > DHT
Hamster	T > DHT	Neither	NI	NI	T (Sialic acid)	Neither	T = DHT	DHT > T
					T (Phospholipid)	Neither		
Monkey	Neither	DHT	NI	NI	Neither	DHT	DHT > T	DHT > T

^{a}The abbreviations are as follows: T, testosterone; DHT, 5α-dihydrotestosterone; NI, not investigated.

cells of the caput epididymidis are exposed to high concentrations of androgens, a finding which is supported by the fact that large amounts of testosterone, DHT, and dehydroepiandrosterone are detectable in this region (152–155). Thus, functions of the caput epididymidis are regulated more by the passage of testicular fluid containing androgens and spermatozoa, whereas the cauda epididymidis is more dependent on the levels of androgens in peripheral circulation than on the presence of spermatozoa (127, 156).

Spermatozoa In any discussion of the factors affecting epididymal physiology, the interaction between spermatozoa and the cells lining them in maintaining homeostasis of the epididymal tubules should not be overlooked. Marked alterations in epididymal function of rats have been reported following efferent duct ligation or selective suppression of spermatogenesis or inhibition of action of androgens by the antiandrogen, cyproterone acetate. The secretory activity of the caput epididymidis was markedly affected by the blockage of testicular fluid-spermatozoa following efferent duct ligation, whereas that of the cauda epididymidis decreased to a greater extent following action of the antiandrogen (127, 156). Analysis of the metabolism of [^{3}H] testosterone in 0.5-cm segments of the epididymis of the adult, intact rhesus monkey (Figure 3) shows that in the first three segments of the epididymis, which have few or no spermatozoa, [^{3}H] testosterone is not metabolized to DHT whereas in other segments of the caput, corpus and cauda epididymides the testosterone is metabolized to DHT (157, 158). In subadult male rhesus monkeys, arrest of spermatogenesis and consequent azoospermia alter selectively the pattern of metabolism of androgens by the caput epididymidis (159, 160). In the monkeys which are azoospermic, the metabolism of testosterone to DHT is markedly inhibited in the caput, but not in the cauda epididymidis (Figures 2 and 3). In monkeys in which spermatogenesis is restored or in untreated monkeys, testosterone is metabolized to DHT in all segments of the epididymidis (157). In addition to the epididymal tissue, the spermatozoa from the caput epididymidis of the rhesus monkey incorporate larger amounts of [^{3}H] testosterone when incubated in vitro than those from other regions of the epididymis (Table 4), the significance of which is not clear.

Action of Androgens on Epididymis

Synthesis of Steroids by Epididymis Histochemical and biochemical investigations show that the epididymis possesses the enzymes for the synthesis of steroids (161, 162). Testosterone and DHT occur in high concentrations in the epididymis and in the luminal fluid (152–155). Androgen biosynthesis occurs in vitro in the epididymis under appropriate conditions of incubation (131). The epididymis may also be capable of synthesizing testosterone in vivo and may modulate the function of the accessory glands through an intact ductus deferens (163).

Androgen Binding by Cytoplasmic and Nuclear Receptors The epididymis concentrates selectively the androgens it receives from testicular fluid or from

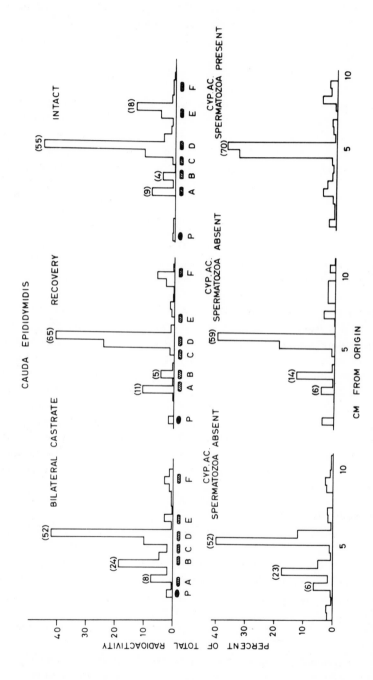

Figure 2. Androgen metabolites formed in slices of the cauda epididymidis of subadult rhesus monkeys after incubation with [³H] testosterone in vitro for 60 min. (For other details, see legend for Figure 1.)

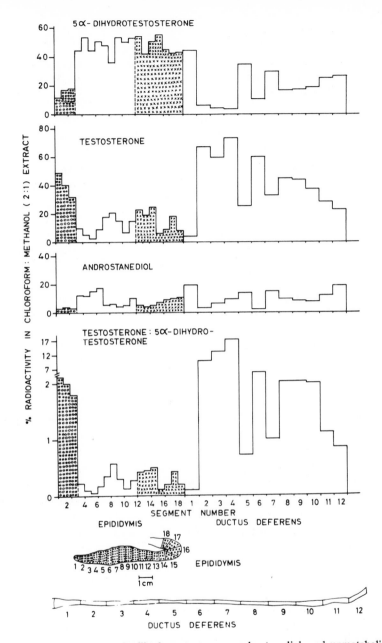

Figure 3. The percentages of 5α-dihydrotestosterone, androstanediol, and unmetabolized testosterone in slices of the epididymis and ductus deferens of a normal mature monkey after incubation with [³H] testosterone in vitro for 60 min. The ratio of unmetabolized testosterone to dihydrotestosterone indicates the extent of 5α reduction. The epididymis was divided into 0.5-cm segments and the ductus deferens into 2.0-cm segments before incubation. In segments 1–3, which were devoid of spermatozoa, the conversion of [³H] testosterone to 5α-dihydrotestosterone and androstanediol was lower than in other subsequent segments in which spermatozoa were present. The differences in pattern of metabolism of [³H] testosterone is comparable to that under conditions in which spermatozoa are present or absent in the epididymis (see Figures 1 and 2).

Table 4. Incorporation of [^{3}H] testosterone by spermatozoa from epididymis and ductus deferens of adult rhesus monkey

Period of incubation[a]	Epididymis			Ductus deferens
	Caput	Corpus	Cauda	
	(cpm/10^6 spermatozoa)			
30 min[b]	1,239	213	406	429
	1,193	219	341	452
60 min	906	136	242	112

From Norman et al. (160a).

[a]Spermatozoa were collected by gentle teasing of the epididymal tubules. Spermatozoa (40–60 million) were incubated at 37°C in a metabolic shaker in Norman-Johnson solution, pH 7.4, containing glucose (0.5%) and 1 μCi of 1,2-[^{3}H] testosterone (9 × 10^{-9} M).

[b]Duplicate samples.

peripheral circulation in vivo and in vitro (146, 164–170). Once testosterone is taken up by the epithelial cells, it is converted principally to DHT and androstenediol (169–171) and is bound to cytoplasmic receptors (151, 166, 172, 173). It is transported to the nucleus and binds with nuclear acceptors on the nuclear chromatin (171, 174–176) and initiates the various phenomena associated with hormone action. The sedimentation coefficient of human epididymal cytosol receptor is similar to that of the rat (J. A. Blaquier, personal communication). The cytoplasmic receptor from the epididymis has physical properties and steroid specificity resembling that from ventral prostate (149, 171, 172, 175–178) and is different from the ABP which binds and transports androgens from the testis to the epididymis. The ABP-androgen complex dissociates rapidly and acts as a carrier protein capable of rapidly releasing its androgen to the surrounding epithelial cells. Another striking difference between the two proteins is in the nature of their affinity for antiandrogens; cyproterone acetate inhibits the binding of [^{3}H] DHT to the epididymal receptors and to the epididymal nuclei to the same extent, but it has no effect on binding of [^{3}H] DHT to ABP (60 177). DHT uptake and binding to nuclei are dependent on cytoplasmic androgen-binding receptors, but not on ABP, which merely increases the amount of androgen entering the cells (60, 174, 177).

Metabolism of Androgens by Spermatozoa

In vitro and in vivo studies of the epididymis have shown that DHT is the major metabolite of testosterone in addition to androstanediol (169–171). However, segmental differences exist in the ability of the epididymis to convert testosterone to DHT. In the rat, maximal amounts of DHT are formed in vitro in the caput epididymidis (170), or in vivo in the cauda epididymidis (179). In the rhesus monkey, the caput and cauda epididymides possess equal abilities to convert testosterone to DHT (157, 162). However, in the hamster, the major metabolite of androgen is androstanediol and not DHT (180). The human epididymal tissue also has limited capacity for DHT formation (181).

Although the pattern of metabolism of androgens has been studied extensively in a number of adult mammals, very little is known about steroid metabolism in immature animals and changes, if any, occurring during the transition from the immature to the adult state. This lacuna appears glaring in view of the well known fact that during the early stages of postnatal life, androstenedione is the principal androgen in the testis and plasma and shifts gradually in favor of testosterone during the transition to the postpubertal state (182). Age-related studies in rats show that there is a concurrent shift in the ability of the epididymis and accessory glands to metabolize steroids in the immature rat; the principal metabolite of testosterone in vivo in the epididymis and accessory glands is androstenedione, whereas formation of DHT is negligible. During the transition from the prepubertal (40-day) to post-pubertal (70-day) condition, a gradual increase in the formation of DHT from testosterone and a corresponding decrease in the production of androstenedione occurs in the sex accessories until the pattern seen in adulthood, namely formation of DHT from testosterone, is established (183) (Table 5). No androphilic proteins or DHT are detectable in the epididymis of the 10-day-old rat; however, by 20 days of age the concentration of epididymal DHT, originally detectable on day 15, shows an increase concurrent with appearance of the 8S epididymal cytosol receptor. During subsequent post-pubertal development, the increase in epididymal DHT concentration is associated with the presence of the 8S cytosol receptor (173). Ejaculated and epididymal spermatozoa bind and metabolize testosterone (184, 185). Ejaculated human spermatozoa convert testosterone principally to androstenedione (186). Epididymal spermatozoa metabolize testosterone to DHT and 5α-androstanediol (166). Spermatozoa from the cauda epididymidis of rat convert testosterone to androstenedione and androstanediol, which is inhibited by prolonged exposure of rats to cyproterone acetate (179); the loss of motility of cauda epididymal spermatozoa in rats following long-term exposure to the antiandrogen (134, 187) may perhaps be due to decreased formation of androstanedione-androstanediol resulting in decreased levels of cAMP and loss of ability of sperm to use glycolyzable substrates (188–190). Levels of intracellular cAMP are correlated with degree of motility; decrease in motility of sperm coincides with loss of cAMP from sperm (188–190). In addition, the motility of spermatozoa is also dependent on their ability to metabolize glycolyzable sugars, which in turn is regulated by the levels of ATP and their breakdown (190). However, the relative roles of the androgen metabolites in maintaining the motility and fertilizing ability of spermatozoa have not been delineated except in the hamster, in which 5α-androstanediol maintains the fertilizing ability of spermatozoa in the cauda epididymidis (191).

Table 5. Relative percentages[a] of three major labeled metabolites in sex accessories of castrate rats on difference postnatal days, following intravenous administration of [³H] testosterone[b]

Postnatal age (days)	Metabolites	Percentage of metabolites recovered						
		Epididymis			Vas deferens	Ventral prostate	Dorsolateral prostate	Seminal vesicles
		Caput	Corpus	Cauda				
30[c]	Testosterone	1.3	2.0	1.4	1.6	1.2	2.0	1.5
	Dihydrotestosterone	9.5	10.5	11.7	17.9	9.2	14.6	19.3
	Androstenedione	74.5	59.4	66.8	49.1	72.9	56.4	53.3
45[c]	Testosterone	4.6	5.5	6.9	11.6	5.5	7.7	7.0
	Dihydrotestosterone	43.4	31.7	39.3	40.7	33.5	28.2	34.1
	Androstenedione	28.1	25.4	19.2	17.0	21.2	14.9	17.7
70[d]	Testosterone	5.2	8.7	7.7	10.9	12.4	6.0	7.6
	Dihydrotestosterone	47.0	41.6	42.2	41.0	45.1	41.6	42.3
	Androstenedione	20.6	26.3	21.5	21.3	21.6	28.6	21.1
80[d]	Testosterone	7.1	6.0	6.1	6.8	4.3	2.7	7.8
	Dihydrotestosterone	66.6	51.0	59.9	41.9	60.6	57.2	61.3
	Androstenedione	11.8	15.6	15.3	14.8	12.6	8.9	10.9
90[d]	Testosterone	9.2	9.0	13.0	28.9	7.0	7.7	9.9
	Dihydrotestosterone	65.6	50.1	61.6	42.7	71.3	61.1	66.7
	Androstenedione	8.6	15.6	14.6	6.4	5.3	5.9	6.3

From Singh et al. (183).

[a]Expressed as percentage of total radioactivity in the organ.
[b]Each animal received 20 μCi of [³H] testosterone/100 g body weight 48 hr after bilateral castration.
[c]Values represent a pool from three animals.
[d]Values represent the mean of three animals.

Sperm Maturation

Transport of spermatozoa through the epididymis is an essential prerequisite for the acquisition of characteristic patterns of motility and fertilizing ability in a number of mammals, including man (130). Rabbit spermatozoa become fertile in the proximal corpus and distal caput epididymidis, but not in the region proximal to the caput flexure, although they are motile (192, 193). Aging alone is sufficient for the development of spermatozoan motility, while fertilizing ability is acquired by interaction with the epididymal environment. The fertilizing capacity of spermatozoa is androgen-dependent; it is lost when the epididymal milieu is altered by hypophysectomy or castration (194–196). Spermatozoa from the rete testis are infertile and do not become fertile by incubation with cauda epididymidis; the rete testis fluid does not decrease the fertilizing ability of the spermatozoa from the cauda epididymidis. The secretions of the cauda epididymidis provide an environment for storage of spermatozoa and maintenance of their viability while fertilizing ability is acquired by contact with the secretions of the corpus epididymidis (197). The spermatozoa contained in an intact epididymal tubule from the corpus epididymidis become fertile in 1–4 days when maintained in organ culture containing testosterone, whereas spermatozoa from the caput epididymidis fail to become fertile when cultured under the same conditions (M. C. Orgebin-Crist, personal communication). Motility of spermatozoa in the cauda epididymidis of rat and hamster is maintained for a longer time than is their fertilizing ability (191, 195) in castrate animals treated with testosterone, whereas in unilaterally castrate rats motility is affected earlier than fertilizing ability in the absence of testicular fluid (198).

Maturation of spermatozoa involves a number of morphological, physiological, and biochemical changes (Table 6). A detailed discussion of these changes is omitted because the subject has been reviewed extensively (131, 199). The role of a few constituents of the epididymis in sperm maturation is discussed. However, no single parameter can be used as a marker for maturation of spermatozoa.

Carnitine The epididymis of rat, boar, bull, monkey, and human contain high concentrations of carnitine, which increase gradually in the epididymal luminal fluid from the caput to cauda epididymidis (200–202). The concentrations of carnitine in the epididymis increase during days 45–60 of postnatal growth (137). The epididymis selectively accumulates carnitine from blood (137), but cannot synthesize it in vitro from trimethylaminobutyrate, its immediate precursor (203). Carnitine may be involved 1) in maintaining the osmotic pressure of the fluid in the cauda epididymidis as a sequel to absorption of ions (137); 2) as a cofactor in the oxidation of fatty acids by the epididymal sperm (200); 3) as an acetyl transfer system in the epididymis and the sperm as a buffer preventing rapid changes in the ratio of acetyl coenzyme A (204), and 4) in generating the acetyl groups as an "energy reservoir" for sperm in a medium poor in substrates.

Table 6. Maturational changes in spermatozoa during transit through epididymis

Biochemical	Morphological	Physiological
Phospholipids[a] ↓		Cold shock susceptibility ↑
Proteins (nonnuclear) ↓	Membranes	Permeability ↑
Sialic acid/sialoprotein ↓	Membranes	Negative charge ↑
Cholesterol ↓		
Water and some electrolytes ↓	?	Specific gravity ↑
−S−S− bonds[b] ↑	Nucleus	?
−S−S− bonds[d] ↑	Outer dense fibers	Motility pattern changes
cAMP ↑	?	
Carnitine ↑		
Rates of glycolysis and fructolysis ↑ in vitro	−	Energy substrate ↑ utilization potential
?	Acrosome	
?	Cytoplasmic droplet migration	?

Modified from Hamilton (131).
[a]The symbols used are as follows: ↓, decrease; ↑, increase; ?, unknown; −, no effect.
[b]Disulfide bonds.

Phospholipids Phospholipids are important constituents of the epididymis in a number of mammals (205). Total lipids and phospholipids occur in higher concentrations in the caput than in the cauda epididymidis of rat (139, 142, 206), sheep (207), and monkey (140, 141). However, in the human, concentration of total lipids and phospholipid phosphorus show very little variation in the caput and cauda epididymides in the young and old age groups (208). The major phospholipids in the caput and cauda epididymides of hamster (206), monkey (140, 141), and human (208) are phosphatidylcholine and phosphatidylethanolamine; the high content of phosphatidylcholine in the caput epididymidis may be related to the synthesis of glycerylphosphoryl-choline (199). The concentration of phospholipids in spermatozoa is high in the caput, but decreases gradually during transport through the corpus and cauda epididymides (209–213). The final steps in the maturation of spermatozoa have been associated with the acquisition of choline plasmalogen during their transit through the epididymis (214–216); however, this has to be viewed with caution since the levels of choline plasmalogen in spermatozoa remain relatively constant (209, 217, 218) or decrease (212) during migration through the male reproductive tract. Prostaglandin $F_{2\alpha}$ is present in the seminal plasma of ram and may have its origin in the testis or epididymis. Prostaglandin, by its antilipolytic activity, may be involved in ensuring the production of spermatozoa less susceptible to changes in seminal plasma. It is likely that degradation or loss of lipid and protein from spermatozoa involving the action of prostaglandin may be under hormonal control (212). Furthermore, maturational changes in the sperm relating to permeability, motility, and resistance to cold shock have been correlated with modifications in the phospholipid levels in the spermatozoa. The contention that intracellular phospholipids serve as a source of oxidizable substrates (219, 220) has been questioned (216, 221).

Sialic Acids Sialic acid is a secretory product of the epididymal epithelium, the levels of which are dependent on androgens (135, 139–142, 222–225). Because sialic acid was estimated in these studies following acid hydrolysis, it is presumably bound to proteins and occurs as sialoproteins (135, 223). The epididymis synthesizes sialic acid (sialoproteins) during the prepubertal period (226); it decreases coincident with the entry of the first wave of spermatozoa into the epididymis and may be related to maturation of spermatozoa. In the rat, hamster, and monkey, the concentration of sialic acid in the spermatozoa decreases sharply during their transit from the caput to the cauda epididymidis. Bound sialic acid in the luminal plasma is maximal in the cauda epididymidis, whereas that bound to the sperm is minimal in this segment of the epididymis (213, 224, 226–228). These data demonstrate the existence of an inverse relationship between the levels of bound sialic acid in the luminal plasma and spermatozoa in the cauda epididymidis. In the rat and hamster, a decrease in levels of testicular androgens following 5 days of treatment with antiserum to lutenizing hormone or an inhibition of effects of androgens by the antiandrogen, cyproterone acetate (227, 228), results in a decrease in sialoproteins in the luminal plasma and those bound to spermatozoa in the cauda epididymidis. These results emphasize the relation between the epididymal environment and the spermatozoa and their dependence on androgens. Alteration of epididymal environment selectively by low doses of antiandrogens and decrease in the levels of sialic acid in the caput and cauda epididymides cause inhibition of sperm maturation (127, 187); the integrity of the plasma and acrosomal membranes of the spermatozoa are impaired along with loss of fertilizing ability (134). Sialoproteins may play a role 1) in the stabilization of the plasma membrane and the acrosome and its membranes; 2) in maintaining spermatozoa in a decapacitated state; 3) in antigenic interaction between sperm and epididymal epithelium; and 4) in maintaining the ionic balance in the epididymal plasma (133, 134).

Proteins The epididymis accumulates proteins selectively from blood (229) or secretes specific proteins to the luminal fluid (230, 231). Differences in the concentration of proteins exist in the different segments of the epididymis. The presence of spermatozoa may play a role in the regulation of the production and accumulation of proteins in the epididymal lumen. Unlike in the testis, barriers may not exist in the epididymis for the movement of large molecular weight proteins (131). Testosterone or DHT increase, in vitro, the incorporation of amino acids into proteins of epididymal tubules (232, 233). The ability of the different segments of the epididymis to synthesize proteins in vivo was demonstrated for the first time by Rajalakshmi and Prasad (234) and was shown to be androgen-dependent. The synthesis of new protein macromolecules in response to the administration of testosterone may be associated with growth and secretory activity of the epididymal cells. Rapid synthesis of new proteins also occurs in the epididymis following decrease in androgen production during experimentally induced cryptorchidism (235) or following inhibition of androgen action by cyproterone acetate (236), and may be associated with the synthesis of lysosomal enzymes as a necessary prerequisite for tissue involution.

Epididymis as Site for Control of Fertility in Male

Considerable attention has been focused in recent years on the epididymis as an extragonadal site for control of fertility in the male. A number of antiandrogens and derivatives of α-chlorohydrin have been used to interfere with processes of maturation and survival of spermatozoa in the epididymis.

α-Chlorohydrin The demonstration by Ericsson and Baker (237) that α-chlorohydrin, a derivative of glycerol, induced functional sterility in the rat by a direct action on the epididymis, evoked considerable interest and has led to its use in inhibition of fertility in a number of mammals (238–243). The rapid onset of infertility in male rats 48 hr after the first administration of α-chlorohydrin (242, 244) and its distribution in the reproductive tract (245, 246) indicate that it acts principally on the cauda epididymidis and vas deferens. The antifertility action of α-chlorohydrin may result from 1) impaired oxygen supply or interference with the absorption and removal of testicular fluid and/or secretory products, resulting in accumulation of metabolic wastes and consequent appearance of epididymal lesions and blockage of spermatozoa (247–249); 2) interference with vasculature of the epididymis (237); 3) interference with normal depletion of zinc from the epididymis (the resulting high concentration of zinc in the seminal plasma may interfere with the expenditure of energy by the spermatozoa (248, 250)); 4) the action of the compound as an alkylating agent (249); 5) reduction (251) or increase (250) in oxygen consumption of epididymal spermatozoa; 6) metabolic antagonism of phospholipid synthesis by competition with glycerol (252); and 7) interference with lipid synthesis and metabolism in the cauda epididymidis (253). α-Chlorohydrin administered to rats (8 mg/kg for 3 days) causes no change in the epididymal epithelium or in levels of sodium, potassium, and glycerylphosphorylcholine, but significantly increases the levels of lactic dehydrogenase and glutamic oxalacetic transaminase (254). These may be associated with loss of fertilizing capacity of the spermatozoa following release of these enzymes from degenerating spermatozoa (255, 256). Low doses of α-chlorohydrin prevent fertilization in the female following mating due to inhibition of sperm transport in the female genital tract, lack of sperm motility, and failure to stimulate oviducal muscles (257). The compound induces a general stress syndrome in rats with loss of body weight, atrophy of the adrenal glands, and increase in total leukocyte count (258). High doses of α-chlorohydrin cause depression of bone marrow and death in monkeys (250). α-Chlorohydrin added to boar and hamster semen in vitro inhibits fertilization (241, 242). The compound inhibits the energy metabolism of ejaculated human spermatozoa, which may be due to limitation of hexose utilization and resultant decrease in motility; the decrease in cAMP may reflect a drop in ATP (259). α-Chlorohydrin may be phosphosphorylated on entry into spermatozoa, and the phosphorylated compound may then competitively inhibit glyceraldehyde-3-phosphate dehydrogenase (260).

Other analogues, namely α-bromohydrin and α-iodohydrin, do not show any antifertility effects; the α-chloro group appears to be essential for the antifertility action (261, 262). α-Amino-3-chloro-2-propanol hydrochloride (aminochlorohydrin) shows promise, because it has a higher efficacy-to-toxicity ratio than α-chlorohydrin in rats and monkeys (263). The compound has rapid antifertility action on sperm in the epididymis and vas deferens, and its mode of action may be similar to that of α-chlorohydrin. The rapid antifertility action and its low antispermatogenic and antihormonal properties are encouraging. The *l* isomer of the amino derivative is effective as an antifertility agent, whereas the *d* form is not (Waites, personal communication). Compounds like chlorohydrin secreted into the epididymis, may be transported through the ejaculate into the female genital tract and may be absorbed by the vaginal mucosa; hence, concurrent studies on the effects of continued exposure of the female to such compounds are necessary before they are recommended for use in humans.

The recent finding of lesions, including glial proliferation, in the central nervous system of rhesus monkeys treated with aminochlorohydrin precludes the possibility of the use of aminochlorohydrin or α-chlorohydrin as a contraceptive agent in man.

Antiandrogens One of the compounds which is now undergoing clinical trials in man as a male contraceptive is the antiandrogen cyproterone acetate. Prasad et al. (187) reported that continuous release of microquantities of cyproterone acetate from subcutaneously implanted silastic capsules, at doses too low to inhibit testicular and accessory gland functions, altered selectively the epididymal milieu, resulting in loss of motility and fertilizing ability of spermatozoa; reversible functional sterility was thus induced without affecting mating behavior. The loss of fertilizing ability of the spermatozoa may be due to impairment of the secretory activity of the epididymis and consequent disruption of the plasma and acrosomal membranes and integrity of the acrosome itself (134,264). These data clearly demonstrated a dissociation of the dose of cyproterone acetate required to alter epididymal functions from that required to inhibit the secretory activity of the accessory glands or mating behavior. Although metabolism of testosterone to dihydrotestosterone is impaired in the accessory glands of rats treated with cyproterone acetate (179), their secretory functions are not altered (187). Schenk et al. (264) were not able to reproduce these results.

Prasad et al. (187) advocated the use of small doses of cyproterone acetate to induce functional sterility in the human male by selective interference with epididymal function without affecting libido or functions of the accessory glands. Clinical trials have been carried out to determine whether low doses of cyproterone acetate administered orally daily would produce similar differential effects in the human. Twenty and 10 mg/day or 10 and 5 mg/day of cyproterone acetate were administered in two clinical trials carried out in West Germany and India. Administration of 5 mg and 10 mg/day of cyproterone acetate orally to normal men over a period of 20 weeks caused a gradual decrease in the count and motility of spermatozoa, concurrent with an increase in the percentage of nonmotile spermatozoa, as well as of abnormal and immature forms (266). The ability of motile spermatozoa to penetrate through the cervical mucus (Kremer test) was markedly inhibited. The levels of acid phosphatase, sialic acid, and glycerylphosphorylcholine in the semen decreased rapidly, whereas the levels of fructose did not change significantly. The volume of semen was not altered. Libido and the frequency of coitus were not significantly altered. The levels of serum glutamic pyruvic transaminase, serum glutamic oxalotransaminase, and serum alkaline phosphatase were not significantly altered; there was no change in blood urea or in hematocrit values in either group (266). A rapid decrease in sperm count, along with increase in nonmotile as well as abnormal and immature sperm, has been reported following a daily dose of 10 or 20 mg of cyproterone acetate (267–268); the ability of ejaculated motile sperm to penetrate the cervical mucus was markedly impaired. Seminal acid phosphatase level was decreased, whereas alkaline phosphatase was elevated. Although the content of seminal fructose was not altered, the rate of fructolysis was decreased. The levels of plasma testosterone decreased marginally, but the levels of FSH and LH remained unaltered.

The human testis appears to be sensitive to action of antiandrogens; the extent of spermatogenic arrest is dose-dependent and may be due to inhibition of the action of androgens required for the maintenance of spermatogenesis. These effects may be mediated by competitive inhibition of progesterone-androgen receptors in the nonflagellate germ cells, resulting in an androgen deprivation effect and arrest of spermatogenesis (269). Cyproterone may act directly on the Sertoli cells to inhibit their function and interfere with spermatogenesis (270). Treatment of adult rhesus monkeys with 200 mg of cyproterone/day caused alterations in the subcellular structure of the Sertoli cells. The number of tight junctions, which are the morphological basis of the blood-testis barrier, was reduced between the Sertoli cells. Microfilaments, mitochondria, Golgi bodies, and endoplasmic reticulum were also reduced. The mitotic activity of type A spermatogonia was reduced. These changes were reversed following cessation of treatment, resulting in resumption of division of spermatogonia (270).

At the dose of cyproterone acetate used in man, the accessory gland functions were normal, although the functions of the epididymis (decrease in levels of glycerylphosphorylcholine, increase in the percentage of nonmotile spermatozoa, and marked inhibition of migration of motile spermatozoa through the cervical mucus) were impaired, similar to results obtained in rats. The results of the clinical trials with cyproterone acetate are of considerable interest and have demonstrated the possibility of selective changes in mechanisms regulating motility of spermatozoa without affecting sexual behavior or other functions of the reproductive system in man. Further clinical studies are necessary to determine whether the motile spermatozoa fail to migrate through the cervical mucus in vivo and whether they have also lost the ability to fertilize ova.

The subcellular mechanisms involved in the expression of the differential action of cyproterone acetate on the epididymis and accessory glands have been investigated in the rhesus monkey. The prostate, seminal vesicles, and epididymis of the monkey metabolize exogenously administered [3H] testosterone to dihydrotestosterone (271); treatment with cyproterone acetate does not alter the pattern of steroid metabolism either in the accessory glands or in the epididymis (157). Cyproterone acetate (10 nM) selectively inhibits in vitro the binding of [3H] testosterone or [3H] dihydrotestosterone to the epididymal cytosol of the monkey, whereas it causes a marginal enhancement of the binding of [3H] dihydrotestosterone to the cytosol of the prostate and seminal vesicles (271). The epididymis, which has a high threshold requirement of androgens (134, 140, 141) is thus affected by androgen deprivation due to impairment of uptake of androgens, but the accessory glands with a low threshold requirement of androgens function normally. These differences in the subcellular action of cyproterone acetate on the epididymis and accessory glands demonstrated in the rhesus monkey may provide the explanation for induction of functional sterility observed in rat, monkey, and perhaps also in man.

In the clinical trials with cyproterone acetate, the drug was administered daily in oral doses. Another form of drug delivery to obviate the necessity of oral administration of a daily pill needs to be devised. Dinakar et al. (160) have shown that cyproterone acetate released from subcutaneously implanted silastic capsules mimics the effects of the antiandrogen in human clinical trials; eight silastic capsules releasing 690 μg/day of cyproterone acetate caused azoospermia in adult rhesus monkeys in 4 months, without interfering with the ejaculate volume or biochemistry or functions of the accessory glands. This is an interesting and promising lead, suggesting an alternate method of drug delivery which could be tried in the human. Further improvements in contraceptive technology and drug delivery systems are necessary to provide continuous release of adequate quantities of the antiandrogen necessary to induce selective antifertility effects in man.

The influence of another antiandrogen, 17α-methyl-B-nortestosterone (SKF 7690), on epididymal sperm maturation has been tested in hamsters; spermatozoa isolated in the epididymis by ligature of the epididymis and exposed to graded doses of the antiandrogen for 12 days exhibited a dose-dependent reduction in the fertilizing ability (272). The compound had no effects on testicular androgen production (273). The histology of the epididymis showed degenerative changes, but the concentration of fructose in the seminal vesicles remained unchanged, indicating that the effect of the antiandrogenic treatment was primarily due to its interference with epididymal function (272).

A new antiandrogen, 6,7-difluoromethylene-4',5'-dehydro-1,2,2-methylene-(17R)-spiro(androst-4-ene-17,2(-3'H)-furan)-3-one, inhibits the histology of the prostate without any effects on the Leydig cells or spermatogenesis (274).

A nonsteroidal antiandrogen 4'-nitro-3'-trifluoro-methylisobutyr-anilide (Flutamide, Sch. 13521) inhibits growth of the rat prostate and retention of the DHT-receptor complex by the prostate cell nuclei (275, 276). The compound has no influence on the reproduction of male rats despite reduced weights of the accessory sex glands (277). R 2796 (17β-hydroxy-2,2β,17-trimethyl-8-estra-4,9,11-triene-3-one) and BOMT (6α-bromo-17β-hydroxy-17-methyl-4-ox-5-androstane-3-one) inhibit the formation of DHT-receptor complexes (278, 279). These antiandrogens have not been used clinically for control of fertility in the male.

Thus, antifertility effects of antiandrogens vary with their potency and the spectrum of endocrine activities. Neumann and Steinbeck (280) conclude that "even if an antiandrogen has a safe antifertility effect, the generalized nature of antiandrogenic activity does not permit its use as a male contraceptive. . . ." However, low doses of cyproterone acetate used in clinical trials have no effects on liver function, hematology, or other physiological responses (266–268). We feel that cyproterone acetate, as an antiandrogen, is promising for fertility regulation in man and has several advantages over the combination of progestogens and androgens now in use for control of fertility in the male. However, its mild progestational activity may cause undesirable side effects. It would, therefore, be desirable and necessary to limit the use of cyproterone acetate for specific periods (2 years), alternating with the female partner who could use a hormonal or nonhormonal contraceptive during the subsequent two years. Such an approach provides a method of contraception that would encourage a real partnership in family planning.

IMMUNOLOGICAL APPROACHES TO CONTROL OF FERTILITY IN THE MALE

Immunological neutralization of endogenous hormones has been achieved in several immature and mature laboratory rodents by passive or active immunization. This approach has been used to study 1) the development of the gonads and reproductive organs; 2) initiation and maintenance of spermatogenesis, and 3) hormonal regulation of epididymal and accessory gland function.

Immunization of rats and rabbits with ovine LH results in suppression of spermatogenesis, cessation of Leydig cell function, and loss of libido (281–283). However, different results were obtained by Talaat and Lawrence (284). Passive immunization with LH antisera obtained from rabbits was more effective than active immunization (285). The inhibition of gonadal function with LH antiserum was reversible in hamsters (286). Active immunization of dogs with LH antiserum caused similar effects; testosterone concentrations were markedly reduced, but libido was not affected (287).

Immunization with ovine FSH does not affect either spermatogenesis or accessory gland function (285, 288). On the other hand, Turner and Johnson (289) reported onset of sterility in rats 1–2 weeks after treatment with FSH antiserum without any effects on spermatogenesis or Leydig cell function; they attributed the quick onset of infertility to changes in the epididymal milieu although they did not explain the rationale for this conclusion. These results highlight the differences in response of different species of animals to the same antigen or immunization procedure; these variations could be due to neutralization of LH also. The role of FSH in influencing specific stages of spermatogenesis needs to be reinvestigated more carefully by use of highly purified FSH antisera free from antibodies reacting with LH or by use of purified FSH used as an antigen. Such studies are necessary in view of the recent demonstration by Means (42, 43) that FSH is involved in maintaining viability of spermatogonia and prevents their degeneration. The role of FSH in spermatogenesis has been discussed earlier in this chapter.

FSH and LH antisera have been administered to rats at different ages from birth to onset of puberty, resulting in varying degrees of interference with spermatogenic function and infertility (61). These results show that gonadotropins are required for initiation of gonadal function even during the early postnatal period and spermatogenesis; however, after spermatogenesis is initiated, hCG or testosterone maintains spermatogenesis in animals deprived of gonadotropins (4, 61). Administration of FSH or LH antiserum during days 5–15 postnatally has no effect on the onset of fertility in treated rats (290).

LH antiserum has been used as a good experimental tool to alter testicular androgen synthesis and to study changes in the physiology of the epididymis in relation to sperm maturation. This is discussed earlier in the chapter (225, 227, 228).

Immunization of rabbits with testosterone leads to loss of sexual activity, atrophy of the accessory glands, increased release of gonadotropins due to a stimulation of the hypothalamo-pituitary axis, resulting in testicular hypertrophy and hyperfunction (291, 292).

The advantages and disadvantages of use of active or passive immunization procedures have been described by Mitchison (293). Eshkol and Lunenfeld (61) have similarly discussed the caution which must be exercised in the use of FSH and LH for active immunization or in the use of gonadotropin antisera in interference with gonadal function and control of fertility in the male.

SUMMARY AND CONCLUSIONS

Recent advances in the control of male reproductive functions have been reviewed. Considerable attention has been paid to developing new approaches to control of fertility in the male, based on elucidation of the ultrastructure of the testis, hormonal regulation of spermatogenesis, and physiology of the epididymis. There are two possible steps in the male reproductive processes which can be modified for regulation of fertility: 1) inhibition of spermatogenesis and 2) inhibition of sperm maturation and their viability in the epididymis. Two basic problems inherent in the application of either of these methods to control fertility in the human male are (a) the long time interval between the initiation of any treatment and the onset of infertility which is usually not evident for 8–10 weeks, even after vasectomy, and (b) difficulty in selective interference with sperm production and sperm maturation without affecting libido or functions of the accessory glands.

The role of FSH in spermatogenesis has been reviewed. FSH secretion may be regulated by "inhibin," which may be of testicular origin. Proteins which cause suppression of FSH and have molecular weights ranging from 25,000–80,000 or 100,000 have been extracted from the rete testis fluid of rams or from bovine testes. Human and bull seminal plasma also contain proteinaceous substances which decrease FSH levels. Control of fertility in the male by lowering levels of FSH by administration of "inhibin" is based on the assumption that deprivation of FSH interferes with spermatogenesis. This theory needs to be tested, particularly in nonhuman primates.

A number of steroidal and nonsteroidal compounds have been tested for their effects on inhibition of spermatogenesis in laboratory animals; however, none has the possibility of use in control of fertility in the human male due to a number of side effects (71). Progestogens, estrogens, or androgens per se may not be of use in the control of fertility. One approach to the control of fertility in the human male is the use of progestational compounds with testosterone for suppression of spermatogenesis and maintenance of libido and accessory gland function, respectively. A number of such combinations have been successfully tested in men. A few questions which need to be answered before this promising approach is cleared for mass use in a family planning program are the long term side effects of progestogens—weight gain, gynecomastia, changes in liver function, and psychological complications following reduction in size of the testis.

The possibility of reversible or irreversible sterilization in human males using modern methods of heat application has been suggested.

Studies on the rete testis and its physiology show that it offers a post-tubular site for action of antifertility agents and regulation of fertility by interference with its luminal physiology and consequent adverse effects on spermatozoa. However, no compound has been found to be effective selectively through the rete testis and its secretions.

The epididymis is an organ in which testicular spermatozoa undergo morphological and functional differentiation leading to the acquisition of characteristic patterns of motility and the ability to fertilize ova. The epididymis has a higher threshold requirement of androgens for the maintenance of its function than the accessory glands. Species differences exist in the different regions of the epididymis for the metabolism of androgens and response to exogenously administered androgens. Epididymal spermatozoa metabolize steroids and also influence the metabolism of steroids by the epididymal tissue. The morphological, biochemical, and physiological changes in spermatozoa during maturation have been reviewed. Androgens stimulate the synthesis of a number of proteins in the epididymis. Changing levels of sialoproteins in the epididymal luminal plasma are related to the sialoproteins bound to the spermatozoa and may play a role in the stabilization of the plasma and acrosomal membranes.

α-Chlorohydrin and the antiandrogen, cyproterone acetate have been used to interfere with sperm maturation and sperm survival in the epididymis. α-Chlorohydrin and aminochlorohydrin induce functional sterility by a direct action on the epididymis and spermatozoa. The toxic effects of these compounds preclude their use as a contraceptive agent in man.

A new approach to induction of functional sterility in the male is selective alteration of epididymal function by a local androgen deprivation effect, without affecting functions of the accessory glands or libido. Clinical trials have been carried out to determine whether low doses of cyproterone acetate administered daily by mouth produce similar differential effects in man.

Administration of 5–10 mg/day of cyproterone acetate orally to men caused a gradual decrease in count and motility of spermatozoa; ability of motile spermatozoa to penetrate through the cervical mucus in vitro was inhibited. Volume of semen, libido, frequency of coitus, and levels of serum glutamic pyruvic transaminase, serum glutamic oxalotransaminase, blood urea, or hematocrit were not altered. An increase in the dose of cyproterone acetate to 20 mg caused similar effects, but decreased sperm counts along with an increase in nonmotile and abnormal, as well as immature spermatozoa. Studies are necessary to determine whether or not such motile spermatozoa fail to migrate through the cervical mucus in vivo and lose the ability to fertilize ova. Induction of functional sterility by the use of small quantities of antiandrogen is promising for fertility regulation in men.

REFERENCES

1. Steinberger, E. (1971). Hormonal control of mammalian spermatogenesis. Physiol. Rev. 51:1.
2. Steinberger, E. (1974). In H. M. Grumbach, G. D. Grave, and F. E. Mayer (eds.), Control of Onset of Puberty. John Wiley & Sons, New York.
3. Berswordt-Wallrabe, R. Von, Steinbeck, H., and Neumann, F. (1968). Effect of FSH on the testicular structure of rats. Endokrinologie 53:35.
4. Kalra, S. P., and Prasad, M. R. N. (1967). Effects of FSH and testosterone propionate in immature rats treated with clomiphene. Endocrinology 81:965.
5. Steinberger, A., Heindel, J. J., Lindsey, J. N., Elkington, J. S. H., Sanborn, B. M., and Steinberger, E. Endocr. Res. Commun. In press.

6. Franchimont, P. (1972). Human gonadotrophin secretion. J. R. Coll. Physicians Lond. 6:283.

7. Johnson, A. D. (1970). Testicular lipids in the testis. *In* A. D. Johnson, W. R. Gomes, and N. L. Vandemark (eds.), Testis, Vol. III, Academic Press, New York.

8. Franchimont, P., Chari, S., and Demoulin, A. (1975). Hypothalamus-pituitary-testis interaction. J. Reprod. Fertil. 44:335.

9. Franchimont, P., Millet, D., Vendrely, E., Letawe, J., Legros, J., and Netter, A. (1972). Relationship between spermatogenesis and serum gonadotrophin levels in azoospermia and oligospermia. J. Clin. Endocrinol. Metab. 34:1003.

10. Paulsen, C. A., Leonard, J. M., deKretser, D. M., and Leach, R. B. (1972). Interrelationship between spermatogenesis and follicle stimulating hormone levels. *In* B. B. Saxena, C. G., Balingand, and H. M. Gandey (eds.), Gonadotrophins, pp. 628–639. Wiley Interscience, New York.

11. deKrester, D. M., Burger, H. G., Fortune, D., Hudson, B., Long, A. R., Paulsen, C. A., and Taft, H. P. (1972). Hormonal histological and chromosomal studies in adult males with testicular disorders. J. Clin. Endocrinol. Metab. 35:392.

12. deKretser, D. M., Burger, H. G., and Hudson, B. (1974). The relationship between germinal cells and serum FSH levels in males with infertility. J. Clin. Endocrinol. Metab. 38:787.

13. Debeljuk, L., Arimura, A., and Schally, A. V. (1973). Pituitary and serum FSH and LH levels after massive and selective depletion of the germinal epithelium in the rat testis. Endocrinology 92:48.

14. de Kretser, D. M., Lee, V. W. K., Keogh, E. J., Burger, H. G., and Hudson, B. (1976). Studies on the relationship between FSH and germ cells: evidence for selective suppression of FSH by testicular extracts. J. Reprod. Fertil. (Suppl.). 24:1.

15. Joshi, L. R., Sheth, A. R., Raghavan, V. P., and Rao, S. S. (1973). Serum FSH and LH levels in fertile and subfertile men. Indian J. Med. Res. 61:1308.

16. McCullagh, D. R. (1932). Dual endocrine activity of the testis. Science 76:19.
    Johnson, S. G. (1964). Studies on the testicular-hypophyseal feedback mechanism in man. Acta Endocrinol. (Kbh.) (Suppl.) 90:99.

18. Keogh, E. J., Lee, V. M. K., Rennie, G. C., Burter, H. G., Hudson, B., and de Kretser, D. M. (1976). Selective suppression of FSH by testicular extracts. J. Reprod. Fertil. 46:7.

19. Nandini, S. G., Lipner, H., and Moudgal, N. R. (1976). A model system for studying inhibin. Endocrinology 98:1460.

20. Franchimont, P., Chari, S., Hagelstein, M. J., and Duraiswami, S. (1975). Existence of a FSH inhibiting factor, 'inhibin,' in bull seminal plasma. Nature 257:402.

21. Bramble, F. J., Houghton, A. L., Eccles, S. S., Murray, M. A. F., and Jacobs, H. S. (1975). Specific control of follicle stimulating hormone in the male: postulated site of action of inhibin. Clin. Endocrinol. 4:443.

22. Setchell, B. P., and Jacks, F. (1974). The effects of injections of ram rete testis fluid on the plasma follicle stimulating and luteinizing hormones in the rat. J. Endocrinol. 62:675.

23. Swederloff, R. S., Walsh, P. C., Jacobs, H. S., and Odell, W. D. (1971). Serum LH and FSH during sexual maturation in the male rat: effect of castration and cryptorchidism. Endocrinology 88:120.

24. Macleod, J., Pazianos, A., and Ray, B. (1966). The restoration of human spermatogenesis and of the reproductive tract with urinary gonadotrophins following hypophysectomy. Fertil Steril. 17:7.

25. Mancini, R. E., Vilar, O., Donini, P., and Perez-Leioret, A. (1971). Effect of human urinary FSH and LH on the recovery of spermatogenesis in hypophysectomised patients. J. Clin. Endocrinol. Metab. 33:888.

26. Gemzell, C., and Kjessler, B. (1964). Treatment of infertility after partial hypophysectomy with human pituitary gonadotrophins. Lancet 1:694.

27. Mellinger, R. C., and Thompson, R. J. (1966). The effect of clomiphene citrate in male infertility. Fertil. Steril. 17:94.

28. Reyes, F. I., and Faiman, C. (1974). Long-term therapy with low-dose cis-clomiphene in male infertility: effects on semen, serum FSH, LH, testosterone and estradiol, and carbohydrate tolerance. Int. J. Fertil. 19:49.

29. Dym, M., and Fawcett, D. W. (1970). The blood-testis barrier in the rat and the physiological compartmentation of the seminiferous epithelium. Biol. Reprod. 3:308.
30. Fawcett, D. W. (1974). Interaction between Sertoli cells and germ cells. *In* R. E. Mancini and L. Martini (eds.), Male Fertility and Sterility, pp. 13–36. Academic Press, New York.
31. Murphy, H. D. (1965). Sertoli cell stimulation following intratesticular injection of FSH in the hypophysectomized rat. Proc. Soc. Exp. Biol. Med. 117:1202.
32. Castro, A. E., Seigeur, A. C., and Mancini, R. E. (1970). Electron microscopic study of the localization of labelled gonadotrophins in the Sertoli and Leydig cells of the rat testis. Proc. Soc. Exp. Biol. Med. 133:582.
33. Means, A. R. (1974). Early sequence of biochemical events in the action of follicle stimulating hormone on the testis. Life Sci. 15:371.
34. Means, A. R., and Vaitukaitis, J. (1972). Peptide hormone acceptors: specific binding of [^{3}H]FSH to testis. Endocrinology 90:39.
35. Bhalla, U. K., and Reichert, L. E. (1974). *In* M. L. Dufau and A. R. Means (eds.), Hormone Binding and Target Cell Activation in the Testis. Plenum Press, New York.
36. Steinberger, A., Thanki, K. H., and Siegal, B. (1974). *In* M. L. Dufau and A. R. Means (eds.), Hormone Binding and Target Cell Activation in the Testis. Plenum Press, New York.
37. Reddy, P. R. K., and Villee, C. (1975). Messenger RNA synthesis in the testis of immature rats: effects of gonadotrophins and cyclic AMP. Biochem. Biophys. Res. Commun. 63:1063.
38. Means, A. R., and Hall, P. (1967). Effect of FSH on protein synthesis of immature rats. Endocrinology 81:1151.
39. Means, A. R., and Hall, P. (1968). Protein biosynthesis in the testis. I. Comparison between stimulation by FSH and glucose. Endocrinology 82:597.
40. Reddi, A. H., Ewing, L. L., and Williams-Ashman, H. G. (1971). Biochem. J. 122:333.
41. Dorrington, J. H., and Fritz, I. B. (1974). Cell-types influenced by FSH in the rat testis. *In* N. R. Moudgal (ed.), Gonadotrophins and Gonadal Function, pp. 500–511. Academic Press, New York.
42. Means, A. R. (1974). Metabolic effects of FSH on cells of the seminiferous epithelium. *In* R. E. Mancini and L. Martini (eds.), Male Fertility and Sterility, pp. 405–422. Academic Press, New York.
43. Means, A. R. (1974). Some effects of FSH on the seminiferous epithelium of mammalian testis. *In* N. R. Moudgal (ed.), Gonadotrophins and Gonadal Function, pp. 485–499. Academic Press, New York.
44. Fawcett, D. W. (1973). Interactions of cell types within the seminiferous epithelium and their implications for control of spermatogenesis. *In* S. J. Segal, R. Crozier, P. A. Corfman, and P. G. Condliffe (eds.), The Regulation of Mammalian Reproduction, pp. 116–138. Charles C Thomas, Illinois.
45. Ritzen, E. M., Dobbins, M. S., French, F. S., and Nayfeh, S. N. (1972). A high-affinity androgen binding protein in rat testis: evidence for its secretion in seminiferous fluid and passage through the epididymis. Excerpta Med. Int. Congr. Series No. 256:79.
46. Vernon, R. G., Kopec, B., and Fritz, I. B. (1973). Studies on the distribution of the high affinity testosterone binding protein (TBP) in rat testis seminiferous tubules. J. Endocrinol. 57:11.
47. Vernon, R. G., Kopec, B., and Fritz, I. B. (1974). Observations on binding of androgens by rat testis seminiferous tubules and testis extracts. Mol. Cell Endocrinol. 1:167.
48. French, F. S., and Ritzen, E. M. (1973). Androgen-binding protein in efferent duct fluid of rat testis. J. Reprod. Fertil. 32:479.
49. Hanson, V., Reusch, E., Ritzen, E. M., French, F. S., Trygstad, O., and Torgersen, O. (1973). FSH stimulation of testicular androgen binding protein. Nature (New Biol.) 246:56.
50. Hansson, V., Trygstad, O., French, F. S., McLean, W. S., Smith, A. A., Tindall, D. J., Weddington, S. C., Petrusz, P., Nayfeh, S. N., and Ritzen, E. M. (1974). Androgen transport and receptor mechanisms in testis and epididymis. Nature 250:387.

51. Hagenas, L., Ritzen, E. M., Ploen, L., Hansson, V., French, F. S., and Nayfeh, S. N. (1976). Sertoli cell origin of testicular androgen binding protein (ABP). Mol. Cell Endocrinol. In press.

52. Fritz, I. B., Kopec, B., Lam, K., and Vernon, G. (1974). In W. L. Dufau and A. R. Means (eds.), Hormone Binding and Target Cell Activation in the Testis. Plenum Press, New York.

53. Sanborn, B., Elkington, J. S. H., Steinberger, A., Steinberger, E., and Meistrich, M. L. (1975), In F. S. French, V. Hansson, S. M. Nayfeh, and E. M. Ritzen (eds.), Hormonal Regulation of Spermatogenesis, pp. 293–309. Plenum Press, New York.

54. Weddington, S. C., Brandtzaeg, P., Sletten, K., Christensen, T., Hanson, V., French, F. S., Petrusz, P., Nayfeh, S. N., and Ritzen, E. M. (1975). Purification and characterization of rabbit testicular androgen binding protein (ABP). In F. S. French, V. Hansson, E. M. Ritzen, and S. N. Nayfeh (eds.), Hormonal Regulation of Spermatogenesis, pp. 433–451. Plenum Press, New York.

55. Weddington, S. C., Brandtzaeg, P., Hansson, V., French, F. S., Petrusz, P., and Nayfeh, S. N. (1975). Immunological cross reactivity between testicular androgen binding protein and serum testosterone-binding globulin. Nature 258:257.

56. Smith, A. A., McLean, W. S., Hansson, V., Nayfeh, S. N., and French, F. S. Androgen receptor in nuclei of rat seminiferous tubules. Steroids. In press.

57. Sanborn, B. S., and Steinberger, E. (1975). Androgen nuclear exchange in rat testis. Endocr. Res. Commun. 2:335.

58. Mulder, E., Peters, M. J., deVries, J., and Van der Molen. (1975). Androgen receptors in the rat testis. Mol. Cell Endocrinol. 25:171.

59. Sar, M., Stumpf, W. E., McLean, W. S., Smith, A. A., Hansson, V., Nayfeh, S. N., and French, F. S. (1975). Localization of androgen target cells in the rat testis: sutoradiographic studies. In F. S. French, V. Hansson, E. M. Ritzen, and S. N. Nayfeh (eds.), Hormonal Regulation of Spermatogenesis. Plenum Press, New York.

60. Hansson, V., Weddington, S. C., French, F. S., McLean, W. S., Smith, A. A., Nayfeh, S. N., Ritzen, E. M., and Hagenas, L. Secretion and role of androgen binding proteins in the testis and epididymis. J. Reprod. Fertil. (Suppl.) In press.

61. Eshkol, A., and Lunenfeld, B. (1975). Use of antisera to gonadotrophins in reproduction research. In E. Neischlag (ed.), Immunization with Hormones in Reproduction Research. North-Holland, Amsterdam.

62. Dym, M., and Clermont, Y. (1970). Role of spermatogonia in the repair of the seminiferous epithelium following X-irradiation of the rat testis. Am. J. Anat. 128: 265.

63. Clermont, Y., and Girard, A. (1973). Existence of a spermatogonial chalone in the rat testis. Anat. Rec. 175:294.

64. Clermont, Y., and Mauger, A. (1974). Existence of spermatogonial chalone in the rat testis. Cell Tissue Kinet. 7:165.

65. Clermont, Y., and Mauger, A. Effect of spermatogonial chalone on the growing rat testis. Cell Tissue Kinet. In press.

66. Gomes, W. R. (1970). Metabolic and regulatory hormones influencing testis function. In A. D. Johnson, W. R. Gomes, and N. L. Vandemark (eds.), The Testis, Vol. 3, pp. 68–138. Academic Press, New York.

67. de Kretser, D. M. (1974). The regulation of male fertility: the state of the art and future possibilities. Contraception 9:561.

68. Ericsson, R. J. (1971). Antispermatogenic properties of 2,3-dehydro-2-Ci-napthyl-4-(1H)-quin-azolinone (U-29, 409). Proc. Soc. Exp. Biol. Med. 137:532.

69. Parvinen, M. (1973). Observations on the freshly isolated and accurately identified spermatogenic cells of the rat: early effects of heat and short-time experimental cryptorchidism. Virchows. Arch. Abs. B. Zellpath. 13:38.

70. Parvinen, M., and Kormano, M. (1974). Early effects of antispermatogenic benzimidazole derivatives, U32,422A and U32,104 on the seminiferous epithelium of rat. Andrologia 6:245.

71. Prasad, M. R. N. (1973). Control of fertility in the male. In Pharmacology and the Future of Man. Proceedings of the 2nd International Congress of Pharmacology, San Francisco, 1972. 1:208. S. Karger, Basel.

Control of Male Reproductive Functions   189

72. Heller, C. G., Laidlaw, W. M., Harvey, H. T., and Nelson, W. O. (1958). Effects of progestational compounds on the reproductive processes of the human male. Ann. N. Y. Acad. Sci. 71:649.
73. Heller, C. G., Moore, D. G., Paulsen, C. A., Nelson, W. O., and Laidlaw, W. M. (1959). The effect of progesterone and synthetic progestins upon the reproductive physiology of normal man. Fed. Proc. 18:454.
74. Dorner, G., Gotz, F., and Mainz, K. (1972). Infertility and maintenance of sexual behaviour in male rats treated with chlormadinone acetate. J. Endocrinol. 52:197.
75. Ericsson, R. J., and Dutt, R. H. (1965). Progesterone and 6α-methyl-17α-hydroxy-progesterone acetate as inhibitors of spermatogenesis and accessory gland function in the ram. Endocrinology 77:203.
76. MacLeod, J. (1965). In C. R. Austin, and J. S. Perry (eds.), Symposium on Agents Affecting Fertility, pp. 93–109. Little Brown & Co., Boston.
77. Singh, J. N., Jehan, Q., Setty, B. S., and Kar, A. B. (1971). Effect of some steroids on spermatogenesis and fertility of rats. Indian J. Exp. Biol. 9:132.
78. Singh, J. N., Setty, B. S., Srivastava, K., and Kar, A. B. (1972). Effect of norgestrel on spermatogenesis and fertility of rats. Indian J. Exp. Biol. 10:159.
79. Setty, B. S., and Kar, A. B. (1969). Antispermatogenic effect of norgestrel in rhesus monkeys. Indian J. Exp. Biol. 7:49.
80. Johansson, E. D. B., and Nygren, K. G. (1973). Depression of plasma testosterone levels in men with norethindrone. Contraception 8:219.
81. Terner, C., and MacLaughlin, J. (1973). Effects of sex hormones on germinal cells of the rat testis: a rationale for the use of progestin and androgen combinations in the control of male fertility. J. Reprod. Fertil. 32:458.
82. Kragt, C. L., Bergstrom, K. K., Kirton, K. T., and Porteus, E. (1975). Male antifertility: an approach. Contraception 11:91.
83. Brenner, P. F., Bernstein, G. S., Roy, S., Jecht, E. W., and Mishell, D. R. (1975). Administration of norethandrolone and testosterone as a contraceptive agent for men. Contraception 11:193.
84. Frick, J. (1973). Control of spermatogenesis in man by combined administration of progestin and androgen. Contraception 8:191.
85. Frick, J., Bartsch, G., and Marberger, G. (1976). Steroidal compounds (injectables and implants) affecting spermatogenesis in man. J. Reprod. Fertil. (Suppl.). 24:35.
86. Coutinho, E. M., and Melo, J. F. (1973). Successful inhibition of spermatogenesis in man without loss of libido: a potential new approach to male contraception. Contraception 8:207.
87. Coutinho, E. M. (1974). Male contraception and the "unisex pill". IPPF Med. Bull. 8:3.
88. Rowley, M. J., and Heller, C. G. (1972). The testosterone rebound phenomenon in the treatment of male infertility. Fertil. Steril. 23:498.
89. Albert, A. (1961). The mammalian testis. In W. C. Young (ed.), Sex and Internal Secretions, Vol. 1, pp. 305–365. Williams & Wilkins Company, Baltimore.
90. Reddy, P. R. K., and Prasad, M. R. N. (1973). Control of fertility in male rats by subcutaneously implanted silastic capsules containing testosterone. Contraception 7:105.
91. Berndtson, W. E., Desjardins, C., and Ewing, L. L. (1974). Inhibition and maintenance of spermatogenesis in rats implanted with polydimethyl siloxane capsules containing various androgens. J. Endocrinol. 62:125.
92. Heller, C. G., Nelson, W. O., Hill, I. B., Hendersen, E., Maddock, W. O., Jungck, E. C., Paulsen, C. A., and Mortimore, G. E. (1950). Improvement of spermatogenesis following depression of the human testis with testosterone. Fertil. Steril. 1:415.
93. Reddy, P. R. K., and Rao, J. M. (1972). Reversible antifertility action of testosterone propionate in human males. Contraception 5:295.
94. Lee, P. A., Jaffee, R. B., Midgley, A. R., Jr., Kohen, F., and Nieswender, G. F. (1972). Regulation of human gonadotropins. VIII. Suppression of serum LH and FSH in adult males following exogenous testosterone administration. J. Clin. Endocrinol. Metab. 35:636.
95. Sherins, R. J., and Loriaux, D. L. (1973). Studies on the role of sex steroids in the feedback control of FSH concentrations in men. J. Clin. Endocrinol. Metab. 36:886.

96. Sherins, R. J., Gaudey, H. M., Thorslund, T. W., and Paulsen, C. A. (1971). Pituitary and testicular function studies. I. Experience with a new gonadal inhibitor, 17α-pregn-4-ene-20-one(2,3-d)isoxazol-17-ol(danazol). J. Clin. Endocrinol. Metab. 32:522.

97. Skoglund, R. D., and Paulsen, C. A. (1973). Danazol-testosterone combination: a potentially effective means for reversible male contraception. A preliminary report. Contraception 7:357.

98. Ulsteur, M., Netto, N., Leonard, J., and Paulsen, C. A. (1975). Changes in sperm morphology in normal men treated with danazol and testosterone. Contraception 12:437.

99. Saxena, S. K., Purvis, K., Cekan, Z., Diczfalusy, E., Giner, J., and Paulsen, C. A. (1976). Endocrine effects of two forms of male fertility control: vasectomy and treatment with danazol-testosterone oenanthate combination. Personal communication.

100. Mauss, J., Borsch, G., Bormacher, K., Richter, E., Leyendecker, G., and Nocke, W. (1975). Effect of long term testosterone oenanthate administration on male reproductive function: clinical evaluation, serum FSH, LH, testosterone and seminal fluid analyses in normal men. Acta Endocrinol. (Kbh.). 78:373.

101. Nieschlag, E., Mauss, J., Coert, A., and Kicovic, P. (1975). Plasma androgen levels in men after oral administration of testosterone or testosterone undecanoate. Acta Endocrinol. (Kbh.). 79:366.

102. Robinson, D., Rock, J., and Menkin, F. (1968). Control of human spermatogenesis by induced changes of intrascrotal temperature. JAMA 204:296.

103. Chaudhury, A. K., and Steinberger, E. (1964). A quantitative study of the effect of heat on germinal epithelium of rat testis. Am. J. Anat. 115:509.

104. Fahim, M. S., Fahim, Z., Der, R., Hall, D. C., and Harman, J. (1975). Heat in male contraception (hot water 60°C, infrared, microwave and ultrasound). Contraception 11:549.

105. Lazarus, B. A., and Zorgniotti, A. W. (1975). Thermoregulation in the human testis. Fertil. Steril. 26:757.

106. Reid, B. L., and Cleland, K. W. (1957). The structure and function of the epididymis. I. The histology of the rat epididymis. Aust. J. Zool. 5:223.

107. Ladman, A. J., and Young, W. C. (1958). An electron microscopic study of the ductuli efferentes and rete testis of the guinea pig. J. Biophys. Biochem. Cytol. 4:219.

108. Leeson, T. S. (1962). Electron microscopy of the rete testis of the rat. Anat. Rec. 144:57.

109. Dym, M. (1972). The mammalian rete testis. Anat. Rec. 172:304.

110. Tuck, R. R., Setchell, B. P., Waites, G. M. H. and Young, J. A. (1970). Composition of fluid collected by micropuncture and catheterization from the seminiferous tubules and rete testis of rats. Pflügers Arch. 318:225.

111. Waites, G. M. H., and Einer-Jensen, N. (1974). Collection and analysis of rete testis fluid from macaque monkeys. J. Reprod. Fertil. 41:505.

112. Voglmayr, J. K., Waites, G. M. H., and Setchell, B. P. (1966). Studies on spermatozoa and fluid collected directly from the testis of the conscious ram. Nature 210:861.

113. Voglmayr, J. K., Larsen, L. H., and White, I. G. (1970). Metabolism of spermatozoa and composition of fluid collected from the rete testis of living bulls. J. Reprod. Fertil. 21:449.

114. Setchell, B. P. (1974). Secretions of the testis and epididymis. J. Reprod. Fertil. 37:165.

115. Levine, N., and Marsh, D. J. (1971). Micropuncture studies of the electrochemical aspects of fluid and electrolyte transport in individual seminiferous tubules, the epididymis and the vas deferens in rats. J. Physiol. (Lond) 213:557.

116. Waites, G. M. H. (1976). Permeability of the seminiferous tubules and the rete testis to natural and synthetic compounds. J. Reprod. Fertil. (Suppl.). 24:49.

117. Setchell, B. P., and Waites, G. M. H. (1975). The blood-testis barrier. In R. O. Greep and E. B. Astwood (eds.), Handbook of Physiology, Endocrinology: Reproductive Biology of Male. American Physiological Society, Washington, D.C.

118. Cooper, T. G., and Waites, G. M. H. (1974). Testosterone in rete testis fluid and blood of rams and rats. J. Endocrinol. 62:619.

119. Cooper, T. G., and Waites, G. M. H. (1975). Steroid entry into rete testis fluid and the blood-testis barrier. J. Endocrinol. 65:195.

120. Waites, G. M. H., Jones, A. R., Main, S. J., and Cooper, T. G. (1973). The entry of antifertility and other compounds into the testis. Adv. Biosci. 10:101.

121. Edwards, E. M., Jones, A. R., and Waites, G. M. H. (1973). The action of α-chlorohydrin(3-chloro-1,2-propane-1,2-diol) on the metabolism of glycerol in male rats. J. Reprod. Fertil. 35:589.

122. Einer-Jensen, N. (1974). Local recirculation of injected (^{3}H)-testosterone from the testis to the epididymal fat pad and the corpus epididymis in the rat. J. Reprod. Fertil. 37:145.

123. Einer-Jensen, N. (1974). Local recirculation of ^{133}xenon and ^{85}krypton to the testes and the caput epididymis in rats. J. Reprod. Fertil. 37:55.

124. Einer-Jensen, N., and Waites, G. M. H. (1976). Testicular blood flow and a study of the spermatic venous to arterial transfer of radioactive krypton and testosterone in rhesus monkey. J. Physiol. (Lond.) In press.

125. Free, M. J., and Jaffe, R. A. (1974). Dyanamics of venous-arterial testosterone transfer in the pampiniform plexus of rat. Seventh Annual Meeting, Society for the Study of Reproduction, August 1974 (Abstr.).

126. Free, M. J., Jaffe, R. A., Jain, S. K., and Gomes, W. R. (1973). Testosterone concentrating mechanism in the reproductive organs of the male rat. Nature (New Biol.) 244:24.

127. Prasad, M. R. N., Rajalakshmi, M., Gupta, G., and Karkun, T. (1973). Control of epididymal function. J. Reprod. Fertil. (Suppl.) 18:215.

128. Crabo, B. (1965). Studies on the composition of epididymal content in bulls and boars. Acta Vet. Scand. (Suppl.) 6:5.

129. Bedford, J. M. (1975). Maturation, transport and fate of spermatozoa in the epididymis. In R. O. Greep and E. B. Astwood (eds.), Handbook of Physiology, Section 7. Endocrinology: Reproductive System—Male, Vol. 3. American Physiological Society, Washington, D.C.

130. Bedford, J. M., Calvin, H., and Cooper, G. W. (1973). The maturation of spermatozoa in the human epididymis. J. Reprod. Fertil. (Suppl.) 18:199.

131. Hamilton, D. W. (1975). Morphology and function of the epithelium lining the ductuli efferentes, epididymis and vas deferens. In R. O. Greep and E. P. Astwood (eds.), Handbook of Physiology, Section 7. Endocrinology: Reproductive System—Male, Vol. 3. American Physiological Society, Washington, D.C.

132. Orgebin-Crist, M. C. (1975). Endocrine control of the development and maintenance of sperm fertilizing ability in the epididymis. In R. O. Greep and E. B. Astwood (eds.), Handbook of Physiology, Section 7. Endocrinology: Reproductive System—Male, Vol. 3. American Physiological Society, Washington, D.C.

133. Prasad, M. R. N., Rajalakshmi, M., Gupta, G., Dinakar, N., Arora, R., and Karkun, T. (1974). Epididymal environment and maturation of spermatozoa. In R. E. Mancini and L. Martini (eds.), Male Fertility and Sterility, pp. 459–478. Academic Press, New York.

134. Rajalakshmi, M., Arora, R., Bose, T., Dinkar, N., Gupta, G., Thampan, T. N. R. V., Prasad, M. R. N., Anand-Kumar, T. C., and Moudgal, N. R. (1976). Physiology of the epididymis and induction of functional sterility in the male. J. Reprod. Fertil. (Suppl.). 24:72.

135. Rajalakshmi, M., and Prasad, M. R. N. (1968). Changes in the sialic acid content of the accessory glands of the male rat. J. Endocrinol. 41:471.

136. Hamilton, D. W. (1972). The mammalian epididymis. In H. Balin and S. Glasser (eds.), Reproductive Biology, pp. 208–237. Excerpta Medica, Amsterdam.

137. Brooks, D. E., Hamilton, D. W., and Mallek, A. H. (1974). Carnitine and glycerylphosphoryl choline in the reproductive tract of the male rat. J. Reprod. Fertil. 36:141.

138. Glover, T. D. (1974). Recent progress in the study of the male reproductive physiology: testis stimulation, sperm formation, transport maturation (epididymal physiology), semen analysis, storage and artificial insemination. In R. O. Greep (ed.), MTP In-

ternational Review of Science, Physiology Series 1. Reproductive Physiology, Vol. 8. University Park Press, Baltimore.

139. Gupta, G., Rajalakshmi, M., and Prasad, M. R. N. (1974). Regional differences in androgen thresholds of the epididymis of the castrated rat. Steroids 24:575.

140. Dinakar, N., Arora, R., and Prasad, M. R. N. (1974). Effects of microquantities of testosterone on the epididymis and accessory glands of the castrated rhesus monkey, *Macaca mulatta*. J. Endocrinol. 60:399.

141. Dinakar, N., Arora, R., and Prasad, M. R. N. (1974). Effects of microquantites of 5α-dihydrotestosterone on the epididymis and accessory glands of the castrated rhesus monkey, *Macaca mulatta*. Int. J. Fertil. 19:133.

142. Karkun, T., Rajalakshmi, M., and Prasad, M. R. N. (1974). Maintenance of the epididymis in the castrated golden hamster by testosterone and dihydrotestosterone. Contraception 9:471.

143. Das, R. P., Roy, S., and Bandopadhya, G. P. (1973). Effect of castration and testosterone on spermatozoa and accessory genital organs of rats. Contraception 8:471.

144. Cavazos, L. F., and Melampy, R. M. (1950). Effects of differential testosterone propionate levels on rat accessory gland activity. Iowa State College J. Sci. 31:19.

145. Jones, R., and Glover, T. D. (1973). The effects of castration on the composition of rabbit epididymal plasma. J. Reprod. Fertil. 34:405.

146. Back, D. J. (1973). The passage of [³H] testosterone into the cauda epididymidis of the rat. J. Reprod. Fertil. 35:586.

147. Hansson, V., Djoseland, O., Reusch, E., Attramadal, A., and Torgersen, O. (1973). An androgen binding protein in the testis cytosol fraction of adult rats: comparison with the androgen binding protein in the epididymis. Steroids 21:457.

148. Danzo, B. J., Eller, B. C., and Orgebin-Crist, M. C. (1974). Studies on the site of origin of the androgen binding protein present in epididymal cytosol from mature intact rabbits. Steroids 24:107.

149. Danzo, B. J., and Eller, B. C. (1975). Androgen binding to cytosol prepared from epididymides of sexually mature castrated rabbits: evidence for a cytoplasmic receptor. Steroids 25:507.

150. Hansson, V., Ritzen, E. M., and French, F. S. (1975). Androgen transport and receptor mechanisms in testis and epididymis. *In* R. O. Greep and E. B. Astwood (eds.), Handbook of Physiology, Section 7. Endocrinology—Male, Vol. 3. American Physiological Society, Washington, D.C.

151. Weddington, S. C., McLean, W. S., Nayfeh, S. N., French, F. S., Hansson, V., and Ritzen, E. M. (1974). Androgen binding protein (ABP) in rabbit testis and epididymis. Steroids 24:123.

152. White, I. G., and Hudson, B. (1968). The testosterone and dehydroepiandrosterone concentration in fluids of the mammalian male reproductive tract. J. Endocrinol. 41:291.

153. Frankel, A. I., and Eik-Ness, K. B. (1970). Testosterone and dehydroepiandrosterone in the epididymis of the rabbit. J. Reprod. Fertil. 23:441.

154. Vreeburg, J. T. M. and Aafjes, J. H. (1971). Dihydrotestosterone (5α-androstan-17β-ol-3-one) in the epididymis of rats. *In* I. Ingelman, A. Sundberg, and M. O. Lunell (eds.), Current Problems in Fertility, pp. 203–206. Plenum Press, New York.

155. Aafjes, J. H., and Vreeburg, J. T. M. (1972). Distribution of 5α-dihydrotestosterone in the epididymis of bull and boar and its concentration in rat epididymis after ligation of efferent testicular ducts, castration and unilateral gonadectomy. J. Endocrinol. 53:85.

156. Rajalakshmi, M., and Prasad, M. R. N. (1971). Alteration in sialic acid in the epididymis of the pubertal rat in response to changes in functional activity of the testis. J. Reprod. Fertil. 24:409.

157. Dinakar, N. (1975). Epididymal function and its regulation in the rhesus monkey, *Macaca mulatta*. Ph.D. Thesis, University of Delhi, Delhi, India.

158. Dinakar, N., Arora, R., and Prasad, M. R. N. (1976). Regional differences in the metabolism of testosterone in the epididymis and ductus deferens of the adult rhesus monkey: an in vitro study. Indian J. Exp. Biol. In press.

159. Dinakar, N., Arora, R., and Prasad, M. R. N. (1975). Influence of spermatozoa on androgen metabolism in the epididymis of the rhesus monkey. Anat. Rec. 181:528.

160. Dinakar, N., Arora, R., and Prasad, M. R. N. (1976). Metabolism of testosterone *in vitro* in the testis, epididymis and the sex accessories of the rhesus monkey: effects of cyproterone acetate on androgen metabolism. Indian J. Exp. Biol. In press.

160a. Norman, C., Goldberg, E., Porterfield, I. D., and Johnson, C. E. (1960). Prolonged survival of human sperm in chemically defined media at room temperature. Nature 188:760.

161. Hamilton, D. W. (1971). Steroid function in the mammalian epididymis. J. Reprod. Fertil. (Suppl.) 13:89.

162. Arora, R. (1975). Hormonal environment and epididymal function in the rhesus monkey, *Macaca mulatta*. Ph.D. Thesis, University of Delhi, Delhi, India.

163. Pierrepoint, C. G., Davies, P., and Wilson, D. W. (1974). The role of the epididymis and ductus deferens in the direct and unilateral control of the prostate and seminal vesicles of the rat. J. Reprod. Fertil. 41:413.

164. Gloyna, R. E., and Wilson, J. D. (1969). A comparative study of the conversion of testosterone to 17β-hydroxy-5-androstan-3-one (DHT) by prostate and epididymis. J. Clin. Endocrinol. Metab. 29:970.

165. Inano, H., Machino, A., and Tamaoki, B. I. (1969). *In vitro* metabolism of steroid hormones by cell free homogenates of epididymidis of adult rats. Endocrinology 84:997.

166. Blaquier, J. A. (1971). Selective uptake and metabolism of androgens by rat epididymis: the presence of a cytoplasmic receptor. Biochem. Biophys. Res. Commun. 45:1076.

167. Calandra, R. S., Blaquier, J. A., Castillo, E. T. D., and Rivarola, M. A. (1975). Androgen dependency of the androgen receptor in rat epididymis. Biochem. Biophys. Res. Commun. 67:97.

168. Ritzen, E. M., Nayfah, S. N., French, F. S., and Dobbins, M. (1971). Demonstration of androgen binding components in rat epididymis cytosol and comparison with binding components in prostate and other tissues. Endocrinology 89:143.

169. Djoseland, O., Hansson, V., and Haugen, H. N. (1973). Androgen metabolism by rat epididymis. 1. Metabolic conversion of [³H] testosterone *in vivo*. Steroids 21:773.

170. Djoseland, O., Hansson, V., and Haugen, H. N. (1974). Androgen metabolism by rat epididymis. 2. Metabolic conversion of [³H] testosterone *in vitro*. Steroids 23:397.

171. Tindall, D. J., French, F. S., and Nayfeh, S. N. (1972). Androgen uptake and binding in rat epididymal nuclei, *in vivo*. Biochem. Biophys. Res. Commun. 49:1391.

172. Hansson, V., Djoseland, O., Reusch, E., Attramadal, A., and Torgersen, O. (1973). Intracellular receptors for 5α-dihydrotestosterone in the epididymis of adult rats: comparison with the androgen binding protein (ABP) in the testicular and epididymal fluid. Steroids 22:19.

173. Calandra, R. S., Podesta, E. J., Rivarola, M. A., and Blaquier, J. A. (1974). Tissue androgens and androphilic proteins in the rat epididymis during sexual development. Steroids 24:507.

174. Tindal, J. D., Hansson, V., Sar, M., Stumpf, W. E., French, F. S., and Nayfeh, S. N. (1974). Further studies on the accumulation and binding of androgen in rat epididymis. Endocrinology 95:119.

175. Blaquier, J. A., and Calandra, R. S. (1973). The intracellular binding of androgens in rat epididymis. Acta Physiol. Latino Am. 21:97.

176. Blaquier, J. A., and Calandra, R. S. (1973). Intranuclear receptor for androgens in rat epididymis. Endocrinology 93:51.

177. Blaquier, J. A. (1974). Androgen receptors from the epididymis of the rhesus monkey. Endocrine Res. Commun. 1:155.

178. Liao, S., Fang, S., Tymoczko, J. L., and Liang, T. (1974). Androgen receptors, antiandrogens and uptake and retention of androgens in male sex accessory organs. *In* D. Brandes (ed.), Male Accessory Sex Organs, pp. 80–105. Academic Press, New York.

179. Rajalakshmi, M., and Prasad, M. R. N. (1976). Metabolism of testosterone by the epididymis and ventral prostate of rat and its inhibition by cyproterone acetate. Steroids. In press.

180. Bose, T., Rajalakshmi, M., and Prasad, M. R. N. (1976). The metabolism of [^{3}H]-testosterone in the epididymis and accessory glands of reproduction in the castrate hamster. Indian J. Exp. Biol. In press.
181. Sulcova, J., and Starka, L. (1973). The metabolism of androgens in normal human testis and epididymis *in vitro*. Endocrinol. Exp. 7:113.
182. Steinberger, E., and Ficher, M. (1971). Formation and metabolism of testosterone in testicular tissue of immature rats. Endocrinology 89:679.
183. Singh, R., Rajalakshmi, M., and Prasad, M. R. N. (1976). Changes in the metabolism of steroids in the epididymis and accessory glands of rat during the pre- and post-pubertal period. Indian J. Exp. Biol. In press.
184. Ericsson, R. J., Cornette, J. C., and Buthala, D. A. (1967). Binding of sex steroids to rabbit sperm. Acta Endocrinol. (Kbh.). 56:424.
185. Seamark, R. F., and White, I. G. (1964). The metabolism of steroid hormones in semen. J. Endocrinol. 30:307.
186. Castaneda, E., Rios, E. P., Perz, A. E., Lichtenberg, R., Cardero, C., Iramain, C. A., and Perez, P. G. (1974). *In vitro* biotransformation of steroid hormones by human semen. Fertil. Steril. 25:261.
187. Prasad, M. R. N., Singh, S. P., and Rajalakshmi, M. (1970). Fertility control in male rats by continuous release of microquantities of cyproterone acetate from subcutaneous silastic capsules. Contraception 2:165.
188. Hoskins, D. D. (1973). Adenine nucleotide mediation of fructolysis and motility in bovine epididymal spermatozoa. J. Biol. Chem. 248:1135.
189. Garbers, D., First, N. L., and Lardy, H. A. (1973). The stimulation of bovine epididymal sperm metabolism by cyclic nucleotide phosphodiesterase inhibitors. Biol. Reprod. 8:589.
190. Hoskins, D. D., Mustermann, D., and Hall, M. L. (1975). The control of bovine sperm glycolysis during epididymal transit. Fertil. Steril. 12:566.
191. Lubicz-Nawrocki, C. M. (1973). The effect of metabolites of testosterone on the viability of hamster epididymal spermatozoa. J. Endocrinol. 58:199.
192. Bedford, J. M. (1967). Efferent of duct ligation on the fertilizing ability of spermatozoa from different regions of the rabbit epididymis. J. Exp. Zool. 166:271.
193. Orgebin-Crist, M. C. (1967). Sperm maturation in the rabbit epididymis. Nature 216:816.
194. Orgebin-Crist, M. C. (1973). Maturation of spermatozoa in the rabbit epididymis: effect of castration and testosterone replacement. J. Exp. Zool. 185:301.
195. Dyson, A. L. M. B., and Orgebin-Crist, M. C. (1973). Effect of hypophysectomy, castration and androgen replacement upon the fertilizing ability of rat epididymal spermatozoa. Endocrinology 93:391.
196. Orgebin-Crist, M. C., and Davis, J. (1974). Functional and morphological effects of hypophysectomy and androgen replacement in the rabbit epididymis. Cell Tissue Res. 148:183.
197. Cooper, T. G., and Orgebin-Crist, M. C. (1975). The effect of epididymal and testicular fluids on the fertilizing capacity of testicular and epididymal spermatozoa. Andrologia 7:85.
198. Gupta, G., Rajalakshmi, M., and Prasad, M. R. N. (1976). Relative importance of androgen in the regulation of epididymal function in the rat. Indian J. Exp. Biol. 14:82.
199. Voglmayr, J. K. (1975). Metabolic changes in the spermatozoa during epididymal transit. *In* R. O. Greep and E. B. Astwood (eds.), Handbook of Physiology, Section 7. Endocrinology—Male, Vol. 3. American Physiological Society, Washington, D.C.
200. Casillas, E. R. (1972). The distribution of carnitine in male reproductive tissues and its effect on palmitate oxidation by spermatozoal particles. Biochim. Biophys. Acta 280:545.
201. Marquis, N. R., and Fritz, I. B. (1965). Effects of testosterone on the distribution of carnitine, acetylcarnitine and carnitine acetyl transferase in the tissues and reproductive system of the male rat. J. Biol. Chem. 240:2197.
202. Frankel, G., Peterson, R. N., Davis, J. E., and Freund, M. (1974). Glycerylphosphorylcholine and carnitine in normal human semen and post-vasectomy semens: differences in concentration. Fertil. Steril. 25:84.

203. Casillas, E. R., and Erickson, B. J. (1975). The role of carnitine in spermatozoan metabolism. Substrate-induced elevation in the acetylation state of carnitine and coenzyme A in bovine and monkey spermatozoa. Biol. Reprod. 12:275.
204. Casillas, E. R. (1973). Accumulation of carnitine by bovine spermatozoa during maturation in the epididymis. J. Biol. Chem. 248:8227.
205. Prasad, M. R. N., and Rajalakshmi, M. (1976). Comparative physiology of the mammalian epididymis. Gen. Comp. Endocrinol. 28:530.
206. Rajalakshmi, M., Reddy, P. R. K., and Prasad, M. R. N. (1973). Distribution of fructose, citric acid, sialic acid and lipids in male accessory glands of the hamster, Mesocricetus auratus. J. Endocrinol. 58:349.
207. Riar, S. S., Setty, B. S., and Kar, A. B. (1973). Studies on the physiology and biochemistry of mammalian epididymis: biochemical composition of epididymis: a comparative study. Fertil. Steril. 24:355.
208. Rajalakshmi, M., Reddy, P. R. K., Prasad, M. R. N., Upadhyaya, P., and Minocha, V. R. (1975). Studies on the physiology of the human epididymis and vas deferens. I. Biochemical constituents. Contraception 12:175.
209. Quinn, P. J., and White, I. G. (1967). Phospholipids and cholesterol content of epididymal and ejaculated ram spermatozoa and seminal plasma in relation to cold shock. Aust. J. Biol. Sci. 20:1205.
210. Poulos, A., Voglmayr, J. K., and White, I. G. (1973). Phospolipid changes in spermatozoa during passage through the genital tract of the bull. Biochem. Biophys. Acta 306:194.
211. Terner, C., Maclaughlin, J., and Smith, B. R. (1975). Changes in lipase and phosphatase activities of rat spermatozoa in transit from the caput to the cauda epididymidis. J. Reprod. Fertil. 45:1.
212. Poulos, A., Brown-Woodman, P. D. C., White, I. G., and Cox, R. I. (1975). Changes in phospholipids of ram spermatozoa during migration through the epididymis and possible origin of prostaglandin $F_{2\alpha}$ in testicular and epididymal fluid. Biochim. Biophys. Acta 388:12.
213. Arora, R., Dinakar, N., and Prasad, M. R. N. (1975). Biochemical changes in the spermatozoa and luminal contents of different regions of the epididymis of the rhesus monkey, Macaca mulatta. Contraception 11:689.
214. Teichman, R. J., Cummins, J. M., and Takei, G. H. (1974). The characterization of a malachite green stainable, gluteraldehyde extractable phospholipid in rabbit spermatozoa. Biol. Reprod. 10:565.
215. Cummins, J. M., and Teichman, R. J. (1974). The accumulation of malachite green stainable phospholipid in rabbit spermatozoa during maturation in the epididymis and its possible role in capacitation. Biol. Reprod. 10:555.
216. Poulos, A., and White, I. G. (1973). The phospolipid composition of human spermatozoa and seminal plasma. J. Reprod. Fertil. 35:265.
217. Scott, T. W., Voglmayr, J. K., and Setchell, B. P. (1967). Lipid composition and metabolism in testicular and ejaculated ram spermatozoa. Biochem. J. 102:456.
218. Lavon, U., Volcani, R. and Danon, D. (1970). The lipid content of bovine spermatozoa during maturation and ageing. J. Reprod. Fertil. 23:215.
219. Mann, T. (1964). The Biochemistry of Semen and of the Male Reproductive Tract, Ed. 2. Methuen, London.
220. Scott, T. W. (1973). Lipid metabolism of spermatozoa. J. Reprod. Fertil. (Suppl.) 18:65.
221. Darin-Bennet, A., Poulos, A., and White, I. G. (1973). A re-examination of the role of phospolipids as energy substrates during incubation of ram spermatozoa. J. Reprod. Fertil. 34:543.
222. Fournier, S. (1966). Distribution of sialic acid in the genital system of adult normal and castrated Wistar rats. C.R. Soc. Biol. 160:1087.
223. Fournier-Delpech, S., Bayard, F., and Boulard, C. L. (1974). Contribution à l'étude de la maturation du sperme; étude d'une proteine acide de l'epididyme chez le rat. Dependance androgène, relation avec l'acide sialique. C. R. Soc. Biol. 167:543.
224. Laporte, P. (1974). Teneur en acides sialiques des spermatozoides epididymaires: estimation indirecte, in vivo chez le rat. C. R. Acad. Sci. Paris 279:761.

225. Gupta, G., Rajalakshmi, M., Prasad, M. R. N., and Moudgal, N. R. (1974). Effects of antiserum to luteinizing hormone (LHAS) on the physiology of the epididymis and accessory glands in albino rat. Contraception 10:491.
226. Rajalakshmi, M., and Prasad, M. R. N. (1969). Changes in sialic acid in the testis and epididymis of the rat during the onset of puberty. J. Endocrinol. 44:379.
227. Gupta, G., Rajalakshmi, M., Prasad, M. R. N., and Moudgal, N. R. (1974). Alteration of epididymal function and its relation to maturation of spermatozoa. Andrologia 6:35.
228. Bose, T. K., Prasad, M. R. N., and Moudgal, N. R. (1974). Changing patterns of sialic acid in the spermatozoa and luminal plasma of the epididymis and vas deferens of the hamster, *Mesocricetus auratus* (Waterhouse). Indian J. Exp. Biol. 13:8.
229. Alumot, E., Lensky, J., and Schindler, H. (1971). Separation of proteins in the epididymal fluid of the ram. J. Reprod. Fertil. 25:349.
230. Huang, H. E. S., and Johnson, A. D. (1975). Amino acid composition of epididymal plasma of mouse, rat, rabbit, and sheep. Comp. Biochem. Physiol. 50B:359.
231. Podesta, E. J., Calandra, R. S., Rivarola, M. A., and Blaquier, J. A. (1975). The effect of castration and testosterone replacement on specific proteins and androgen levels of the epididymis. Endocrinology 97:399.
232. Blaquier, J. A. (1973). An *in vitro* action of androgens on protein synthesis by epididymal tubules maintained in organ culture. Biochem. Biophys. Res. Commun. 52:1177.
233. Blaquier, J. A. (1975). The influence of androgens on protein synthesis by cultured rat epididymal tubules. Acta Endocrinol. (Kbh.) 79:403.
234. Rajalakshmi, M., and Prasad, M. R. N. (1976). Stimulation of nucleic acid and protein synthesis in the epididymis and accessory organs of rat by testosterone. J. Endocrinol. 70:263.
235. Rajalakshmi, M., and Prasad, M. R. N. (1974). Changes in the epididymis of adult rats following experimentally induced cryptorchidism. Andrologia 6:293.
236. Rajalakshmi, M., and Prasad, M. R. N. (1975). Action of cyproterone acetate on the accessory organs of reproduction in prepubertal and sexually mature rats. Fertil. Steril. 26:137.
237. Ericsson, R. J., and Baker, V. F. (1970). Male antifertility compounds: biological properties of U-5897, and U-15,646. J. Reprod. Fertil. 21:267.
238. Ericsson, R. J., and Youngdale, G. A. (1970). Male antifertility compounds: structure and activity relationships of U-5897, U-15,646 and related substances. J. Reprod. Fertil. 21:263.
239. Kirton, K. T., Ericsson, R. J., Ray, J. A., and Forbes, A. D. (1973). Male antifertility compounds: efficacy of U-5897 in primates (*Macaca mulatta*). J. Reprod. Fertil. 21:275.
240. Kreider, J. L., and Dutt, R. N. (1970). Induction of temporary infertility in rams with an orally administered chlorohydrin. J. Anim. Sci. 37:95.
241. Johnson, L. A., and Pursel, V. G. (1972). Oviductal and cervical insemination of α-chlorohydrin treated porcine semen. Presented at the Fifth Annual Meeting of the Society for Study of Reproduction (USA).
242. Lubicz-Nawrocki, C. M., and Chang, M. C. (1974). The onset and duration of infertility in hamsters treated with α-chlorohydrin. J. Reprod. Fertil. 39:291.
243. Brown-Woodman, P. D. C., Salomon, S., and White, I. G. (1974). Effects of α-chlorohydrin on the fertility of rams. Acta Eur. Fertil. 5:193.
244. Turner, M. A. (1971). Effects of α-chlorohydrin upon the fertility of spermatozoa of the cauda epididymidis of the rat. J. Reprod. Fertil. 24:267.
245. Crabo, B., and Appelgren, L. A. (1972). Distribution of [^{14}C]α-chlorohydrin in mice and rats. J. Reprod. Fertil. 30:161.
246. Edwards, E. M., Jones, A. R., and Waites, G. M. H. (1973). The action of α-chlorohydrin(3-chloro-1,2-propane-1,2-diol) on the metabolism of glycerol in male rats. J. Reprod. Fertil. 35:589.
247. Ericsson, R. J. (1970). Progesterone and related compounds as inhibitors of spermatogenesis and accessory gland function in the rabbit and ram. Discuss. Abstr. Int. 31B:1619.
248. Gunn, S. A., Gould, T. C., and Anderson, W. R. D. (1969). Possible mechanism of

post-testicular antifertility action of 3-chloro-1,2-propanediol. Proc. Soc. Exp. Biol. Med. 132:656.

249. Cooper, E. R. A., Jones, A. R., and Jackson, M. (1974). Effects of α-chlorohydrin and related compounds on the reproductive organs and fertility of the male rat. J. Reprod. Fertil. 38:379.

250. Setty, B. S., Kar, A. B., Roy, S. K., and Chaudhury, S. R. (1970). Studies with subtoxic doses of α-chlorohydrin in the male monkey (*Macaca mulatta*). Contraception 10:279.

251. Samojlik, E., and Chang, M. C. (1970). Antifertility activity of 3-chloro-1,2-propanediol on male rats. Biol. Reprod. 2:299.

252. Coppola, J. A. (1969). An extragonadal male antifertility agent. Life Sci. 8:43.

253. Voglmayr, J. K. (1974). Alpha chlorohydrin induced changes in the distribution of free myo-inositol and prostaglandin F_2 synthesis of phosphatidyl inositol in the rat epididymis. Biol. Reprod. 2:593.

254. Glover, T. D. (1976). Investigations into the physiology of the epididymis in relation to male contraception. J. Reprod. Fertil. (Suppl.). 24:95.

255. Crabo, B. G., Bower, R. E., Brown, K. I., Graham, E. F., and Pace, M. M. (1971). Extra-cellular glutamic-oxaloacetic transaminase as a measure of membrane injury in spermatozoa during treatment. In A. Ingelmann-Sundberg, and N. O. Lunell (eds.), Current Problems in Fertility, pp. 33–60. Plenum Press, New York.

256. Jones, R., and Glover, T. D. (1973). The effects of castration on composition of rabbit epididymal plasma. J. Reprod. Fertil. 34:405.

257. Vickery, B. H., Erickson, G. I., and Bennet, J. P. (1974). Mechanism of antifertility action of low doses of α-chlorohydrin in the male rat. J. Reprod. Fertil. 38:1.

258. Brown-Woodman, P. D. C., and White, I. G. (1974). Effect of the male antifertility agent α-chlorohydrin on epididymal physiology of the rat. Aust. Physiol. Pharmacol. Soc. 5:47.

259. Homonnai, Z. T., Paz, G., Sofer, A., and Yeelwab, G. A. (1975). A direct effect of α chlorohydrin on motility and metabolism of ejaculated human spermatozoa. Contraception 12:579.

260. Mori, H., Suter, D. A. I., Brown-Woodman, P. D. C., White, I. G., and Ridley, D. D. (1975). Identification of the biochemical lesion produced by α-chlorohydrin in spermatozoa. Nature 255:75.

261. Banik, U. K., Tanikella, T., and Rakhit, S. (1972). Oral antifertility effects of halo-propanediol derivatives in male rats. J. Reprod. Fertil. 30:117.

262. Cooper, E. R. A., Jones, A. R., and Jackson, H. (1974). Effects of chlorhydrin and selected compounds on the reproductive organs and fertility of the male rat. J. Reprod. Fertil. 38:379.

263. Coppola, J. A., and Saldarini, R. J. (1974). A new orally active male antifertility agent. Contraception 9:459.

264. Rajalakshmi, M., Singh, S. P., and Prasad, M. R. N. (1971). Effects of micro quantities of cyproterone acetate released through silastic capsules on the histology of the epididymis of rat. Contraception 3:335.

265. Schenck, B., Elger, W. Schopflin, G., and Neumann, F. (1975). Failure to induce sterility in male rats with microdoses of cyproterone acetate. Contraception 12:517.

266. Roy, S., Chatterjee, S. L., Prasad, M. R. N., Poddar, A. K., and Pandey, D. C. (1976). Effects of cyproterone acetate on reproductive functions in normal human males. Contraception 14:403.

267. Koch, U. J., Lorenz, F., Danehl, K., and Hammerstein, J. (1974). Uber die Verwendbarkeit von Cyproteron Acetate aus Fertilitatstemmung beim Mann. Morphologische Veranderungen und Einflusse und die spermien Motilitat. 40, Tagung der Deutschen Gessellschaft fur Gynaekologie und Geburtshelfe, Weisbaden, Sept. 1974.

268. Lorenz, F., Koch, U. J., Danechl, K., Lubke, K., and Hammerstein, J. (1974). Uber die Verwendbarkeit von Cyproterone Acetate zur Fertilitatstemmung beim Mann. Physiologische und Biochemische Veranderungen. 40, Tagung der Deutschen Gesellschaft fur Gynaekologie und Geburtschelfe, Weisbaden, Sept. 1974.

269. Galena, H. J., Pillai, A. K., and Terner, C. (1974). Progesterone and androgen receptors in non-flagellate germ cells of the testis. J. Endocrinol. 63:223.

270. Aumuller, G., Schenck, B., and Neumann, F. (1975). Fine structure of monkey (*Macaca mulatta*) Sertoli cells after treatment with cyproterone. Andrologia 7:317.

271. Thampan, T. N. R. V., Dinakar, N., Arora, R., Prasad, M. R. N., and Duraiswami, S. (1974). Androgen receptors in the epididymis of the rhesus monkey, *Macaca mulatta*. Anat. Rec. 178:524.

272. Lubicz-Nawrocki, C. M., and Glover, T. D. (1973). The influence of 17α-methyl-B-nortestosterone (SK & F 7690) on the fertilizing ability of spermatozoa in hamster. J. Reprod. Fertil. 34:331.

273. Cummins, J. M., and Orgebin-Crist, M. C. (1974). Effects of the antiandrogen, SK & F 7690 on the fertility of epididymal spermatozoa in the rabbit. Biol. Reprod. 11:56.

274. Brooks, J. R., Bush, R. D., Patanelli, D. J., and Steelman, S. L. (1973). A study of the effects of a new antiandrogen on the hyperplastic dog prostate. Proc. Soc. Exp. Biol. Med. 143:647.

275. Peets, E. A., Henson, M. F., and Neri, R. (1974). On the mechanism of the antiandrogenic action of flutamide (α-α-α-trifluoro-2-methyl-4'-nitro-M-propionotoluidide) in the rat. Endocrinology 94:532.

276. Liao, S., Howell, D. K., and Chang, T. M. (1974). Action of a non-steroidal antiandrogen, flutamide, on the receptor binding and nuclear retention of 5α-dihydrotestosterone in rat ventral prostate. Endocrinology 94:1205.

277. Neri, R., Florance, K., Koziol, P., and Cleave, S. Van. (1972). A biological profile of a non-steroidal antiandrogen, Sch. 13521(4'-nitro-3'-trifluoromethylisobutyranitide). Endocrinology 91:427.

278. Baulieu, E. E., and Jung, I. (1970). A prostatic cytosol receptor. Biochem. Biophys. Res. Commun. 38:588.

279. Mangon, F. R., and Mainwaring, W. I. P. (1972). An explanation of the antiandrogenic properties of 6α-bromo-17β-hydroxy-17α-methyl-oxa-5α-androstane-3-one. Steroids 20:331.

280. Neumann, F., and Steinbeck, H. (1974). Antiandrogens. *In* O. Eichler, A. F. Rensselaer, H. Herken, and A. D. Welch (eds.), Androgens and Antiandrogens, Vol. II, pp. 235–484. Springer Verlag, Berlin.

281. Hayshida, T. (1966). Immunological reactions of pituitary hormones. *In* G. W. Harris and B. T. Donovan (eds.), The Pituitary. II, p. 62. Butterworth, London.

282. Qadri, S. K., Harbers, L. H., and Spies, H. (1966). Inhibition of spermatogenesis and ovulation in rabbits with antiovine LH rabbit serum. Proc. Soc. Exp. Biol. Med. 123:809.

283. Pineda, K. H., Luecker, D. C., Faulkner, L. C., and Hopwood, M. L. (1967). Atrophy of rabbit testis associated with production of antiserum to bovine luteinizing hormone. Proc. Soc. Exp. Biol. Med. 125:605.

284. Talaat, M., and Lawrence, K. A. (1971). Impairment of spermatogenesis and libido through antibodies to luteinizing hormone. Fertil. Steril. 22:113.

285. Lawrence, K. A., and Hassouna, H. (1972). Antihormones and their use in studies of reproduction. Biol. Reprod. 6:422.

286. Moudgal, N. R. (1974). Hormonal antibodies: an appraisal of their use in reproductive endocrinology. *In* A. C. Guyton and D. F. Horrobin (eds.), MTP International Review of Science, Vol. 8, pp. 33–62. Butterworths, London.

287. Lunnen, J. E., Faulkner, L. C., Hopwood, M. L., and Pickett, B. W. (1974). Long term effects of immunization with LH on gonadal, adrenal, and thyroid function in dogs. Biol. Reprod. 6:42.

288. Monastirsky, R., Lawrence, K. A., and Tovan, E. (1971). The effects of gonadotrophin immunization of prepubertal rabbits on gonadal development. Fertil. Steril. 22:318.

289. Turner, P. C., and Johnson, A. D. (1971). The effect of anti-FSH serum on the reproductive organs of the male rat. Int. J. Fertil. 16:169.

290. Anderson, C. H., Schwartz, N. B., Nequin, L. G., and Ely, C. A. (1976). Effects of early treatment with antiserum to ovine follicle stimulating hormone and luteinizing hormone on gonadal development in the rat. Fertil. Steril. 27:47.

291. Nieschlag, E., Usadel, K. H., Schwedes, U., Kley, H. K., Schöfflung, K., and Kruskemper, H. L. (1973). Alterations in testicular morphology and functions in rabbits following active immunization with testosterone. Endocrinology 92:1142.

292. Nieschlag, E., Usadel, K. H., Wickings, E. J., Kley, H. K., and Wuttke, W. (1975). Effects of active immunization with steroids on endocrine and reproductive functions in male animals. *In* E. Nieschlag (ed.), Immunization with Hormones in Reproduction Research, pp. 155–172. North Holland, Amsterdam.
293. Mitchison, N. A. (1974). Long-term hazards in immunological methods of fertility control. *In* E. Diczfalusy (ed.), Immunological Approaches to Fertility Control. Karolinska Symposia on Research Methods in Reproductive Endocrinology 7:405.

International Review of Physiology
Reproductive Physiology II, Volume 13
Edited by Roy O. Greep
Copyright 1977   University Park Press   Baltimore

6
Hormonal Interplay in the Establishment of Pregnancy

K. YOSHINAGA

Laboratory of Human Reproduction
and Reproductive Biology
Harvard Medical School
Boston, Massachusetts

INITIATION OF PREGNANCY: ROLE OF NEUROHUMORAL SYSTEM

Estrous Behavior and Ovarian Steroids

In human and subhuman primates, females accept males at any stage of the menstrual cycle. In most other species of animals, however, females accept males only at estrus, usually shortly before ovulation. This estrous behavior is initiated by a combination of estrogen and progesterone (or other progestins such as 20α-hydroxypregn-4-en-3-one (20α-OHP)). Although estrogen alone can induce estrous behavior in ovariectomized animals (1), it is more effectively induced by a combination of estrogen and progestin (in guinea pigs (2), rats (3), mice (4), and hamsters (5)). In these species, a marked increase in ovarian progestin secretion is observed during the preovulatory surge of gonadotropin. Progestins appear to act directly on the brain, because estrous behavior was induced in ovariectomized estrogen-treated hamsters by progesterone administration into the lateral brain ventricle (6). Sawyer and his associates (7, 8) demonstrated that progesterone induced changes in electroencephalographic activity related to estrous behavior in ovariectomized estrogen-treated rabbits. Meyerson (9) showed that progestins act specifically on the central nervous system to induce estrous behavior, this effect being different from the general anesthetic effect of progestins, because the effectiveness of various progestins in inducing estrous behavior was not directly related to their anesthetic effectiveness. Barfield and Lisk (10) advanced the onset of estrous behavior of regular 4-day cycling rats by injecting progesterone between late diestrus and early proestrus. Progesterone injection into the mesencephalic reticular formation effectively induced estrous behavior in ovariectomized rats, as did estrogen injection into the medial preoptic area, as well as into the mesencephalic reticular formation. Thus, two ovarian steroids appear to have two separate loci for the induction of sexual behavior.

Roles of Progestin in Ovulation

In the rabbit, an extremely large amount of 20α-OHP is produced by ovarian interstitial tissue soon after mating or following the administration of an ovulatory dose of luteinizing hormone (LH) (11–13). In ovariectomized rabbits which had been primed with ovarian steroids to induce heat, coitus initiated a transient discharge of LH by the pituitary. Plasma LH largely disappeared from circulation within 2 hr, but if 20α-OHP was injected subcutaneously immediately after mating, plasma levels of LH were maintained. These results, obtained by Hilliard et al. (14), clearly indicate that this progestin acts as a positive feedback agent to prolong and heighten LH discharge in the mated rabbit.

In the rat, a preovulatory increase in progestin secretion occurs during the critical period in the afternoon of the proestrous day (15–17). Hashimoto and Wiest (18), in their elegant experiments, showed that at the preovulatory stage progesterone is secreted mainly by the follicles and 20α-OHP by the corpora lutea. The continual and accelerated secretion of progesterone during the preovulatory stage appears to be essential for the final step of follicular rupture. Rondell (19) reported that progesterone increased distensibility of the rabbit follicular wall and suggested that progesterone may increase collagenolytic enzyme activity in the follicular wall, resulting in ovulation. Lipner and Greep (20) prevented ovulation in pregnant mare's serum (PMS)-treated immature rats by inhibitors of steroidogenesis and proposed that ovarian steroidogenesis induced by LH might play an obligatory role in the ovulatory process. Takahashi et al. (21) hypophysectomized adult rats at various times during the critical period in the afternoon of proestrus and found that hypophysectomy at 1550 hr consistently blocked ovulation. However, when the hypophysectomized rats were given large doses of progesterone, ovulation was restored.

"Hypothalamic Pseudopregnancy"

The mating stimulus applied to the uterine cervix induces an important change in the hypothalamo-pituitary axis in incomplete cyclers and reflex ovulators. Incomplete cyclers such as rats, mice, and hamsters have short estrous cycles. Their cycle lacks the true luteal phase and the corpora lutea of the estrous cycle are "non-functional." Progesterone secretion by the corpora lutea increases on the day of metestrus, but declines on the following day (15, 16). Mating or equivalent stimuli applied to the uterine cervix (22) change the function of the hypothalamo-pituitary-ovarian axis from the cyclic type to the pregnancy type. The stimulus applied to the uterine cervix is transmitted to the hypothalamus. In the hypothalamus, the stimulus appears to reduce prolactin-inhibiting factor (PIF) and perhaps increases prolactin-releasing hormone (PRH). Thus, the pituitary initiates prolactin secretion, which in turn converts the corpora lutea (CL) from a nonfunctional to a functional state. The mechanical stimulus applied to the uterine cervix is transmitted by the pelvic nerve, bilateral resection of the pelvic nerve preventing both pregnancy and pseudopregnancy (23). Because reserpine injection can induce pseudopregnancy in pelvic neuroectomized rats (24), resection of the pelvic nerve does not seem to interfere with the function of the hypothalamo-pituitary-ovarian axis. According to Spies and Niswender (25), prolactin, LH, and follicle-stimulating hormone (FSH) increased significantly 20 min postcoitum (PC) in both intact and pelvic-neuroectomized rats. These three hormones declined and stayed low at 8, 24, and 48 hr PC in pelvic-neuroectomized rats, but prolactin alone was significantly higher at 8–24 hr PC in intact rats than in pelvic-neuroectomized rats. Because luteal activation occurs in intact rats, but not in pelvic-neuroectomized rats, the initial post-coital rise in LH, FSH, and prolactin is not important in luteal activation. The marked increase in serum prolactin at 8–24 hr PC in intact rats is considered important. The stimulus which influences prolactin secretion by the pituitary appears to reach the hypothalamus through the preoptic-hypothalamic connections. Carrer and Taleisnik (26) showed that, when the connections had been severed, the operated rats could not be made pseudopregnant by such procedures as stimulation of the uterine cervix or a single injection of progesterone or reserpine.

Stimulation of the uterine cervix at times far from ovulation time can also result in pseudopregnancy (27, 28). Such stimulation is somehow remembered and remains effective until the next cycle. Therefore, if the set of corpora lutea present at the time of stimulation fails to become functional, the next set of corpora lutea becomes functional. Thus, a "delayed pseudopregnancy" can be induced. Zeilmaker (29) showed that a spontaneous pseudopregnancy was initiated if a new set of corpora lutea was formed within a 6-day period following cervical stimulation, but not after the 7th day. Freeman et al. (30) observed that cervical stimulation in the rat instituted a pattern of prolactin secretion consisting of two daily surges, diurnal and nocturnal. In rats ovariectomized the day after cervical stimulation (CS), the diurnal prolactin surges disappeared no later than 6 days after CS, whereas the nocturnal surges were present at 7 days after CS but disappeared by 11 days after CS. Based on these results, these investigators postulated that cervical stimulation activates a hypothalamic "mnemonic system" which retains and expresses this information repeatedly even in the absence of the ovaries.

Thus, when corpora lutea are formed during the period of the "hypothalamic pseudopregnancy," they can be activated by repeated surges of prolactin. The activated corpora lutea secrete progesterone, which may, in turn, inhibit recurrence of the cycle by preventing cyclic secretion of gonadotropin. The mechanism involved in the activation of the "mnemonic system" remains to be solved in future work. Pertinent information is scarce at the present time. The hypothalamic content of gonadotropin-releasing hormone (GnRH) rose markedly immediately after cervical stimulation (31). GnRH reached a high level 2 hr after cervical stimulation and was comparable to the GnRH levels on day 5 of pseudopregnancy. This change may be one of the features of "hypothalamic pseudopregnancy." Information on the fluctuation of PIF and PRH in different nuclei of the hypothalamus would clarify this problem.

Activation of "Nonfunctional" Corpora Lutea

Smith et al. (32) studied the hormonal factors associated with converting nonfunctional corpora lutea of the estrous cycle to functional corpora lutea of pseudopregnancy by measuring LH, FSH, prolactin, estradiol, and progesterone. In contrast to a fall of progesterone in cycling rats, progesterone levels in pseudopregnant rats continued increasing in the morning 2 days after ovulation. It was only the pattern of prolactin secretion which could account for the marked divergence in progesterone secretion between cycling and pseudopregnant rats. Thus, prolactin is the major luteotropic stimulus which rescues corpora lutea of the estrous cycle and transforms them into functional corpora lutea of pseudopregnancy.

Once the pattern of ovarian hormone secretion of the estrous cycle is altered to that of pseudopregnancy, an equilibrium seems to be established between the ovary and the pituitary, namely, the secretion of pituitary gonadotropin is suppressed by sustained secretion of ovarian hormones. Chatterjee and Harper (33) showed that hemiovariectomy on days 1, 2, and 3 after the appearance of leukocytes in vaginal smears in pseudopregnant rats caused a return of proestrus between 2 and 3 days after operation. From this result and that of Freeman et al. (30), it appears that continued progesterone secretion is necessary for the suppression of gonadotropin secretion, rather than for the maintenance of prolactin secretion.

HORMONAL CONTROL OF OVUM TRANSPORT THROUGH OVIDUCT

The development of culture technique has permitted fertilized ova to develop from the one cell stage to the blastocyst stage in vitro. Because ovarian steroid hormones are dispensable in culture media, the development of fertilized ova appears to be independent of ovarian control. However, the in vivo milieu of fertilized ova in the oviduct is influenced by ovarian steroid hormones. The volume of oviduct fluid is increased by estrogen, but this estrogen effect can be suppressed by progesterone. According to Greenwald (34–36), mucin is released by the action of progesterone from the rabbit oviduct epithelium following coitus; injection of estrogen at this time inhibits the secretory discharge causing a marked reduction in the thickness of the mucin layer of fertilized ova. When the ova with a thin mucin layer were transferred into the uterus of pseudopregnant rabbits, their development was comparable to that of normal ova. Adams (37) found that, although the morulae developed in vitro for 24 hr had a thin mucin layer, they nevertheless implanted successfully when transferred into the recipients. Thus, the amount of mucin layer deposited on the fertilized ova in the oviduct does not seem to influence the development of the ova.

Although ovariectomy after fertilization of ova does not affect the speed of ovum transport through the oviduct (38, 39), exogenous steroid hormones cause alterations in the ovum transport. There seems, however, to exist no simple rule that relates the action of steroids to the responses of the oviduct. In the rabbit, low doses of estrogen accelerated tubal passage of ova and high doses caused retention of ova (40). On the other hand, in rats and mice, massive doses of estrogen accelerated the passage of ova, but moderate doses caused their retention in the tube (tube-locking) (41–44). In the rat, Alden showed that progesterone accelerates ovum transport; in the rabbit, however, progesterone injection caused a delay in egg transport (38).

A recent report (45) suggests that the effect of estrogen and progesterone on ovum transport through the oviduct may be partially mediated through adrenergic processes. Namely, depletion of endogenous sympathetic neurotransmitter from the oviduct failed to disrupt normal ovum transport. However, estrogen-induced "tube-locking" and progesterone-induced acceleration of ovum transport were partially antagonized by depletion of neurotransmitter from the intrinsic adrenergic nerves of the rabbit oviduct. Pauerstein et al. (46) think that estrogen enhances α receptors adrenergic activity and thus isthmic constriction, and progesterone enhances β receptor activity sufficiently to relax the isthmic constriction and allow the ova to be transported to the uterus.

Clear-cut clues are not available at the present time as to what the crucial role of endogenous ovarian steroids is in the period of tubal passage of ova. Possibly, ovarian hormones may act as coordinators to synchronize the timely arrival of fertilized ova into the uterus with the uterine preparation for the spacing and subsequent implantation of ova. Artificial retention of fertilized ova in the oviduct by ligation of the utero-tubal junction results in blastocyst formation in the oviduct. These blastocysts, however, are not capable of differentiating into embryos when they are transferred under the kidney capsule (47). Therefore, the timely arrival of the ova into the uterus appears to be crucial for the normal development of the ova.

HORMONAL REQUIREMENTS FOR OVUM IMPLANTATION

Species Variation

A close examination of the ovarian steroid secretion pattern during the fertile cycle and the subsequent period of early pregnancy reveals the following characteristics in various species of animals: 1) progesterone secretion gradually increases after corpus luteum formation, and 2) estrogen secretion increases prior to ovulation and, in many species of animals, prior to ovum implantation. In most mammals, progesterone is indispensable in preparing the uterine endometrium for ovum implantation. The amount and duration of progesterone secretion varies depending on the animal species. The extract of corpus luteum was shown to induce implantation of the fertilized ova and maintain pregnancy in ovariectomized rabbits (48, 49). Although progesterone alone has been reported to induce implantation and maintain pregnancy in rabbits ovariectomized on day 5 PC (50), estrogen is also necessary for ovum implantation in rabbits ovariectomized at an earlier stage of pregnancy (51).

In the rhesus monkey, Meyer et al. (52) showed that progesterone treatment of monkeys ovariectomized between the 2nd and 6th day after ovulation was able to induce implantation and maintain pregnancy. Ovum implantation takes place on the 9th day after fertilization in this species (53). According to Hotchkiss et al. (54), most animals showed no luteal phase peaks in serum estradiol concentration. However, estrogen level, which was lowest immediately after ovulation, recovered about 8 days later. Because no alteration in circulating estrogen level is seen prior to ovum implantation (55), the importance of estrogen in uterine preparation for ovum implantation is not clear in this species. Even in species of animals such as the hamster, with a clear-cut preimplantation estrogen surge (56), it has been shown that progesterone alone can induce implantation after ovariectomy (57). Besides the hamster, it is well known that in the guinea pig very small amounts of progesterone are required for implantation (58).

In the rat, mouse, and Mongolian gerbil, both progesterone and estrogen are required for induction of implantation; the species difference in steroid requirements for implantation may be explained, in part, by the difference in the genetic ability of the uterine cells to respond to a steroid. For example, progesterone markedly stimulates endometrial carbonic anhydrase in the rabbit (59); the same enzyme is stimulated by estrogen in the rat and mouse and, indeed, progesterone reduces uterine carbonic anhydrase in these species (60). Thus, the species difference in hormonal requirements for ovum implantation appears to involve a difference in activation of enzymes by steroids at the uterine level.

Uterine Sensitivity for Ovum Implantation

Psychoyos (61) demonstrated in the rat that the uterus must be exposed to progesterone for 2 days and to estrogen at the end of the 2-day progesterone action to acquire "receptivity" for ovum implantation. Once the uterus is made receptive, this organ stays at this stage for about half a day or so and is automatically rendered refractory. The refractory state of the uterus can be maintained by continued progesterone treatment, and this state can be abolished by an interruption of progesterone treatment for more than 2 days (62). Maintenance of the refractory state for a prolonged period of time appears to require both progesterone and estrogen, and the refractory state can be effectively abolished with estrogen treatment (63). In the rat, the level of progesterone secretion during the first 2 days of pseudopregnancy or pregnancy is relatively low, but is increased by day 5 (15, 16). An estrogen surge is observed on day 4 (64). Thus, the uterus becomes receptive in the latter half of day 5 and refractory after day 6. This temporal aspect of the uterine sensitivity is also observed in pseudopregnant rats, reflecting a similar pattern of ovarian hormone secretion. Daily injection of progesterone to pseudopregnant rats from the day of proestrus and a single injection of estrogen on day 3 of pseudopregnancy could induce deciduoma formation after uterine traumatization on day 3. Because the uterine traumatization on day 3 does not produce deciduoma in untreated pseudopregnant rats, the temporal pattern of the uterine sensitivity was shown to be advanced by exogenous hormones (65). Progesterone was injected in pregnant rats daily subcutaneously from the day of mating or 1 day earlier, and a small amount of estrogen was injected locally into the mesometrial fat tissue on day 3 of pregnancy. This treatment made a limited portion of the uterus receptive while the ova were in the oviduct, and by the time the ova arrived into the uterus, that portion of the uterus became refractory. Thus implantation of ova was locally inhibited (66). A similar advancement of the uterine sensitivity within a limited area of the uterus was shown by deciduoma formation (67). Namely, a group of unilaterally ovariectomized rats were mated, and the remaining ovary was removed on day 3 of pregnancy (Figure 1). The rats were injected subcutaneously with 5 mg of progesterone daily from days 3–13.

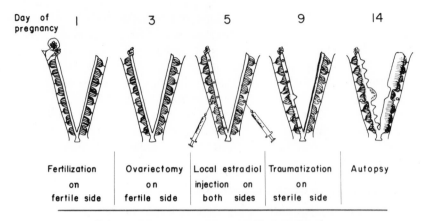

Figure 1. Localized sensitization and desensitization of the rat uterus for ovum implantation and decidualization. Unilaterally ovariectomized rats were mated (day 1) and the
remaining ovary was removed on day 3. Progesterone (5 mg/day) was injected subcutaneously daily from day 3 to day 13. On day 5, estradiol (5 ng) was injected locally into the
mesometrial fat tissue on both sides. On day 9, the endometrium of the sterile horn was
traumatized and a single injection of estradiol (0.2 μg) was given subcutaneously. On day
14, the fertile horn was found to carry two sets of embryos at different developmental
stages. The sterile horn had decidual tissue along the entire horn, except in the limited area
which was influenced by the locally injected estradiol. (See text for further explanation.)

On day 5 of pregnancy, 5 ng of estradiol were injected locally in one spot into
the mesometrial fat tissue of each horn according to the method described in a
previous publication (68). This locally given estrogen sensitized the limited areas
of both horns and induced implantation of blastocysts which happened to be
located in that area, but in the sterile horn, the injection made that portion of
the uterus sensitive and subsequently refractory. On day 9, a single injection of
0.2 μg of estradiol was given subcutaneously to induce implantation of the
blastocysts which had still been free in the area of the uterus beyond the limited
area affected by the locally given estradiol (5 ng). The antimesometrial side of
the sterile horn was traumatized with a curved needle along the entire horn to
produce deciduoma. The uterus obtained on day 14 is shown in Figure 1. In the
figure, the fertile horn carries two sets of embryos at different developmental
stages: the large swellings contain embryos which implanted following locally
administered estrogen (on day 5) and the small swellings contain embryos which
implanted following systemic estrogen (on day 9). The sterile horns show
deciduoma formation except in the area which was influenced by the locally
administered estrogen. These results clearly demonstrate that the chronological
sequence of progesterone and estrogen secretion is very closely related to the
temporal changes in the uterine receptivity. Morphological and biochemical
characterizations of the uterus at various states of the sensitivity for ovum implan-

tation have been reviewed (69–73). Although biochemical changes in the uterus produced by estrogen are observed in the receptive state, the uterus influenced by estrogen alone is not receptive (74, 75). Further studies are needed to characterize receptive and refractory states of the uterus. The refractory condition of the uterus produced by the physiological hormones will provide a good tool for contraceptive purposes.

Gonadotropins and Ovum Implantation

If the pituitary is removed from pregnant rats on days 1–4, implantation is prevented (76). Treatment of pregnant rats with chlorpromazine or reserpine resulted in a delay in ovum implantation (77, 78). These drugs are known to block gonadotropin secretion (79, 80). It is a well-known phenomenon that ovum implantation is delayed during lactation in the rat and mouse (81) due to a decreased estrogen secretion (82). Whitten (83) postulated that the delay in implantation in lactating mice is due to suppression of pituitary gonadotropin release. The question as to which pituitary hormone is responsible for induction of ovum implantation has not yet been resolved.

Administration of LH to rats hypophysectomized after mating induced implantation (84), and antiserum to LH, but not to FSH, prevented ovum implantation in the rat (85, 86) and in the mouse (87). The plasma level of LH, but not FSH, was high prior to ovum implantation in the mouse (88). These results indicate that LH plays an important role in stimulating the ovary to produce the steroids necessary for ovum implantation. On the other hand, the importance of FSH in initiating implantation was evidenced by the following studies in normal rats and mice. Pituitary content of FSH fell significantly on day 3 of pregnancy in the rat and mouse (89). In the mouse and rat, lactational delay in implantation was abolished by a single injection of FSH or PMS (87, 90). According to Raud (90), the peripheral serum level of FSH was significantly lower in pregnant lactating rats than in normal pregnant rats, and the effect of FSH in initiating implantation appears to be reduced by prolactin. A supplement with FSH did abolish the delay in implantation, but FSH was not effective in the rats whose ovaries were irradiated to destroy follicles. Thus, estrogen which is necessary for induction of implantation appears to be produced by the follicles. A delay in implantation by lactation depends on the intensity of the suckling stimulus. When the suckling litter is small, there is no delay in implantation (91). Moudgal et al. (92) injected prolactin (100 µg/day) to pregnant rats nursing two pups and found that implantation was delayed. Although it still remains to be clarified whether FSH or LH is responsible for the preimplantation increase in estrogen secretion, Bindon (87) thinks both FSH and LH are reduced during lactation. He showed that both FSH and PMS are effective in abolishing the delay, but that PMS is more effective. We have also found that human chorionic gonadotropin (hCG) was 100% effective in abolishing the delay in implantation in lactating rats, but that LH was not always effective and FSH was ineffective (93). It may be that a particular combination of FSH and LH is necessary for estrogen secretion by an ovary occupied by young functional corpora lutea producing a large quantity of progesterone. Low levels of gonadotropin during lactation are under the influence of the hypothalamus, which may be considered to produce little PIF and/or more PRH. This state of the hypothalamus during lactation in the mouse can be upset by the odor of strange males, resulting in an increased gonadotropin secretion (Bruce effect). Thus, Bloch (94) was able to abolish a delay in ovum implantation in lactating mice by placing males near the pregnant lactating mice. The gonadotropin release for the preimplantation estrogen surge in the rat was shown to be regulated by a catecholaminergic pathway in the hypothalamus. Labhsetwar (95) caused a delay in implantation by a neuroleptic agent, haloperidol, which is known to interfere selectively with dopaminergic transmission.

RECOGNITION OF PREGNANCY BY MATERNAL ENDOCRINE SYSTEM

Primates

The cyclic activity of the female endocrine system (the hypothalamo-pituitary-ovarian axis) associated with periodic ovulation ceases during pregnancy. This endocrine alteration is accompanied by increased progesterone levels in blood, which can be attributed to the presence of the conceptus in the uterus and to the physiological modification of the uterus resulting from implantation of the ovum. The influence of the conceptus on the maternal endocrine system varies among different species. The variation is due to the steroidogenic ability of the placenta and to the type, time, and quantity of gonadotropic hormones produced by the trophoblast cells of the placenta.

In the primate, the mechanism by which the conceptus alters the maternal endocrine system involves the secretion of chorionic gonadotropin by the trophoblast cells. In the human, this hormone, human chorionic gonadotropin, is a glycoprotein composed of α and β subunits. Various aspects of this hormone have recently been reviewed (96, 97). During the menstrual cycle, progesterone secretion by the corpus luteum reaches a peak about 8 days after its formation and declines thereafter to a low level prior to the onset of the next menses (98). In the pregnancy cycle, the progesterone level does not decline after reaching a peak of the cycle. The maintenance of progesterone levels coincides with an increase in hCG secretion. Therefore, hCG is considered to rescue the corpus luteum, which was to regress should fertilization have failed, and to stimulate it for further production of progesterone. The ability of a corpus luteum to maintain high levels of progesterone appears to be limited to 6–8 weeks, even with continuous stimulation by endogenous hCG. The corpus luteum secretes not only progesterone, but also 17α-hydroxyprogesterone (17α-OHP). Yoshimi et al. (99) followed plasma levels of progesterone, 17α-OHP, and hCG during early pregnancy and found that the progesterone level reached a nadir at 6–8 weeks after ovulation and then increased again, whereas 17α-OHP continued to decline. Because the secretory capacity of 17α-OHP by the placenta is limited (100, 101), the first peaks of progesterone and 17α-OHP are attributable to the corpus luteum, but the second peak of progesterone without the concomitant rise of 17α-OHP is attributable to progesterone secretion by the placenta. Thus, the placenta itself produces progesterone at later stages and secures maintenance of pregnancy in the human, subhuman primates, cow, ewe, and some other animal species. Besides the rescue of the corpus luteum, hCG was suggested to have a role in suppressing the maternal immune response to the fetus. Adock et al. (102) showed that hCG completely inhibited the response of lymphocytes to phytohemagglutinin. The effect was both reversible and non-cytotoxic. Thus, hCG is considered to represent trophoblastic surface antigen and to block the action of maternal lymphocytes.

Compared with the human, less chorionic gonadotropin is secreted in quantity and in duration in the rhesus monkey (103). Tullner (104) found by a bioassay that monkey chorionic gonadotropin (mCG) was detectable in urine as early as the 13th day of pregnancy in some monkeys, but in others, the hormone was not detectable until the 18th or 20th day. A sensitive radioimmunoassay method detected serum mCG first at about 10 or 11 days after ovulation (105). Histological study of the early embryonic development showed that the penetration of the epithelium by trophoblastic cells had begun at a few sites along the area of attachment on the 9th day after ovulation. Broad microvilli from the trophoblastic cells extended into the epithelial surface, but the basement membrane was not reached. These events preceded measurable increments in circulating chorionic gonadotropin and the rescue of the corpus luteum (55). Plasma concentration of progesterone starts declining after peaking at the mid-luteal phase in the nonfertile cycle. In the fertile cycle, a marked but short-lived increase in progesterone secretion takes place after ovum implantation. Neill and Knobil (106) showed that daily injections of hCG, starting on day 22 of the cycle, followed a precipitous rise in plasma progesterone concentration in the rhesus monkey. Despite daily hCG injection, progesterone levels could not be sustained. An ephemeral progesterone rise was also observed by Atkinson et al. (105) at the outset of mCG secretion during early pregnancy. The serum progesterone level fell and that of estrogen rose when mCG secretion was at its maximum; progesterone then increased again when mCG secretion fell. These results indicate that mCG initially stimulates progesterone secretion, but the precipitous increase in mCG secretion results in an increase in estrogen, which acts as a luteolytic factor. Progesterone levels rose again around days 25–30 of pregnancy. Ovariectomy on day 23 of pregnancy did not affect progesterone levels (105). mCG was not detectable by bioassay or radioimmunoassay after the 40th day of pregnancy (104, 105, 107). Placental progesterone in rhesus monkeys must be adequate for maintenance of pregnancy by the 21st postmating day, because ovariectomy at this time did not alter the normal course of pregnancy (108).

Rabbit

Fuchs and Beling (109) showed in the rabbit that plasma levels of progesterone were significantly higher in pregnancy than in pseudopregnancy from 5–8 days after mating. Because attachment of blastocysts to the uterine wall does not take place until 7 days after mating in this species, the presence of free blastocysts in the uterus appears to be recognized by the maternal endocrine system. Haour et al. (110, 111) claimed that 6–7 days after mating, rabbit blastocyst fluid contained a substance which competes with iodinated hCG for binding to bovine corpus luteum plasma membrane and that the blastocyst fluid stimulated morphological luteinization and progesterone secretion when added to monkey granulosa cell cultures. Although the chemical nature of the active principle in rabbit blastocyst fluid remains to be characterized, these observations not only demonstrate that the rabbit blastocyst secretes a gonadotropic substance which stimulates progesterone secretion, but also provide evidence to support the results obtained by Fuchs and Beling (109). Contrary to the results obtained by Haour et al., Sundaram et al. (112) failed to demonstrate any hCG-like activity in 100 μl of the day-6 blastocyst fluid. Their assay method was based on the production of testosterone during incubation with mouse testes, which had a sensitivity of detecting 0.4 mI.U. of hCG.

No matter when the gonadotropic hormone becomes detectable, the requirement of the placenta for the sustained secretion of progesterone by the ovary above and beyond that of pseudopregnancy has been well documented (113–116). In pseudopregnant rabbits, ovarian progesterone secretion reaches a peak (20–30 μg/g of ovary/hr) around 8–14 days after sterile mating, then it declines to premating levels by day 21; in pregnant rabbits, ovarian progesterone secretion increases to a higher peak (70–80 μg/g of ovary/hr) on days 15–20, then gradually declines (117). Because ovariectomy causes termination of pregnancy at any stage of gestation (118, 119), the principal source of progesterone is the corpora lutea during pregnancy. It has been reported that the conversion rate of pregnenolone to progesterone by the mid-term rabbit placenta is only one-one hundredth that of the ovary (120). Holt and Ewing (116) showed that systemic plasma and ovarian release of progesterone and 20α-dihydroprogesterone remained high when placentae were left in situ 24 hr after fetectomy on day 21 of gestation. However, when the placentae were also removed, progestin levels declined markedly 24 hr after operation. Therefore, it is apparent that the luteal function in the rabbit pregnancy is dependent on the placental luteotropin.

Hormonal requirements for ovo-implantation and deciduoma formation in the rabbit were studied by Kehl and Chambon (121, 122). They found that in the ovariectomized adult rabbit, the deciduoma is not fully formed with progesterone alone and concomitant administration of estrogen is needed for full development. Because progesterone alone induced implantation in the rabbit ovariectomized 2 days after mating, Kehl and Chambon (123) assumed that the very small amount of estrogen necessary for implantation is derived from the developing trophoblast. Rabbit blastocyst fluid was found to contain 17β-estradiol (124). Whether the amount of estrogen detected in the blastocyst fluid is derived from the trophoblast or from the genital tract fluid is not clear. Further study is needed on the role of the trophoblast in estrogen secretion in relation to ovum implantation.

Although it has been a long time since an organ-specific glycoprotein, uteroglobin, or blastokinin was found in the rabbit uterus and the secretion was progesterone-dependent, the physiological significance of this protein has not yet been clarified.

Rodents and Others

Rat Placental Luteotropin Astwood and Greep (125) first showed that rat placental extract has luteotropic activity because administration of the extract to hypophysectomized pseudopregnant rats allowed the development of deciduoma following traumatization of the endometrium, the evidence of progesterone secretion. Secretion of rat placental luteotropin begins soon after implantation of blastocysts in the uterus. Alloiteau (126, 127) hypophysectomized pregnant rats just before or soon after ovum implantation and treated the rats with steroids to maintain pregnancy. He found that the corpora lutea in the rats hypophysectomized before ovum implantation regressed despite successful maintenance of pregnancy with exogenous steroids; on the other hand, corpora lutea proliferated in those rats hypophysectomized shortly after ovum implantation. These results indicate that hypophysectomy prior to ovum implantation interrupted the continuous supply of luteotropic hormones, and removal of the pituitary gland after ovum implantation permitted a smooth transfer of the luteotropic role from pituitary prolactin to the placental luteotropin. The luteotropic activity of the young conceptus appeared soon after implantation, gradually increased in proportion to the developmental stage of the embryos, and reached a peak at mid-pregnancy (128). After mid-pregnancy, placental luteotropic activity becomes sufficient to maintain luteal function for maintaining pregnancy, because hypophysectomy after mid-pregnancy did not terminate pregnancy in both the rat (129) and the mouse (130, 131).

Kirby (132) showed in the mouse that the trophoblast itself was unable to stimulate corpora lutea, but development of the mammary gland was observed in mice bearing trophoblast transplants. Zeilmaker (133), on the other hand, found that ectopic trophoblast tissue, which developed from a blastocyst transplant under the kidney capsule, produced a luteotropic substance and cyclic mice were made pseudopregnant. These transplantation experiments show that under certain conditions the ectopic trophoblast can secrete a luteotropic hormone and activate the corpora lutea of the estrous cycle.

Using the rate of survival of embryos supported by implanting placental tissue to pregnant rats hypophysectomized on day 6, Averill et al. (134) assayed luteotropic potency of rat placental tissue at various stages of pregnancy. They found that the placentae obtained on day 12 of pregnancy had the highest potency. The placental luteotropin has not only luteotropic, but also mammotropic capability. Thus, Matthies (135) proposed that the hormone be designated rat chorionic mammotropin (RCM). A detailed account of this hormone is referred to in a chapter by Matthies (136). Linkie and Niswender (137) confirmed that the hormone secretion is highest on day 12 of pregnancy and characterized the hormone as a heat-labile protein of approximately 25,000–50,000 molecular weight. Further study of the hormone by Kelly et al. (138) revealed that there are two peaks of hormone activity, the first between days 11 and 13 and the second between days 17 and 21 of pregnancy. The extract of day-12 placentae contained two different substances with lactogenic activity. On the other hand, that of day 17 contained only one. Chatterton et al. (139) claimed that the unimplanted blastocyst stimulated ovarian progestin biosynthesis because ovaries from pregnant rats with viable blastocysts in the uterine lumen synthesized more progesterone than those from control pseudopregnant rats.

Pregnant Mare Serum Gonadotropin   A gonadotropin found in the blood serum of pregnant mares (PMSG) has biological properties similar to a mixture of pituitary FSH and LH, with more FSH activity than LH activity (140). Its appearance in the blood corresponds to the time of implantation of the blastocyst (days 37–42 of pregnancy). The level of PMSG remains high until about day 120 and becomes nondetectable thereafter (140, 141). When PMSG appears in the blood, the corpus luteum formed at the initiation of pregnancy regresses. An increase in PMSG secretion stimulates the ovary, and some follicles ovulate to form corpora lutea (accessory corpora lutea). This new set of corpora lutea persists until about day 180 of pregnancy. Thus, in this species, the luteotropic role of the pituitary is taken over by the placenta, which is marked by formation of a new set of corpora lutea. The latter ensures further maintenance of pregnancy by progesterone secretion.

The origin of the equine endometrial cups, from which PMSG is secreted, had long been considered to be exclusively maternal (142, 143). However, recent studies (144–146) clarified that the endometrial cup cells originate from the fetal chorionic cells and that modified chorionic cells ("girdle cells") invade and phagocytose the endometrial epithelium and migrate through the basal lamina into the stroma to develop into endometrial cup cells.

Steroidogenic Ability of Blastocyst   Evidence has accumulated that blastocysts secrete steroids. A series of studies by Dickmann and Dey (147, 148) demonstrated that Δ^3-3β-hydroxysteroid dehydrogenase activity appears in the embryo, temporarily coinciding with the time of transformation from morula to blastocyst and with that of implantation of the blastocyst. Because the enzyme activity was observed in a culture condition, the appearance of enzyme activity is autonomous in blastocysts and independent of the maternal pituitary function. The steroidogenic ability of blastocysts suggests that the embryos at their implantation sites exert a local hormonal effect on the endometrium which is critical for implantation. Chew and Sherman (149) showed that this enzyme is also present in mouse trophoblasts, and pregnenolone was converted in vitro into progesterone by trophoblast cells collected after implantation, but not, however, by free blastocysts. The temporal appearance of steroid-metabolizing enzymes is also reported in pig blastocysts (150). The steroidogenic ability of pig embryos is age-dependent. Perry et al. (151) found that day-16 pig embryos could synthesize estrogens, but day-10 embryos could not. Robertson and King (152) suggested that the synthesis of estrogens by blastocysts at about days 10–12 of pregnancy and the subsequent rise in the level of maternal plasma estrogen sulfate may be a requirement for initiating blastocyst implantation locally in the uterus, and at the same time, informing the maternal endocrine system that conception has occurred.

Embryos and Uterine Luteolytic Factors   In some species of mammals, such as the guinea pig, pig, sheep, and cow, circulating progesterone levels decline sharply at the end of the luteal phase of the cycle. When the uterus is removed, progesterone levels do not fall at the expected time, but are maintained at a high level for an extended period of time (see a review article by Anderson (153)). The result has been attributed to the removal of luteolytic factors produced by the uterus at the end of the cycle. Moor and Rowson (154, 155) showed in the sheep that the presence of embryos in the uterus between the 12th and 13th day after estrus is essential to prevent the regression of the corpus luteum and that homogenates prepared from frozen 14- or 15-day sheep embryos prevented luteal regression when infused daily into the uteri of nonpregnant ewes from the 12th day after estrus (156). A similar result has been reported in the pig (157). Because the homogenate of frozen embryos was effective in preventing the luteal regression, the embryos must contain some substance(s) which prevent the production or release of uterine luteolytic factor(s) or antagonize the action of the latter. Although the chemical nature of the embryonic luteotropic (or antiluteolytic) substance remains to be clarified, there is some evidence that prostaglandin $F_{2\alpha}$ (158) and its precursor, arachidonic acid (159), have luteolytic activity and are produced in the uterus. Even if these fatty acids were not the sole luteolytic factors responsible for the luteal regression, clarification of the mode of action of luteolytic factors would lead to understanding of the role of embryos in preventing luteal regression.

The influence of the uterus on the luteal function is also observed in the rat. Decidualization of the uterine endometrium prolongs the duration of pseudopregnancy. The degree of prolongation depends on the amount of deciduoma formed (160). The mechanism by which decidual tissue influences the luteal function has not yet been clarified. DeJongh and Wolthius (161) thought that progesterone is "spared" by decidual tissue. Wiest (162) showed that uterine progesterone concentration in decidualized horns was significantly higher than in control horns and suggested that decidual tissue specifically "retains" progesterone. Hashimoto et al. (16) found that the progesterone secretion rate by the ovary of rats bearing deciduoma in the uterus was almost twice that of normal pseudopregnant rats. The progesterone secretion rate was found to be elevated during the prolonged period of pseudopregnancy. Their data show that the ovarian blood flow rate is increased twice as much as that of the controls. Therefore, the elevation of ovarian progesterone output may be partly due to the increased ovarian blood flow. An elevated progesterone secretion during the prolonged period of pseudopregnancy, however, cannot be explained by a mere increase in ovarian blood flow.

Recently, Gibori et al. (163) examined the ability of decidual tissue to maintain progesterone secretion by observing the serum progesterone level after suppression of prolactin release by ergocornine treatment. In pseudopregnant rats, estrus returned 2 days after ergocornine injection. In pseudopregnant rats bearing deciduoma, estrus returned on an average of 6.8 days after ergocornine injection, and the serum progesterone level in these rats was approximately twice that in pseudopregnant rats without deciduoma. Based on these data, they postulated that decidual tissue produces a hormone which can sustain progesterone secretion in the relative absence of prolactin. A recent report by Castracane and Shaikh (164) indicated that prolongation of pseudopregnancy in the rat with decidualized uterus is not due to a decreased production of uterine prostaglandins, supporting the suggestion that decidual tissue has a luteotropic effect. Another possible role of decidual tissue in prolonging pseudopregnancy was suggested by Lin (165), who showed that decidual cells in vitro responded to a combination of LH and prolactin, but not to LH or prolactin alone, and that these cells produced progesterone. It is not certain whether this "endocrine function" of decidual cells can be directly applied to the situation in vivo. Decidual tissue which is produced by transformation of stromal cells at the site of ovum implantation appears to have multiple roles in the establishment of pregnancy. Their clarification is awaited.

ACKNOWLEDGMENTS

The author wishes to express his gratitude to Dr. R. O. Greep for reviewing the manuscript and his valuable comments. The typing of the manuscript by Mrs. S. Nieland is also appreciated.

REFERENCES

1. Davidson, J. M., Rodgers, C. H., Smith, E. R., and Bloch, G. J. (1968). Stimulation of female sex behavior in adrenalectomized rats with estrogen alone. Endocrinology 82:193.
2. Dempsey, E. W., Hertz, R., and Young, W. C. (1936). The experimental induction of oestrus (sexual receptivity) in the normal and ovariectomized guinea pig. Am. J. Physiol. 116:201.
3. Boling, J. L., and Blandau, R. J. (1939). The estrogen-progesterone induction of mating responses in the spayed female rat. Endocrinology 25:359.
4. Ring, J. R. (1944). The estrogen-progesterone induction of sexual receptivity in the spayed female mouse. Endocrinology 34:269.
5. Frank, A. H., and Fraps, R. M. (1945). Induction of estrus in the ovariectomized golden hamster. Endocrinology 37:357.
6. Kent, G. C., and Lieberman, M. J. (1949). Induction of psychic estrus in the hamster with progesterone administered via the lateral brain ventricle. Endocrinology 45:29.
7. Sawyer, C. H., and Kawakami, M. (1961). Interactions between the central nervous system and hormones influencing ovulation. In C. A. Villee (ed.), Control of Ovulation, pp. 79–100. Pergamon Press, New York.
8. Sawyer, C. H., Kawakami, M., and Kanematsu, S. (1966). Neuroendocrine aspects of reproduction. In R. Levine (ed.), Endocrines and the Central Nervous System,

Association for Research in Nervous and Mental Diseases, Vol. 43, pp. 59–85. Williams and Wilkins, Baltimore.

9. Meyerson, B. J. (1967). Relationship between the anesthetic and gestagenic action and estrous behavior-inducing activity of different progestins. Endocrinology 81:369.

10. Barfield, M. A., and Lisk, R. D. (1970). Advancement of behavioral estrus by subcutaneous injection of progesterone in the four-day cyclic rat. Endocrinology 87:1096.

11. Forbes, T. R. (1953). Preovulatory progesterone in the peripheral blood of the rabbit. Endocrinology 53:79.

12. Hilliard, J., Archbald, D., and Sawyer, C. H. (1963). Gonadotropic activation of preovulatory synthesis and release of progestin in the rabbit. Endocrinology 72:59.

13. Hilliard, J., Spies, H. G., and Sawyer, C. H. (1968). Cholesterol storage and progestin secretion during pregnancy and pseudopregnancy in the rabbit. Endocrinology 82:157.

14. Hilliard, J., Penardi, R., and Sawyer, C. H. (1967). A functional role for 20α-hydroxy-pregn-4-en-3-one in the rabbit. Endocrinology 80:901.

15. Eto, T., Masuda, H., Suzuki, Y., and Hosi, T. (1962). Progesterone and pregn-4-en-20α-ol-3-one in rat ovarian venous blood at different stages in reproductive cycle. Jpn. J. Anim. Reprod. 8:34.

16. Hashimoto, I., Henricks, D. M., Anderson, L. L., and Melampy, R. M. (1968). Progesterone and pregn-4-en-20α-ol=3=one in ovarian venous blood during various reproductive states in the rat. Endocrinology 82:333.

17. Miyake, T. (1968). Interrelationship between the release of pituitary luteinizing hormone and the secretions of ovarian estrogen and progestin during estrous cycle of the rat: integrative mechanism of neuroendocrine system. Hokkaido Univ. Med. Library Ser. 1; 139.

18. Hashimoto, I., and Wiest, W. G. (1969). Correlation of the secretion of ovarian steroids with function of a single generation of corpora lutea in the immature rat. Endocrinology 84:873.

19. Rondell, P. (1970). Follicular processes in ovulation. Fed. Proc. 29:1875.

20. Lipner, H., and Greep, R. O. (1971). Inhibition of steroidogenesis at various sites in the biosynthetic pathway in relation to induced ovulation. Endocrinology 88:602.

21. Takahashi, M., Ford, J. J., Yoshinaga, K., and Greep, R. O. (1974). Induction of ovulation in hypophysectomized rats by progesterone. Endocrinology 95:1322.

22. Long, J. A., and Evans, H. M. (1922). The oestrous cycle in the rat and its associated phenomena. Mem. Univ. Calif. 6:1.

23. Kollar, E. J. (1953). Reproduction in the female rat after pelvic nerve neurectomy. Anat. Rec. 115:641.

24. Carlson, R. R., and DeFeo, V. J. (1965). Role of the pelvic nerve vs. the abdominal sympathetic nerves in the reproductive function of the female rat. Endocrinology 77:1014.

25. Spies, H. G., and Niswender, G. D. (1971). Levels of prolactin, LH and FSH in the serum of intact and pelvic-neurectomized rats. Endocrinology 88:937.

26. Carrer, H. F., and Taleisnik, S. (1970). Induction and maintenance of pseudopregnancy after interruption of preoptic hypothalamic connections. Endocrinology 86:231.

27. Greep, R. O., and Hisaw, F. L. (1938). Pseudopregnancies from electrical stimulation of the cervix in the diestrum. Proc. Soc. Exp. Biol. Med. 39:359.

28. Everett, J. W. (1967). Provoked ovulation or long-delayed pseudopregnancy from coital stimuli in barbiturate-blocked rats. Endocrinology 80:145.

29. Zeilmaker, G. H. (1965). Normal and delayed pseudopregnancy in the rat. Acta Endocrinol. (Kbh.) 49:558.

30. Freeman, M. E., Smith, M. S., Nazian, S. J., and Neill, J. D. (1974). Ovarian and hypothalamic control of the daily surges of prolactin secretion during pseudopregnancy in the rat. Endocrinology 94:875.

31. Takahashi, M., Ford, J. J., Yoshinaga, K., and Greep, R. O. (1975). Effects of cervical stimulation and anti-LH releasing hormone serum on LH releasing hormone content in the hypothalamus. Endocrinology 96:453.

32. Smith, M. S., Freeman, M. E., and Neill, J. D. (1975). The control of progesterone secretion during the estrous cycle and early pseudopregnancy in the rat: prolactin, gonadotropin and steroid levels associated with rescue of the corpus luteum of pseudopregnancy. Endocrinology 96:219.
33. Chatterjee, A., and Harper, M. J. K. (1970). Interruption of pseudopregnancy in rats by hemiovariectomy. Endocrinology 87:173.
34. Greenwald, G. S. (1957). Interruption of pregnancy in the rabbit by the administration of estrogen. J. Exp. Zool. 135:461.
35. Greenwald, G. S. (1958). Endocrine regulation of the secretion of mucin in the tubal epithelium of the rabbit. Anat. Rec. 130:477.
36. Greenwald, G. S. (1962). The role of the mucin layer in development of the rabbit blastocyst. Anat. Rec. 142:407.
37. Adams, C. E. (1960). Development of the rabbit eggs with special reference to the mucin layer. Advance Abstracts of short communications, First International Congress of Endocrinology, Copenhagen. Abstr. No. 345, p. 687. Acta Endocrinol. Suppl. 51.
38. Alden, R. H. (1942). Aspects of egg-ovary-oviduct relationship in the albino rat. I. Egg passage and development following ovariectomy. J. Exp. Zool. 90:159.
39. Adams, C. E. (1958). Egg development in the rabbit: the influence of post-coital ligation of the uterine tube and of ovariectomy. J. Endocrinol. 16:283.
40. Greenwald, G. S. (1963). Interruption of early pregnancy in the rabbit by a single injection of oestradiol cyclopentylpropionate. J. Endocrinol. 26:133.
41. Burdick, H. O., and Pincus, G. (1935). The effect of oestrin injections upon the developing ova of mice and rabbits. Am. J. Physiol. 111:201.
42. Burdick, H. O., Whitney, R., and Pincus, G. (1937). The fate of mouse ova tube-locked by injections of oestrogenic substances. Anat. Rec. 67:513.
43. Burdick, H. O., and Whitney, R. (1937). Acceleration of the rate of passage of fertilized ova through the Fallopian tubes of mice by massive injections of an estrogenic substance. Endocrinology 21:637.
44. Whitney, R., and Burdick, H. O. (1937). Acceleration of the rate of passage of fertilized ova through the Fallopian tubes of rabbits by massive injections of progynon-B. Endocrinology 22:639.
45. Pauerstein, C. J., Hodgson, B. J., Fremming, B. D., and Martin, J. E. (1974). Effects of sympathetic denervation of the rabbit oviduct on normal ovum transport and on transport modified by estrogen and progesterone. Gynecol. Invest. 5:121.
46. Pauerstein, C. J., Hodgson, B. J., and Kramer, M. A. (1974). The anatomy and physiology of the oviduct. In R. M. Wynn (ed.), Obstetrics and Gynecology Annual, 1974, pp. 134–201. Appleton-Century-Crofts, New York.
47. Kirby, D. R. S. (1965). The role of the uterus in the early stages of mouse development. In G. E. W. Wolstenholme and M. O'Connor (eds.), Ciba Found. Symp. Preimplantation Stages of Pregnancy, pp. 325–339. J. and A. Churchill, London.
48. Allen, W. M., and Corner, G. W. (1929). Physiology of corpus luteum. III. Normal growth and implantation of embryos after early ablation of ovaries, under the influence of extracts of the corpus luteum. Am. J. Physiol. 88:340.
49. Allen, W. M., and Corner, G. W. (1930). Physiology of corpus luteum. VII. Maintenance of pregnancy in rabbit after very early castration, by corpus luteum extracts. Proc. Soc. Exp. Biol. Med. 27:403.
50. Pincus, G., and Werthessen, N. T. (1938). The maintenance of embryo life in ovariectomized rabbits. Am. J. Physiol. 124:484.
51. Chambon, Y. (1949). Besoins endocriniens qualitatifs et quantitatifs de l'ovoimplantation chez la lapine. C. R. Soc. Biol. 143:1172.
52. Meyer, R. K., Wolf, R. C., and Arslan, M. (1969). Implantation and maintenance of pregnancy in progesterone-treated ovariectomized monkeys (Macaca mulatta). In Proceedings of the Second International Congress on Primates, Vol. 2, pp. 30–35. Karger, New York.
53. Wislocki, G. B., and Streeter, G. L. (1938). On the placentation of the macaque (Macaca mulatta) from the time of implantation until the formation of the definitive placenta. Contrib. Embryol. Carnegie Inst. (Wash.) 27:1.

54. Hotchkiss, J., Atkinson, L. E., and Knobil, E. (1971). Time course of serum estrogen and luteinizing hormone (LH) concentrations during the menstrual cycle of the rhesus monkey. Endocrinology 89:177.
55. Reinius, S., Fritz, G. R., and Knobil, E. (1973). Ultrastructure and endocrinological correlates of an early implantation site in the rhesus monkey. J. Reprod. Fertil. 32:171.
56. Joshi, H. S., and Labhsetwar, A. P. (1972). The pattern of ovarian secretion of oestradiol and oestrone during pregnancy and the post-partum period in the hamster. J. Reprod. Fertil. 31:299.
57. Orsini, M. W., and Meyer, R. K. (1962). Effect of varying doses of progesterone on implantation in the ovariectomized hamster. Proc. Soc. Exp. Biol. Med. 110:713.
58. Deanesly, R. (1960). Implantation and early pregnancy in ovariectomized guinea pigs. J. Reprod. Fertil. 1:242.
59. Lutwak-Mann, C. (1955). Carbonic anhydrase in the female reproductive tract: occurrence, distribution and hormonal dependence. J. Endocrinol. 13:26.
60. Pincus, G., and Bialy, G. (1963). Carbonic anhydrase in steroid-responsive tissues. Recent Prog. Horm. Res. 19:201.
61. Psychoyos, A. (1963). Précisions sur l'état de "non-réceptivité" de l'utérus. C. R. Acad. Sci. 257:1153.
62. Meyers, K. (1970). Hormonal requirements for the maintenance of oestradiol-induced inhibition of uterine sensitivity in the ovariectomized rat. J. Endocrinol. 46:341.
63. Yoshinaga, K., and Greep, R. O. (1974). Uterine sensitivity with regard to the ovo-implantation. In S. M. Husain and A. F. Guttmacher (eds.), Progress in Reproduction Research and Population Control, pp. 137–146. Publications International, Quebec.
64. Yoshinaga, K., Hawkins, R. A., and Stocker, J. F. (1969). Estrogen secretion by the rat ovary in vivo during the estrous cycle and pregnancy. Endocrinology 85:103.
65. Yoshinaga, K., and Greep, R. O. (1970). Precocious sensitization of the uterus in pseudopregnant rats. Proc. Soc. Exp. Biol. Med. 134:725.
66. Yoshinaga, K., and Greep, R. O. (1971). Local inhibition of ovo-implantation in the rat. Endocrinology 88:627.
67. Shaikh, A., and Yoshinaga, K. (1969). Uterine sensitivity for deciduoma formation and ovo-implantation in the rat. Anat. Rec. 163:260 (Abstr.).
68. Yoshinaga, K. (1961). Effect of local application of ovarian hormones on the delay in implantation in lactating rats. J. Reprod. Fertil. 2:35.
69. Psychoyos, A. (1973). Endocrine control of egg implantation. In R. O. Greep and E. B. Astwood (eds.), Handbook of Physiology, Section 7. Endocrinology, Vol. II, Part 2, pp. 187–215. American Physiological Society, Washington, D.C.
70. Schlafke, S., and Enders, A. C. (1975). Cellular basis of interaction between trophoblast and uterus at implantation. Biol. Reprod. 12:41.
71. Nilsson, O. (1974). The morphology of blastocyst implantation. J. Reprod. Fertil. 38:187.
72. Glasser, S. R., and Clark, J. H. (1975). A determinant role for progesterone in the development of uterine sensitivity to decidualization and ovo-implantation. In C. L. Markert and J. Papaconstantinou (eds.), The Developmental Biology of Reproduction, pp. 311–345. Academic Press, New York.
73. Yochim, J. M. (1975). Development of the progestational uterus: metabolic aspects. Biol. Reprod. 12:231.
74. Yoshinaga, K. (1972). Biological and biochemical characteristics of the rat uterus at various phases of uterine sensitivity with regard to ovo-implantation. Fed. Proc. 31:812 (Abstr.).
75. Mester, I., Martel, D., Psychoyos, A., and Baulieu, E. E. (1974). Hormonal control of oestrogen receptor in uterus and receptivity for ovo-implantation in the rat. Nature 250:776.
76. Pencharz, R. I., and Long, J. A. (1931). The effect of hypophysectomy on gestation in the rat. Science 74:206.
77. Psychoyos, A. (1963). A study of the hormonal requirements for ovum implantation

in the rat, by means of delayed nidation-inducing substances (chlorpromazine, trifluo-perazine). J. Endocrinol. 27:337.

78. Mayer, G. (1963). The experimental control of ovum implantation. *In* P. Eckstein and F. Knowles (eds.), Techniques in Endocrine Research, pp. 245–259. Academic Press, New York.

79. Barraclough, C. A., and Sawyer, C. H. (1959). Induction of pseudopregnancy in the rat by reserpine and chlorpromazine. Endocrinology 65:563.

80. Ben-David, M., Danon, A., and Sulman, F. G. (1971). Evidence of antagonism between prolactin and gonadotrophin secretion: effect of methallibure on perphena-zine induced prolactin secretion in ovariectomized rats. J. Endocrinol. 51:719.

81. Lataste, M. F. (1891). Des variations de durée de la gestation chez les mammifères et des circonstances qui déterminent ces variations: théorie de la gestation retardèe. C. R. Soc. Biol. 9:21.

82. Krehbiel, R. H. (1941). The effects of Theelin on delayed implantation in the pregnant lactating rat. Anat. Rec. 81:381.

83. Whitten, W. K. (1955). Endocrine studies on delayed implantation in lactating mice. J. Endocrinol. 13:1.

84. Macdonald, G. J., Armstrong, D. T., and Greep, R. O. (1967). Initiation of blastocyst implantation by luteinizing hormone. Endocrinology 80:172.

85. Hayashida, T., and Young, W. P. (1963). Interruption of pregnancy in rats with antiserum. Anat. Rec. 145:323.

86. Madhwa Raj, H. G., Sairam, M. R., and Moudgal, N. R. (1968). Involvement of luteinizing hormone in the implantation process of the rat. J. Reprod. Fertil. 17:335.

87. Bindon, B. M. (1971). Gonadotrophin requirements for implantation in the mouse. J. Endocrinol. 50:19.

88. Murr, S. M., Bradford, G. E., and Geschwind, I. I. (1974). Plasma luteinizing hormone, follicle-stimulating hormone and prolactin during pregnancy in the mouse. Endocrinology 94:112.

89. Bindon, B. M. (1969). Follicle-stimulating hormone content of the pituitary gland before implantation in the mouse and rat. J. Endocrinol. 44:349.

90. Raud, H. R. (1974). The regulation of ovum implantation in the rat by endogenous and exogenous FSH and prolactin: possible role of ovarian follicles. Biol. Reprod. 10:327.

91. Enzmann, E. V., Saphir, N. R., and Pincus, G. (1932). Delayed pregnancy in mice. Anat. Rec. 54:325.

92. Moudgal, N. R., Rao, A. J., Maneckjee, R., Muralidhar, K., Mukku, V., and Rani, C. S. S. (1974). Gonadotropins and their antibodies. Recent Prog. Horm. Res. 30:47.

93. Yoshinaga, K. (1976). Ovarian hormone secretion and ovum implantation. *In* K. Yoshinaga, R. K. Meyer, and R. O. Greep (eds.), Implantation of the Ovum. Harvard University Press, Cambridge, Massachusetts, pp. 3–17.

94. Bloch, S. (1973). Nidation induced in mice during the lactational delay by the presence of strange males. J. Endocrinol. 57:185.

95. Labhsetwar, A. P. (1973). Evidence for the involvement of a catecholaminergic pathway in gonadotrophin secretion for implantation in rats. J. Reprod. Fertil. 33:545.

96. Friesen, H. G. (1973). Placental protein and polypeptide hormones. *In* R. O. Greep and E. B. Astwood (eds.), Handbook of Physiology, Section 7, Vol. II, Part 2, pp. 295–309. American Physiological Society, Washington, D.C.

97. Morgan, F. J., Kammerman, S., and Canfield, R. E. (1973). Chemistry of human chorionic gonadotrophin. *In* R. O. Greep and E. B. Astwood (eds.), Handbook of Physiology, Section 7, Vol. II, Part 2, pp. 311–311. American Physiological Society, Washington, D. C.

98. Ross, G. T., Cargill, C. M., Lipsett, M. B., Rayford, P. L., Marshall, J. R., Strott, C. A., and Rodbard, D. (1970). Pituitary and gonadal hormones in women during sponta-neous and induced ovulatory cycles. Recent Prog. Horm. Res. 26:1.

99. Yoshimi, T., Strott, C. A., Marshall, J. R., and Lipsett, M. B. (1969). Corpus luteum function in early pregnancy. J. Clin. Endocrinol. Metab. 29:225.

100. Oertel, G. W., Weiss, S. P., and Eik-Nes, K. B. (1959). Determination of progesterone in human blood plasma. J. Clin. Endocrinol. Metab. 19:213.
101. Palmer, R., Blair, J. A., Eriksson, G., and Diczfalusy, E. (1966). Studies on the metabolism of C-21 steroids in the human foetoplacental unit. 3. Metabolism of progesterone and 20α- and 20β-dihydroprogesterone by mid-term placentas perfused in situ. Acta Endocrinol. (Kbh.) 53:407.
102. Adock, E. W., III, Teasdale, F., August, C. S., Cox, S., Meschia, G., Battaglia, F. C., and Naughton, M. A. (1973). Human chorionic gonadotropin: its possible role in maternal lymphocyte suppression. Science 181:845.
103. Hobson, B. M. (1971). Production of gonadotropin, oestrogens and progesterone by the primate placenta. In M. W. H. Bishop (ed.), Advances in Reproductive Physiology, Vol. 5, pp. 67–102. Academic Press, New York.
104. Tullner, W. W. (1968). Urinary chorionic gonadotropin excretion in the monkey (Macaca mulatta)–early phase. Endocrinology 82:874.
105. Atkinson, L. E., Hotchkiss, J., Fritz, G. R., Surve, A. H., Neill, J. D., and Knobil, E. (1975). Circulating levels of steroids and chorionic gonadotropin during pregnancy in the rhesus monkey, with special attention to the rescue of the corpus luteum in early pregnancy. Biol. Reprod. 12:335.
106. Neill, J. D., and Knobil, E. (1972). On the nature of the initial luteotropic stimulus of pregnancy in the rhesus monkey. Endocrinology 90:34.
107. Hodgen, G. D., Niemann, W. H., and Tullner, W. W. (1975). Duration of chorionic gonadotropin production by the placenta of the rhesus monkey. Endocrinology 96:789.
108. Tullner, W. W., and Hertz, R. (1966). Normal gestation and chorionic gonadotropin levels in the monkey after ovariectomy in early pregnancy. Endocrinology 78:1076.
109. Fuchs, A.-R., and Beling, C. (1974). Evidence for early ovarian recognition of blastocysts in rabbits. Endocrinology 95:1054.
110. Haour, F., and Saxena, B. B. (1974). Detection of gonadotropin in rabbit blastocyst before implantation. Science 185:444.
111. Haour, F., Channing, C. P., and Saxena, B. B. (1975). A stimulatory effect of preimplantation rabbit blastocyst fluid upon luteinization of monkey granulosa cell cultures. Presented at the 57th Annual Meeting of the Endocrine Society, June 18–20, New York.
112. Sundaram, K., Connel, G., and Passautino, T. (1975). Control of corpus luteum function in the pregnant rabbit: absence of HCG-like material in the blastocyst. Presented at the 57th Annual Meeting of the Endocrine Society, June 18–20, New York.
113. Greep, R. O. (1941). Effects of hysterectomy and of estrogen treatment on volume changes in the corpora lutea of pregnant rabbits. Anat. Rec. 80:465.
114. Chu, J. P., Lee, C. C., and You, S. S. (1946). Functional relations between the uterus and the corpus luteum. J. Endocrinol. 4:392.
115. Scholfield, B. M. (1960). Hormonal control of pregnancy by the ovary and placenta in the rabbit. J. Physiol. (London) 151:578.
116. Holt, J. A., and Ewing, L. L. (1974). Acute dependence of ovarian progesterone output on the presence of placentas in 21-day pregnant rabbit. Endocrinology 94:1438.
117. Hilliard, J., Spies, H. G., and Sawyer, C. H. (1969). Hormonal factors regulating ovarian cholesterol mobilization and progestin secretion in intact and hypophysectomized rabbits. In K. W. McKerns (ed.), The Gonads, pp. 55–92. Appleton-Century-Crofts, New York.
118. Allen, W. M., and Corner, G. W. (1929). Physiology of the corpus luteum. III. Normal growth and implantation of embryos after very early ablation of the ovaries, under the influence of extracts of the corpus luteum. Am. J. Physiol. 88:340.
119. Zarrow, M. X., and Neher, G. M. (1955). Concentration of progestin in the serum of the rabbit during pregnancy, the puerperium and following castration. Endocrinology 56:1.
120. Matsumoto, K., Yamane, G., Endo, H., Kotoh, K., and Okano, K. (1969). Conversion

of pregnenolone to progesterone by rabbit placenta *in vitro*. Acta Endocrinol. 61:577.

121. Kehl, R., and Chambon, Y. (1949). Sur le conditionnement lutéinique du déciduome traumatique chez la lapine. C. R. Soc. Biol. (Paris) 143:698.

122. Chambon, Y. (1949). Besoins progestéro-folliculiniques quantitatifs du déciduome traumatique chez la lapine. C. R. Soc. Biol. (Paris) 143:701.

123. Kehl, R., and Chambon, Y. (1949). Synergie progestéro-folliculinique d'ovo-implantation chez la lapine. C. R. Soc. Biol. (Paris) 143:1169.

124. Dickmann, Z., Dey, S. K., and Gupta, J. S. (1975). Steroidogenesis in rabbit preimplantation embryos. Proc. Natl. Acad. Sci. U.S.A. 72:298.

125. Astwood, E. B., and Greep, R. O. (1938). A corpus-stimulating substance in the rat placenta. Proc. Soc. Exp. Biol. Med. 38:713.

126. Alloiteau, J.-J. (1957). Evolution normale des corps jaunes gestatifs chez la ratte hypophysectomesée au moment de la nidation. C. R. Soc. Biol. (Paris) 151:2009.

127. Alloiteau, J.-J. (1958). Destruction active des corps jaunes chez la ratte gestante hypophysectomisée avant la nidation. C. R. Soc. Biol. (Paris) 152:707.

128. Yoshinaga, K., and Adams, C. E. (1967). Luteotrophic activity of the young conceptus in the rat. J. Reprod. Fertil. 13:505.

129. Pencharz, R. I., and Long, J. A. (1933). Hypophysectomy in the pregnant rat. Am. J. Anat. 53:117.

130. Newton, W. H., and Beck, N. (1939). Placental activity in the mouse in the absence of the pituitary gland. J. Endocrinol. 1:65.

131. Deansley, R., and Newton, W. H. (1940). The influence of the placenta on the corpus luteum of pregnancy in the mouse. J. Endocrinol. 2:317.

132. Kirby, D. R. S. (1965). Endocrinological effects of experimentally induced extra-uterine pregnancies in virgin mice. J. Reprod. Fertil. 10:403.

133. Zeilmaker, G. H. (1968). Luteotrophic activity of the ectopic mouse trophoblast. J. Endocrinol. 41:455.

134. Averill, S. C., Ray, E. W., and Lyons, W. R. (1950). Maintenance of pregnancy in hypophysectomized rats with placental implants. Proc. Soc. Exp. Biol. Med. 75:3.

135. Matthies, D. L. (1967). Studies of the luteotropic and mammotropic factor found in trophoblast and maternal peripheral blood of the rat at mid-pregnancy. Anat. Rec. 159:55.

136. Matthies, D. L. (1974). Placental peptide hormones affecting fetal nutrition and lactation: effects of rodent chorionic mammotrophin. *In* J. B. Josimovich, M. Reynolds, and C. Cobo (eds.), Lactogenic Hormones, Fetal Nutrition and Lactation, pp. 297–334. John Wiley and Sons, New York.

137. Linkie, D. M., and Niswender, G. D. (1973). Characterization of rat placental luteotropin: physiological and physicochemical properties. Biol. Reprod. 8:48.

138. Kelly, P. A., Shiu, R. P. C., Robertson, M. C., and Friesen, H. G. (1975). Characterization of rat chorionic mammotropin. Endocrinology 96:1187.

139. Chatterton, R. T., Jr., Macdonald, G. J., and Ward, D. A. (1975). Effect of blastocysts on rat ovarian steroidogenesis in early pregnancy. Biol. Reprod. 13:77.

140. Cole, H. H., and Hart, G. H. (1930). The potency of blood serum of mares in progressive stages of pregnancy in effecting the sexual maturity of the immature rat. Am. J. Physiol. 93:57.

141. Wide, M., and Wide, L. (1963). Diagnosis of pregnancy in mares by an immunological method. Nature 198:1017.

142. Clegg, M. T., Boda, J. M., and Cole, H. H. (1954). The endometrial cups and allanto-chorionic pouches in the mare with emphasis on the source of equine gonadotropin. Endocrinology 54:448.

143. Amoroso, E. C. (1955). Endocrinology of pregnancy. Br. Med. Bull. 11:117.

144. Allen, W. R., and Moor, R. M. (1972). The origin of the equine endometrial cups. I. Production of PMSG by fetal trophoblast cells. J. Reprod. Fertil. 29:313.

145. Allen, W. R., Hamilton, D. W., and Moor, R. M. (1973). The origin of equine endometrial cups. II. Invasion of the endometrium by trophoblast. Anat. Rec. 177:485.

146. Hamilton, D. W., Allen, W. R., and Moor, R. M. (1973). The origin of equine

endometrial cups. III. Light and electron microscopic study of fully developed equine endometrial cups. Anat. Rec. 177:503.

147. Dickmann, Z., and Dey, S. K. (1974). Steroidogenesis in the preimplantation rat embryo and its possible influence on morula-blastocyst transformation and implantation. J. Reprod. Fertil. 37:91.

148. Dey, S. K., and Dickman, Z. (1974). Δ^5-3β-hydroxysteroid dehydrogenase activity in rat embryos on days 1 through 7 of pregnancy. Endocrinology 95:321.

149. Chew, N. J., and Sherman, M. I. (1975). Biochemistry of differentiation of mouse trophoblast Δ^5,3β-hydroxysteroid dehydrogenase. Biol. Reprod. 12:351.

150. Flood, P. F. (1974). Steroid-metabolizing enzymes in the early pig conceptus and in the related endometrium. J. Endocrinol. 63:413.

151. Perry, J. S., Heap, R. B., and Amoroso, E. C. (1973). Steroid hormone production by pig blastocysts. Nature 145:45.

152. Robertson, H. A., and King, G. J. (1974). Plasma concentrations of progesterone, oestrone, oestradiol-17β and of oestrone sulfate in the pig at implantation, during pregnancy and at parturition. J. Reprod Fertil. 40:133.

153. Anderson, L. L. (1973). Effects of hysterectomy and other factors on luteal function. In R. O. Greep and E. B. Astwood (eds.), Handbook of Physiology, Section 7, Vol. II, Part 2, pp. 69–86. American Physiological Society, Washington, D.C.

154. Moor, R. M., and Rowson, L. E. A. (1966). The corpus luteum of the sheep: functional relationship between the embryo and the corpus luteum. J. Endocrinol. 34:233.

155. Moor, R. M., and Rowson, L. E. A. (1966). The corpus luteum of the sheep: effect of the removal of embryos on luteal function. J. Endocrinol. 34:497.

156. Rowson, L. E. A., and Moor, R. M. (1967). The influence of embryonic tissue homogenate infused into the uterus, on the life-span of the corpus luteum in the sheep. J. Reprod. Fertil. 13:511.

157. Longnecker, D. E., and Day, B. N. (1972). Maintenance of corpora lutea and pregnancy in unilaterally pregnant gilts by intrauterine infusion of embryonic tissue. J. Reprod. Fertil. 31:171.

158. McCracken, J. A., Baird, D. T., and Goding, J. R. (1971). Factors affecting the secretion of steroids from the transplanted ovary in the sheep. Recent Prog. Horm. Res. 27:532.

159. Hansel, W., Shemesh, M., Hixon, J., and Lukaszewska, J. (1975). Extraction, isolation and identification of a luteolytic substance from bovine endometrium. Biol. Reprod. 13:30.

160. Velardo, J. T., Olsen, A. G., Hisaw, F. L., and Dawson, A. B. (1953). The influence of decidual tissue upon pseudopregnancy. Endocrinology 53:216.

161. De Jongh, S. E., and Wolthuis, O. L. (1964). Factors determining cessation of corpus luteum function: the possible role of oestradiol and progesterone. Acta Endocrinol. Suppl. 90: 125.

162. Wiest, W. G. (1973). DCR sensitivity related to uterine estradiol (E_2), progesterone (P) and 20α-OH-P. Presented at the 55th Annual Meeting of the Endocrine Society, June 20–22, Chicago, Illinois.

163. Gibori, G., Rothchild, I., Pepe, G. J., Morishige, W. K., and Lam, P. (1974). Luteotrophic action of decidual tissue. Endocrinology 95:1113.

164. Castracane, V. D., and Shaikh, A. A. (1976). Effect of decidual tissue on the uterine production of prostaglandins in pseudopregnant rats. J. Reprod. Fertil. 46:101.

165. Lin, Y. C. (1975). Progestin output by rat decidual cells in vitro. Presented at the 8th Annual Meeting of the Society for the Study of Reproduction, July 22–25, Fort Collins, Colorado.

International Review of Physiology
Reproductive Physiology II, Volume 13
Edited by Roy O. Greep
Copyright 1977   University Park Press   Baltimore

7
Neurohormonal Bases of Male Sexual Behavior

J. M. DAVIDSON

Department of Physiology
Stanford University
Stanford, California

This work was supported by Grants MH 21178 and HD 00778 from the National
Institutes of Health, United States Public Health Service.

The primary objective of this chapter is to analyze issues regarding the role of androgen in male sexual behavior and its interplay with the nervous system. The presentation will be constructed of three overlapping levels of analysis which correspond to three conceptual-experimental approaches. The first is that of biobehavioral correlation, involving careful measurement and classification of behavioral responses and their relationships to eliciting stimuli, together with the attempt to elucidate biological correlates of the behavioral phenomena. The second approach is anatomical. Here one asks where relevant events are occurring, and what the relative importance of changes at different locations is. Finally, we shall proceed to the psychological level of analysis. The third approach involves the interpretation of anatomical, physiological, and behavioral information (and in the human, subjective reports), in terms of possible psychobiological processes which intervene between stimulus and response.

There is much overlap between these three points of view. Although the first approach asks "what?" the second "where?" and only the third asks "how?" it is not in fact possible to avoid "how" questions in any of them insofar as they are all concerned with mechanisms. But it is only the third approach that refers to the principal "how" of biobehavioral science: how do physical phenomena relate to behavior? By following a series of increasingly complex questions, this chapter develops a kind of cascading argument which might give the reader a better picture of the state of this field than would be conveyed by the more usual systematic approach. In the last part of this chapter, the relevance to men of the findings from animal research is discussed in the context of a consideration of the nature of human sexuality.

DESCRIPTIVE-CORRELATIVE APPROACH

Before embarking on a description of selected findings from the animal research laboratory, the restricted nature of our data base should be noted. The use of selected, vigorously mating laboratory (or farm-reared) animals, housed in small, often crowded cages and tested in arenas where they are conditioned to prompt copulation, may produce results at variance with what would be seen in the wild, and experimental observations in the field are rare. The laboratory situation removes or at least modifies the influence of social and other environmental factors. Among resulting discrepancies, one would expect decreases in courtship behavior in the laboratory, and overall increases in copulatory responses. There is, however, reason to believe that the discrepancies are quantitative rather than qualitative (1).

Behavior Patterns

With this caveat in mind, consider first the mating pattern of the rat, which is the most studied species. (Except when otherwise specified, the species under discussion will be the rat, until the final section of the chapter.) When placed with a receptive female, a brief period of exploration of the immediate environment and of the female partner ensues. Within seconds or minutes there follows a series of mounts, mounts with intromission (referred to hereafter as "intromissions"), separated by interintromission intervals of the order of 1 min in duration, during which the male usually grooms himself. Ejaculation follows about 5–15 of these events, interspersed with occasional mounts without intromission (referred to as "mounts"). These three events can be clearly distinguished by the experienced observer by watching the activity of the whole animal (the penis itself is generally not seen).

This behavior is relatively invariant and stereotyped and is subject to experimental manipulation. It can be quantified (2) in terms of intervals to the first event, mount latency (ML) or intromission latency (IL), time from first intromission to ejaculation, ejaculation latency (EL), and number of events before ejaculation, mount frequency (MF) and intromission frequency (IF). It should be noted that although the final consummatory event includes seminal emission, ejaculation, as used in this context, refers not to this, but to a pattern of behavior which is characterized by a long intromission and slow dismount. It is followed by a postejaculatory interval (PEI) of about 5 min, 75% of which comprises an absolute refractory period during which time the male shows sleep-like electroencephalogram (EEG) patterns and emits vocalization in the supersonic range (3). The animal is not arousable during the remaining 25% of the PEI. The whole series is repeated around five to ten times, albeit with predictable changes in the IF, EL, and PEI, after which a state of sexual exhaustion ensues.

From the comparative viewpoint, Dewsbury (1) has constructed a classification scheme for male copulatory behavior among mammals in general. He differentiates mating patterns according to the presence or absence of a genital lock, thrusting during intromissions, multiple intromissions, and multiple ejaculations. The number of possible forms of copulatory behavior in terms of this classification is $2^4 = 16$, and although not all may exist in nature, 7 of the 16 patterns have been found in different species of muroid rodents. In addition to these characteristics, other items of importance which vary among species are the conditions under which copulation occurs, precopulatory or courtship behaviors, postures, quantitative detail regarding frequency and temporal characteristics, and other behavioral patterns which may be found during a copulatory episode (1).

Role of Testosterone

A most important biological correlate of male sexual behavior in all mammalian species is the presence of adequate testicular androgen (primarily testosterone (T)) in the circulation. The dependence of male copulation on androgen appears from laboratory studies to be almost absolute with the qualification that a lag period is found between removal of testicular androgen from the blood (by castration) and the decline in behavioral capacity. This lag is highly variable within and between species, but greater or lesser residues of behavior (including ejaculatory patterns) may persist for prolonged periods of time in many species (4). The suspicion that adrenal androgen may be involved in this phenomenon has not been substantiated in any species studied (5, 6).

Given adequate androgen, copulatory behavior may persist after removal of large portions of the nervous system (7) and even in the absence of a penis (8). But what is an adequate blood level of androgen, and can the normal inter- and intra-individual variations in quantity and/or quality of sexual behavior be correlated with variations in androgen level?

Earlier studies on guinea pigs and rats demonstrated that there are constitutional factors determining the probability and level of male copulation (9), which may be conceived of as individual differences in sensitivity to the behavioral actions of androgen. Furthermore, normal guinea pigs were thought to be functioning at their maximal possible androgen-dependent level of behavior, although there is some evidence that androgen can raise behavioral performance of rats above the normal level (9). These studies do not, however, directly address themselves to the issue of the extent to which normal variations in testicular androgen production may contribute to inter- and intra-individual behavioral differences. Only with the advent, in recent years, of practical, reliable, and sensitive radioimmunoassays for steroids has it become feasible to provide direct answers to this question. Studies in the rhesus monkey have so far failed to find such correlations (10, 11), but insufficient data have been reported to date to settle this question.

Recently, mean plasma T levels were obtained in this laboratory from 70 rats by four blood samplings over a period of 11 days, and behavioral data were obtained from the same animals by three interspersed mating tests during the same period (12). No correlations between the probability of occurrence of complete (ejaculatory) behavior or any other derived measure of the mating pattern could be observed. These same animals were subsequently castrated and received subcutaneous implants of T in constant release (silastic) capsules, which produce extremely stable circulating hormone levels over periods of several weeks, these levels being a highly predictable function of capsule surface area. Repeated behavioral testing and blood sampling for T analysis showed a clearly increasing trend of IL (i.e., slowing of responses) with decreasing T levels but commencing only well below the normal mean (Figure 1). The surprising outcome of these studies (12), however, was the high level of behavior in animals with extremely low plasma T. Thus, in the group with the smallest capsules, full ejaculatory responses were observed in over 80% of the tests, although the plasma T level was less than 10% of that found in the normal control group. (It should be noted that restoration of sexual behavior following its disappearance in castrates requires considerably more androgen than does maintenance (13).) This is impressive, even allowing for possible enhancement of the behavioral effect of a given mean plasma T level when artificially held constant as compared with the usual situation of episodic release (14). These data provide compelling evidence for a considerable degree of redundancy or margin for error in the

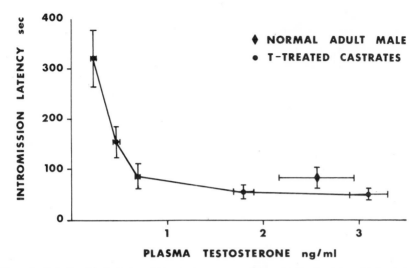

Figure 1. Relationship between plasma testosterone and sexual behavior (mean intromission latency over repeated tests) during 3 weeks after castration and simultaneous implantation of subcutaneous constant release capsules containing T. Rats of the Long-Evans strain were used in this and subsequent figures. Reprinted from Ref. 12.

normal circulating level of T. Most of the variance in behaviorally normal, experienced male rats does not seem related to androgen levels, and dose-response relationships are probably to be found mostly at the lowest level of behavioral performance.

What about the changes which accompany normal maturation? A recent study fom this laboratory (15) has shown that the normal onset of mating behavior in male rats coincides or precedes the peripubertal increase in circulating T (Figure 2). In view of the considerable period of time required before administration of T can influence copulatory and especially ejaculatory behavior, it is clear that the pubertal appearance of behavioral capacity, although requiring this presence of androgen, seems to depend on maturational processes within the organism rather than simply on being elicited by increased hormone titers. Another related maturational factor, that adult rats are more sensitive to the behavior-activating effect of androgen than prepubertal animals, was also demonstrated (15).

Finally, the interplay between level of exposure to T and sensitivity or responsiveness to the hormone should be considered in relation to the seasonal variations in reproductive and behavioral phenomena. When, as in many wild or photoperiodically responsive species, testicular activity is profoundly depressed in nonbreeding seasons (16), T level and sexual performance will co-vary, but the correlation may be less clear in species like the rhesus monkey, in which the change in T level is not so profound (10, 11). In the hamster, which belongs to the former category, a decreased sensitivity to behavioral activation by T has

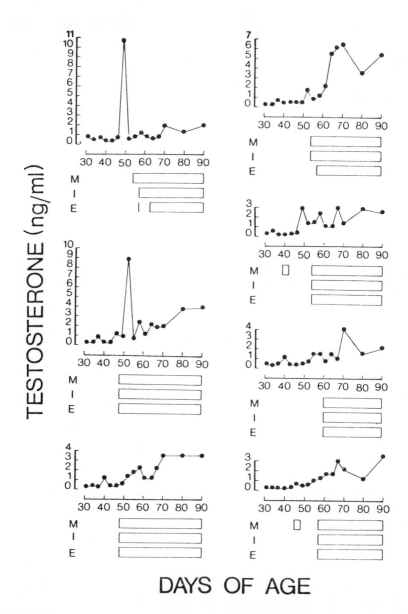

Figure 2. Plasma testosterone in developing male rats, in relation to the onset of mounts (M), intromissions (I), and ejaculations (E). Blood sampling (performed shortly after each test) was shown to have no effect on the behavior. Reprinted from Ref. 15.

been reported on short photoperiods (as in the nonbreeding season) (17), although it is not clear to what extent the change was related merely to passage of time after decline in T level.

Other Biochemical Correlates

In recent years, work on biological correlates of sexual behavior has been extended to the cellular-molecular level with the opening up of three major areas of active investigation. First is the search for neurotransmitters, particularly those which might mediate effects of hormones on sexual behavior. As in many other branches of neurobehavioral science, a plethora of research activity is being devoted to the role of biogenic amines. Considerable data from females, as well as males, support an inhibitory role for serotonin and a facilitatory one for catecholamines (18). Agonists and antagonists of central (and peripheral) biogenic amines are new and powerful agents for the experimentalist interested in modifying the behavioral responses of animals. Both L-dihydroxyphenylalanine (L-dopa, precursor of dopamine) and p-chlorophenylalanine (a serotonin synthesis inhibitor) have been proposed for clinical use in treating depressed male sexuality, although to date the hopes of finding universal aphrodisiacs, or cures for impotent Parkinsonism patients, are far from fulfilment (18). Because the same amines are presumably involved in many physiological and behavioral systems, the unequivocal demonstration of the specific role of a given neurotransmitter will remain beyond our grasp until neurons related to sexual behavior are identified.

The search for hormone receptors inside the brain cell is presently being pursued with vigor. Progress in this endeavor, as compared with peripheral receptor studies, is slow but sure (19). Not so sure is the relevance to behavior. Here again, interpretations depend upon clarification of the problems of anatomical specificity. Until then, the findings on steroid receptors in brain could relate to a variety of possible cerebral actions of these hormones. Certainly, the apparent distribution of steroid receptors in brain tissue is wide enough to cover areas postulated to be involved in a variety of behavioral and other endocrine (e.g., feedback) effects.

As to the metabolic conversion of androgen within the cell, recent attention has centered on the hypothesis that aromatization is a necessary prerequisite for behavioral action, i.e., the "active androgen" is in fact an estrogen (20). This would clearly differentiate the behavioral mechanism from androgen action on the reproductive tract which apparently depends upon 5α-reduction to dihydrotestosterone (DHT). In fact, apart from the finding that the hypothalamus seems to have the necessary enzymatic machinery (20), the major original support for the aromatization hypothesis was the observation that DHT has no effect on the sexual behavior of rats (21). Interestingly, this androgen is even more active than T on the hypothalamic-pituitary androgen feedback receptors regulating gonadotropin secretion.

The aromatization hypothesis has engendered a flurry of research which has resulted in some interesting supportive findings. These include the already well-established observation that small doses of estrogen synergize with DHT to elicit male sex behavior in castrated males (22) and that intracranial administration of aromatization inhibitors has been found to suppress mounting behavior elicited by intracranial T but not estrogen (23). However, there are major problems with the hypothesis, such as recent observations that this androgen is quite effective in stimulating sex behavior in some species and very effective in others (23). Even in the rat, it has recently been found that a nonaromatizable steroid, 6α-fluorotestosterone is as behaviorally effective as T (24). One problem in this area is that most information on the aromatizability of different steroids comes from in vitro studies on placenta. Whether the relevant cells perform the postulated conversion as part of their behavioral function cannot again be determined until these cells are identified.

Another area of activity in the molecular sphere, involving mostly females at present, is the investigation of the possibility that steroids activate sex behavior mechanisms by altering genomic expression. Among difficulties in this area of investigation is the fact that the protein synthesis inhibitors used are strongly cytotoxic drugs which, on systemic administration at least, produce nonspecific defects, making interpretation of behavioral findings difficult. This problem has been partially alleviated by using local intra-cerebral implants of actinomycin, which produce reversible suppression of lordosis behavior (25). Again, however, progress in this area depends to a great extent on the solution of the anatomical questions which will now be discussed.

ANATOMY OF BEHAVIORAL REGULATION

If the role of androgen in sex behavior is specific, i.e., not, as Kinsey suggested, merely that of a general maintainance of health and behavioral capacity, then it must act via specific neural mechanisms. These mechanisms may be represented at the level of the brain, spinal cord, and/or peripheral nervous system. The quest for localization of the action of T thus proceeds logically via the prior demonstration of the location of these neural mechanisms.

Brain

The evidence for the specific localization of male sexual functions in the brain is derived from stimulation, lesion, and neurochemical experiments. Electrical stimulation of the brain can elicit or facilitate copulatory behavior when the electrodes are in the medial preoptic area or the posterior lateral hypothalamus (26). Such facilitatory effects from electrical stimulation can be found to extend to more posterior sites (the areas probably being linked via the medial forebrain bundle), all the way back to the mid-brain (26). However, serious doubt attaches to the application of brain stimulation experiments to the localization of specific motivated behaviors. Thus, a given behavior may result from different electrode placements in different animals and stimulation of the same site can produce different results, depending on the motivational state of the subject or simply the passage of time in one subject (27, 28). Although it is not yet clear whether such observations reflect great neural plasticity or are merely forms of experimental artifact (28), they do show that electrical stimulation data provide an uncertain basis for demonstrating neuroanatomical localization of sexual functions.

Brain lesion experiments seem to be more reliable and are only in part supportive of the findings from electrical stimulation. This approach has provided ample evidence for both excitatory and inhibitory mechanisms. The existence of the latter is inferred from the fact that facilitation of male sexual behavior can result from destruction of portions of the limbic-hypothalamic system. Although the original report on male rats was based on rather large lesions (29), the major finding of striking decreases in the postejaculatory interval has recently been confirmed using smaller, more focal destruction in the mid-brain (30).

The effects of large lesions can focus attention on relevant mechanisms but can also seriously mislead, as is shown by literature dating back to observations on the "Klüver-Bucy syndrome" in monkeys (31), which first indicated the possibility of limbic system inhibition of male sexuality. The interpretation of experiments on temporal lobe resection or amygdalectomy with resulting hypersexuality is, however, by no means unequivocal; behavioral facilitation may, for instance, relate to such nonspecific factors as more rapid habituation to novel stimuli, resulting in lack of discrimination of sexual partners (28). In rats, in which striking hypersexuality does not result from limbic system lesions, less dramatic effects on behavior have been reported from dorsal hippocampal and amygdaloid lesions (32), suggesting modulatory influences from the hippocampus (inhibitory) and amygdala (excitatory). The latter may relate to input from the olfactory system, parts of which apparently play important roles in male sex behavior (32, 33).

The most convincing evidence for cerebral localization of male sexual behavior centers on the medial preoptic area (MPO). The finding of virtually complete cessation of copulation in a high percentage of rats with lesions in this region (34) and in cats, dogs (35), and monkeys (J. Slimp, B. Hart, and R. W. Goy, unpublished observations) suggests the phylogenetically widespread role of this brain region in mammals. The effects of these lesions are not mediated by a decrease in testicular function, and T treatment does not restore the behavior (34). Bearing on the specificity of the behavioral role of the MPO are Hart's observations that aggressive-dominance behavior was not impaired in lesioned dogs, although another T-related behavior—urine marking—was affected (35).

The MPO control of male sexual behavior is apparently linked to descending and/or ascending pathways in the medial forebrain bundle (MFB), because lesions in parts of this pathway effectively reduce the behavior. Various views have been expressed on the nature of these connections. First, MPO activation (or disinhibition) of descending pathways in the MFB may produce facilitation of brainstem and spinal cord systems subserving mounting, intromission, and ejaculation (35, 36). Because the MPO does not seem to contribute directly to the formation of the MFB, the influences may be relayed through the anterior hypothalamus. This is supported by the finding that parasagittal knife cuts which impair male rat behavior are more effective when placed posterior to the MPO (37). The idea that the pathway involved uses monoamine transmission is supported by neurochemical experiments (38). It has also been proposed that the MPO is a final common path of a response system encompassing much of the limbic system (39).

Extracerebral Mechanisms

The spinal cord has an obvious function in sexual behavior relating to information transmission between the brain and periphery. More interesting is its importance in the integration of sexual reflexes. A variety of older studies and the more recent and detailed observations of Hart (40) have clearly demonstrated the presence, in spinally sectioned dogs and rats, of sexual reflex responses which can be elicited by mechanical stimulation. Much thought has been devoted to the attempt to explain the role of the brain in terms of its inhibitory influence on these reflexes (7). As to the peripheral nervous system, its obvious role in perception of the sexual partner and execution of copulatory acts need not be discussed here. Of major relevance to this discussion, however, are functions of the penile innervation relating to the fact that, in the behavioral context, the phallus plays a part far greater than that of a conduit for semen. Genital denervation, by section of penile or pudendal nerves or local anesthesia of the penis, is inconsistent with normal intromission and ejaculatory behavior in rats (41), cats (42), and monkeys (43), although mounting behavior continues.

Site of Action of Testosterone

The cerebral, spinal, and peripheral systems briefly outlined above represent the possible sites of the behavioral action of androgen. Influenced by the data on restoration of male sexual behavior by intracerebral T implants (see below), earlier reviews have stressed the importance of androgen's action on the preoptic-hypothalamic region (13, 44). As more information becomes available, however, this is seen to be an oversimplification, for the situation is more complex and less easily resolved than was previously thought. Unfortunately, there have still been no intracerebral androgen implantation studies on males of mammalian species other than the rat, and for this and other reasons the periphery-center debate is not yet closed. The following discussion reverses the order of consideration followed in the previous section and proceeds from periphery to brain in considering possible sites of androgen action.

Periphery   Because there is no evidence of extragenital peripheral sensory actions of androgen, discussion of the periphery will be limited to the penis. It is well known that penile structure depends on androgen, but the effect of castration is a function of the age of the animal at the time of surgery. Prepubertal castration results in an infantile penis, but when performed in adulthood, regression is much slower and does not revert to the infantile condition (45). T administration restores penile structure in the adult castrate, but prepubertal castration is not reversible in adulthood, particularly if surgery was performed in the neonatal period (46). Among the structures subject to the organizing influence of androgen during the early critical period in rodents are the penile bone and the papillae or spines on the epithelium of the glans (46), which are exquisitely sensitive to the action of androgen (47). The function of these structures is not, however, known, and many castrates continue to show complete mating behavior for long periods after they have regressed.

Because no known androgen dependent structure of the penis has been shown to be necessary for sexual behavior, the question is whether the behavioral decrement after denervation or anesthesia relates to an androgen-dependent function. In other words, is T necessary for the reception and transmission of behaviorally relevant information?

A process which might reflect the functional role of T in sex behavior is that of erection. However, although it is often assumed that the androgen dependence of this function is mediated by a direct action of T on the penis, it could equally be mediated by the central nervous system (CNS): the well-known studies of MacLean and Ploog in squirrel monkeys showed that erection could be elicited by stimulation of various brain sites (48). Unfortunately, the androgen dependence of erection has been very little studied. Rodgers and Alheid (49) found that castration did not reduce the frequency of spontaneous erections in the intact, lightly anesthetized supine rat, although the duration of the individual erections was shortened. Unpublished data from both Rodgers' and our laboratories suggest, however, that erection frequency decreased in castrates can be increased by T treatment.

Interesting evidence in favor of the peripheral action of T is the behavioral restoration obtained with combined treatment of castrates with DHT, an effective stimulator of peripheral androgen-dependent tissues, which apparently does not act on the brain to stimulate male sexual behavior (21, 22), and estradiol, which does not stimulate the periphery but is believed to have behavioral action on the brain (see under "Other Biochemical Correlates"). However, more recent reports suggest the possibility of a synergistic action of these two agents at the cerebral level (50), so that there is no real need to postulate that the peripheral action of DHT is essential for the success of these treatments.

An indication of a peripheral action of T in ejaculatory behavior comes from experiments with a potent nonsteroidal antiandrogen, flutamide (α-α-α-trifluoro-2-methyl-4-nitro-m-propiono-toluidide), which successfully countered the action of T on the peripheral organs (including penile spines) and gonadotropin suppression, but did not inhibit sexual behavior in intact adult males (51). Although the findings suggested that this antiandrogen, like the steroidal one, cyproterone, specifically lacks behavioral antiandrogenicity, these intact animals had very high circulating levels of testosterone due to increased gonadotropin secretion. Given the fact that mating behavior can be maintained with very low T levels (see under "Role of Testosterone"), the ratio of antiandrogen to androgen may not have been high enough for competitive inhibition of sexual behavior. When given together with T to castrates, there was no significant effect of flutamide on intromission or mounting behavior, but it reduced ejaculatory frequency quite markedly. It is tempting to conclude that this was due to its interference with T action on the penis, particularly because when the penis was anesthetized by local application of tetracaine, which prevents intromissions or ejaculation, flutamide did not inhibit the effects of T on mounting. In fact, there was a slight stimulation (52) reminiscent of the slight behavioral stimulation with the steroidal antiandrogen cyproterone (47). The important point in the present context, however, is not this anomalous effect of antiandrogens on sexual behavior mechanisms, but that significant (though incomplete) suppression of ejaculatory behavior was obtained with a drug which antagonizes androgenic effects on penile morphology, but not on the presumably central mechanisms regulating mounting.

Testosterone Action at Spinal Level   The best published evidence for a noncerebral action of androgen on erection is the study of Hart on spinally sectioned castrate rats and dogs (40, 53). T propionate treatment of spinal rats significantly facilitated erections and other sexual reflexes in response to genital stimulation. In dogs, no effect on copulatory skeletal movement or erection was found, but several genital reflexes were facilitated by T (53). Hart interprets these results (40, 53) as indicating a spinal action of T and supports this conclusion with another study in which DHT was less effective than T propionate (in a higher dose) in castrated spinal rats (54). However, it is not established that DHT acts at the spinal level, and the results of both studies are consistent with androgenic action on all segments of the reflex arc, including peripheral nerves and sensory receptors. The most direct evidence for a strictly spinal action comes from Hart's study on intraspinal implants of T (55). This procedure significantly activated only one of the genital reflexes examined ("long flips"), however, and, although sexual accessory glands were examined, T release into the circulation cannot be excluded (see below).

Brain as Site of Testosterone Action   The major evidence for a central effect of T in male sexual behavior comes from the well substantiated observation that hypothalamic implants of crystalline T restore complete patterns of sexual behavior in a high proportion of rats which have ceased to mate by virtue of long-term castration; similar findings have been obtained in several avian species (13, 44). Because fairly large implants (200 μg) had to be used to obtain good restoration of behavior in rats (56, 57), it is important to determine whether significant amounts of T are released into the peripheral circulation, thereby affecting areas other than those near the site of implantation. Earlier efforts to exclude circulating T as an explanation for the effects of these implants were not conclusive. This is because they depended on a single observation of the histology of the sexual accessory glands (or penile epithelium) at the time of autopsy, and this would not have reflected earlier release of the hormone from the implant, which might have significantly affected the behavioral results (56, 57).

Although implants in the cortex and other extra-hypothalamic areas were relatively ineffective, this could be due to unequal rates of release of T into the circulation from different areas of the brain. Recent experiments (58) have in fact substantiated the possibility of differential release. However, the areas of T implantation from which leakage was greatest are not those from which the best effects are obtained. Thus, the highest rates of release were from implants in the medial basal hypothalamus and pituitary (58), whereas anterior hypothalamic-medial preoptic (AHPO) implants resulted in circulating levels at the low end of the normal range for the first week after implantation (Figure 3). Implants in the posterior hypothalamus released as much T to the circulation as those in the AHPO, but were slightly less behaviorally effective (Figure 4); in other studies the behavioral difference was greater (56, 59). Most convincingly, implants in the cortex stimulated few behavioral responses (Figure 4), but were not distinguishable in plasma T levels from the AHPO group. Thus, although the possibility remains of synergistic effects resulting from the small amounts of T in the blood acting together with a direct brain effect on the AHPO, little doubt is left about the existence and presumed importance of the behavioral action of androgen on this area.   ·

It is necessary to emphasize, however, that these studies do not eliminate from consideration the peripheral and/or spinal action of T as possibly contributing to the total effect on sexual behavior. Intra-hypothalamic implants result in complete mating patterns only in about 50% of the tests performed, and not more than 80% of any group of implanted rats have ever shown positive results in our studies (56–58). It is not excluded that the failure to reach 100% performance (which is obtained with systemic injections) is due to the lack of androgenic exposure of the peripheral tissues, although it is equally possible that the artificiality of the method of T propionate (TP) implantation is responsible. (It should be noted, however, that the incompleteness of behavioral restoration is not due to the production of a lesion by the implant, because it has been found that single or even double implants of cholesterol do not retard the rate of behavioral restoration resulting from daily injections of TP (47)). Experiments in which administration of DHT did not bring the performance of rats with AHPO implants up to normal seem to counter the possibility that peripheral hypoandrogenicity is responsible for the subnormal performance of these animals (60). They do not, however, eliminate this possibility, because the morphological effects of DHT need not reflect correction of a possible deficit in peripheral function.

It remains true, also, that intracerebral implantation of TP is a very unphysiological procedure. Other less disruptive approaches should still be sought before definitive conclusions are drawn about the relative roles of central versus

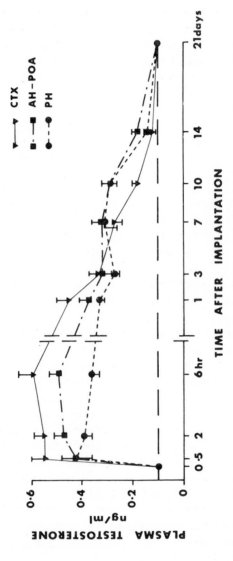

Figure 3. Plasma testosterone resulting from intracerebral implantation of TP. *Broken line*, minimal T level detectable in assay; *CTX*, cortex; *AH-POA*, anterior hypothalamic medial preoptic; *PH*, posterior hypothalamus. Reprinted from Ref. 58.

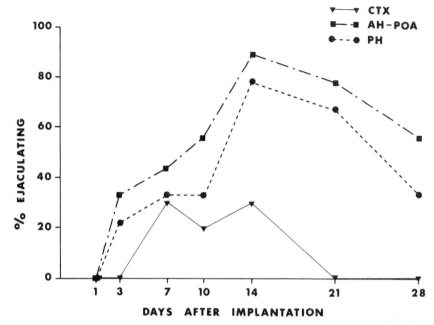

Figure 4. Sexual behavior of rats whose blood testosterone levels are shown in Figure 3. *CTX,* cortex; *AH-POA,* anterior hypothalamic medial preoptic; *PH,* posterior hypothalamus.

peripheral receptors for the behavioral effects of androgen. It should, for example, be noted that the implantation approach has only been useful to study male sex behavior in long-term castrates. Attempts in our laboratory to duplicate the results by using animals implanted on the day of castration failed to produce significant results, because it often takes several weeks for most of the control rats to cease copulating, and androgen implants are only effective for a limited period of time, even when long acting T esters are used (J. M. Davidson, unpublished data).

The conclusion which appears to emerge from these findings is that, in the presence of an established penis (i.e., not in a prepubertal castrate), withdrawal of T from the periphery while maintaining brain levels limits sex behavior only in quantitative terms; some subjects will perform normally and some not at all. To what extent quantitatively and qualitatively optimal behavioral performance depends on peripheral androgen is unclear, as is the nature of the contribution of DHT to behavioral restoration when given together with estrogen. The reduction in androgen-induced ejaculatory but not mounting behavior caused by flutamide is presumptive evidence of a peripheral effect, which is hard to reconcile with the rather high incidence of ejaculation resulting from intracerebral T implantation. Conceivably, flutamide could be acting on an extra penile mechanism regulating ejaculatory behavior.

PSYCHOLOGICAL APPROACH

Beyond the search for biological correlates, mechanisms, and sites of action, lies another level of enquiry. The failure of naive behaviorism to construct a science out of stimuli, responses, and black boxes is evident to any reader of the recent psychological literature. Even if we had very specific and precise knowledge of the cellular and subcellular mechanisms involved in the behavioral action of androgen, a basic question would remain: What does it mean to say that testosterone stimulates male sexual behavior by acting on specific areas of the nervous system? This section is an attempt to look at how the behavior is organized and to uncover phenomena which intervene between hormonal, nervous, and other physical manifestations on the one hand and overt events in the realm of behavior and experience on the other.

Processes Underlying Male Copulation

At the simplest level of analysis, afferent information from the environment can be related to various sexual responses (erection, mounting, etc.) on the output or efferent side. With increasing depth of analysis, however, it soon appears necessary to introduce a more nebulous function, which is usually called "arousal" (61). This term, which can be roughly translated as "readiness to respond" or "level of activation of the organism," has been created by psychologists because mechanistic explanations for behavior which did not include some such postulated intervening variable seemed inadequate for the understanding of behavioral phenomena. General arousal as a nonspecific entity which fuels all behavior is a difficult, controversial concept whose ontological status and even heuristic value has often been questioned (62). It is safer to refer only to specific sexual arousal and, moreover, define it in terms of actual behavioral measures.

The first detailed and authoritative analysis of psychological structures underlying male copulation was that of Beach (63). He postulated a dichotomy between an arousal mechanism (AM) and an intromission-ejaculatory mechanism or copulatory mechanism (CM). Thus, mount or intromission latency and postejaculatory interval were functions of the AM and ejaculation latency, and intromission frequency reflected the CM (for definitions of behavioral measures, see under "Behavior Patterns"). It should be clear that behavioral acts themselves cannot be identified with AM or CM, but behavioral latencies and frequencies are related to them. In general, activity of the AM is negatively correlated with the onset of behavior, and the numbers of events and the time required for ejaculation are negatively correlated with the CM activity. The dual mechanism can account for the observation that the initiation of copulatory behavior may vary independently of the rates and efficiency of copulatory events leading to ejaculation. A variety of situations is known today in which there seems to be a block to the onset of behavior (e.g., certain brain manipulations, drug treatments, and noncopulator rats), although once commenced or elicited by unusual means (e.g., electric shock), the copulatory pattern can proceed without major impairment (64). The separate control of different behavioral components is also seen in situations in which, although mating is absent, various sexual reflexes can be elicited by the experimenter (see under "Anatomy of Behavioral Regulation").

Although this postulated duality has provided a useful guide for conceptualization over the years, it has become abundantly clear that a sharp temporal distinction between the two mechanisms is not acceptable, and the systems operating to control ML, IL, or PEI (i.e., initiation latencies) must also continue to operate during the further progress of copulation. In a recent review, Sachs and Barfield (65) cite many examples of experimental situations (drug treatments, shock, stimulus variations, etc.) which indicate a positive relationship between arousal or arousability and copulatory rate. Furthermore, the CM can in no way be considered a unitary mechanism, but as a recognized concept, it provides a useful shorthand term to cover the plurality of processes which occur from the initiation of copulation in the male to its final consummation. Its two major components, as used here, are sexual reflexes such as erection and ejaculation and the neural events which finally lead to ejaculation.

Localization of Arousal and Copulatory Mechanisms

Clearly, the elucidation of the physiological underpinnings of sexual behavior would be greatly furthered were it possible to locate the neural representation of the AM and CM. Although, despite the efforts of various authors in this direction, no clear-cut conclusions can be derived from the existing data, a discussion of the evidence will be instructive. First, it should be pointed out that surgical penile denervation or local anesthesia does not prevent sexual arousal in intact rats or cats, in that these animals may show high mount frequencies for a period of time (41–43). The import of this is that the arousal function, although undoubtedly affected by genital input, is not dependent in any absolute sense on the penis, but rather presumably depends on sensory input of different modalities.

As to the central mechanisms, the first area to consider is the MPO. Malsbury and Pfaff (32) conclude from a review of the literature on various mammalian species that the MPO is mainly involved in mediating the CM rather than the AM. This was based primarily on the evidence that electrical stimulation in that area elicits erection in squirrel monkeys (48). However, in the opossum as well as the squirrel monkey and rat, stimulation of other areas can also activate sexual reflexes (32). Furthermore, there is the problem of relating such reflexes to events during normal copulation (in this species, erection is part of the social behavior of genital display), and also the general problem of specificity in relation to electrical stimulation of the brain alluded to earlier. The rats with MPO lesions studied by Heimer and Larsson (34) and Giantonio et al. (39) showed various pursuit, sniffing, and other precopulatory behaviors in the absence of intromissions or even mounts with thrusts. This was interpreted as indicating retention of sexual arousal in the absence of the MPO. However, because sexual arousal is generally measured by latency to the first true copulatory event—mount or intromission—it could be argued that these animals did not show true sexual arousal.

Other authors have, in fact, concluded that the MPO is primarily involved in arousal (38, 66). Thus, in the few instances in which copulation occurs in rats with MPO lesions, the behavioral patterns seem normal except for initiation latencies. Similar reasoning was used by Hitt et al. (36) in relation to lesions in the medial forebrain bundle, which presumably has a close functional relationship to the MPO, probably via catecholamine pathways. Because an alternative interpretation—variability in effective completeness of the lesions—may account for these partial deficiencies, it is hard to use this argument to exclude involvement of the MPO in the CM. Of course, the basic problem is that there is no way to demonstrate the maintainence of either one of these two mechanisms in the total absence of the other, because each depends on the display of at least one true copulatory event. Nevertheless, the striking lack of "interest" of MPO-lesioned animals in stimulus females, even following powerful arousing stimuli (38), is convincing evidence that an essential substrate of the AM has been removed.

A recent study on rhesus monkeys with MPO lesions has produced the surprising finding that these males continue to masturbate normally with ejaculation despite the marked suppression of copulation (J. Slimp, B. Hart, and R. W. Goy, unpublished observations). This could be taken as evidence for a specific involvement of MPO in sexual arousal, although only if that term were limited to events leading to copulation per se, excluding autoerotic behavior. Finally, there is no evidence that a sensory deficit is the basis of the MPO lesion effect on the male sexual behavior (32).

One must conclude that while removal of the MPO produces a major deficit in the AM, it is premature to draw conclusions about the specific location of the CM. Similarly, the most posteriorly located lesions which facilitate male sexual behavior presumably remove inhibitory influences on the AM, as indicated by shortening of the PEI (29, 30). The latter lesions can, however, also produce facilitation of ejaculatory reflexes (decreased IF and EL) (29, 67). Many areas may well be involved in the multicomponent CM. Recent data (68) on specific inhibition of ejaculation with corticomedial amygdala lesions point to one of these.

Androgens and Arousal and Copulatory Mechanisms

If it is accepted that the MPO is necessary for sexual arousal, the fact that this area is the most sensitive to the behavior-activating effect of directly applied androgen seems to indicate that T functions via stimulation of the AM. Moreover, the essential role of androgen in arousal follows from the conclusion that castration immediately reduces, and eventually eliminates, arousal. It is also a clear consequence of the fact that, of all the measures of male sexual behavior, IL (and PEI to a lesser extent) shows the most direct dose-response relationship to blood T level in castrates with replacement therapy (12) (see under "Role of Testosterone"). Finally, it has been reported that when sexual behavior is restored by T treatment in castrates, anogenital exploration, i.e., "courtship" behavior, is restored well before copulatory behavior per se (69), and it is often found that mounting returns before ejaculatory patterns appear (22).

At the same time, there is considerable evidence for androgenic regulation of the CM, whether or not this can be related to specific anatomical structures. First, there is the evidence (discussed earlier) for androgenic involvement in penile reflexes, which we are including in the CM. Although there is very little information on possible dose-response relationships in erection, androgen plays at least a "permissive" role.

The behavior of rats which continue to copulate following castration demonstrates other aspects of androgenic involvement in the CM. Thus, IF is low for a number of weeks after castration and drops with the cessation of TP injections in castrates (5). Similar, although less consistent and maintained decreases are found in EL (70). Recently, McGill et al. (71) demonstrated a dose-dependent effect of T in increasing EL in prepubertally or post-pubertally castrated mice. In a strain with long-term retention of sexual behavior after castration, these authors (71) previously noted a period of decreased EL. In our experiments (12) cited under "Descriptive-Correlative Approach," rats with low levels of circulating T from constant release capsules showed low IF.

Thus, it appears that the CM is affected by androgen in at least two ways: the reflex component has an androgen requirement whereas the hormone retards the occurrence of the ejaculatory pattern, so that erection is facilitated while ejaculation is delayed. Is there any conceivable connection between these two effects of androgen on the CM and between either of them and its arousal-enhancing ation? This question is hard to approach in the light of our minimal understanding of the neurophysiology of copulation. Thus, next to nothing is known of hormonal effects on activity in the nerves supplying the penis. In the only published account on this subject, five castrated cats did not differ from intact controls in electrophysiological responses of the pudendal nerve to various tactile stimuli applied to the penis (72). Changes in cortical potentials evoked by genital stimulation indicated, however, possible central effects of androgen on sensory information derived from the periphery or that relevant effects on peripheral nerve function were overlooked.

It is, in part, the difficulty of pinpointing a single essential anatomical structure for the CM (like the MPO for the AM) which constrains attempts to elucidate the role(s) of androgen in this aspect of male sex behavior. The reflex component, as has been seen, undoubtedly involves organization at the spinal level—no doubt with subservience to facilitative and inhibitory mechanisms from the brain—and these spinal mechanisms can be influenced by T. For intromission, penile innervation is essential, due to the necessity for orientation and/or erection. The mechanism which summates the effects of intromissions to trigger the ejaculatory mechanism may depend on a peripheral androgen-dependent input. This is suggested, although not proven, by the evidence of experiments with flutamide in T-treated castrates. As mentioned above, this antiandrogen does not suppress arousal as indicated by normal mount frequencies, but does significantly reduce ejaculatory behavior, without impairing intromissions (51). Furthermore, in rats whose penes are anesthetized with tetracaine, flutamide even slightly stimulates mounting behavior (52) which strengthens the conclusion that any antiandrogenic effects of this compound on ejaculation cannot be mediated by inhibition of arousal.

The conclusion as to CM, therefore, would be that multiple sites of action are probably involved: the brain, spinal cord, and penis. It will be a difficult task to separate the relative importance of these three components, if such separation is feasible. It is worth reiterating that the results of the experiments on hypothalamic implantation of T mean that a high degree of behavioral performance results from chronic exposure of only a small area of the brain to high concentrations of androgen, with only minimal exposure of the periphery. The force of this observation is such as to lower our estimation of the importance of genital or any other peripheral effects of androgen, although it is premature to adopt any dogmatic stance in light of the apparent contradictions between the results of different experimental approaches.

We shall return to the question of relationships between the various effects of androgen in the next section, which takes up the consideration of these problems from the vantage point of human sexuality.

SEXUALITY IN MEN AND ITS NEUROENDOCRINE COMPONENT

Animals, Humans, and Conscious Experience

In this section, it will be our goal to assess the relevance of concepts derived from animal research to problems of sexual behavior in men. Because authoritative experimental data on the neuroendocrine basis of human sexual behavior are virtually nonexistent and the available clinical information is not very helpful for elucidation of underlying mechanisms, any discussion in this area has to be quite speculative. Moreover, because it is futile to discuss the role of hormones or other biological factors in human behavior without giving due consideration to the change of context involved in the extrapolation from animal research, some comments are in order on the nature of human sexuality.

First, it is necessary to clarify a major disparity of perceived goals which has separated most biological and medical scientists from those involved in dealing with problems of human sexuality. The aim of animal researchers is to determine the conditions—physiological and environmental—which determine the occurrence and quality of copulatory behavior in a given situation. The animal is often regarded as a machine which, given adequate functioning of its internal parts and appropriate stimuli from the outside, will "emit" certain behavioral responses. In the human context, however, the desired end point (other than procreation) is a cognitive state: the conscious perception of sexual satisfaction. This state is not merely dependent on the accomplishment of certain mechanical responses: it rather consists of a curious amalgam of perceptions of personal experiences and those of the sexual partner. Contributing to the individual variance in sexual experience are differences in the relative importance of different stages of arousal and of orgasm, as well as differences in expectations relating to the partner (or the absence of these in the case of autoeroticism).

Thus, it can be seen that the emphasis in any broad consideration of human sexuality must be the primary importance of those intrapsychic processes which are determined by cultural and social factors and whose possible biological correlates we are far from understanding. This difference in emphasis is responsible for the disparity of goals which tends to keep the experimentalist poles away from human problems; and it is what makes extrapolation of animal data to humans an enterprise to be embarked upon only with extreme caution.

The juxtaposition of conscious experience as the important variable in humans to that of behavioral responses in animals is not meant to imply that animals do not have minds, but, if they do, we have no knowledge of them. The processing of the sensory input which leads to sexual behavior is relatively inaccessible to the investigator of animal behavior, but much more accessible to humans studying human sexuality. Through the media of introspection and verbal communication, something can be learned about the processing of sexual sensory input by a cognitive apparatus, which manifest as a set of individual attitudes, expectations, capacities for experience, and so forth. Of course, in animal research we are also accustomed to thinking of the relationship between sensory input and behavior as depending on processing by a central apparatus informed in some mysterious way by genetic and experiential information. In the human context, however, it becomes more blatantly obvious in that the cognitive apparatus is more than just the means whereby sensory input is transduced to behavior. The same input can, in fact, be converted—between and within individuals—into vastly different experiences. Extreme examples are seen in reports on women who could be brought to orgasm by stroking the eyebrows or applying pressure to the teeth (73), although the influence of cognition is most strikingly revealed in the case of individuals who can achieve orgasms without any obvious sensory stimulation (74). But recourse to extreme examples is unnecessary to show that processes such as selective attention can influence the extent and quality of the reception of sensory input. Relevant examples are the sensory loss in orgasm (73) and the transformation of painful to pleasureable stimuli at the height of sexual arousal. Sexual (as other) experience must therefore be viewed as a continuous transaction between the sensory and the cognitive.

Arousal and Copulatory Mechanisms in Human Context

Now in returning to the initial question, how could the concepts derived from experimental manipulation and behavioristic analysis on animals be applied to this complex psychophysiological process? Human sexual behavior, like that of animals, can be considered to involve both an AM and a CM. The human AM can be defined as those conscious processes which lead to copulatory activity. It is seen as the conscious result of the transaction between sensory input and cognitive pedisposition. From the vantage point of introspection, it seems clear that this process is not limited to initiation, but must continue into copulation itself, confirming the more recent theorizing in animal behavior (65, 75). Furthermore, copulatory events must provide further fuel for the AM, primarily by the resulting tactile stimulation.

The CM, defined as those physiological-behavioral mechanisms which lead to ejaculation and orgasm, is, on the other hand, neither voluntary nor conscious. It does not refer to most of the neuromuscular activities involved in coitus, which are voluntary acts that any man can perform whether or not he is in possession of his testes. Excluding these does not, however, imply that the CM is basically different in animals and men. In both, the CM should be conceived as a catch-all term comprising the sum of the truly reflex copulatory events and the neural mechanisms which summate sensory information derived from copulatory events and from sexual arousal per se to trigger further sexual reflexes, particularly ejaculation and orgasm. What does seem different in subhuman mammalian species is the apparently more reflex nature of mounting and intromissions. Thus, in the rat, mounts including intromissions occur in inborn patterns of "bouts," which are repeated at relatively constant intervals (65). For humans, copulation is obviously much less automatic and much more flexible and varied.

There are, however, copulatory events involving skeletal muscle which can be viewed as involuntary also in humans: these include the contractions of the anal sphincter and various intense and apparently automatic spasmodic contractions of the extremities during orgasm. More important reflex components of relevance in this context are erection and ejaculation. Both these events will supply sensory information which contributes importantly to the sexual experience. Thus, erection is at the center of a positive feedback loop; it contributes to sexual arousal and, in turn, is stimulated by it. This differs from the situation in animal species such as rats and macaques, in which multiple intromissions with brief erections are separated by longer periods of detumescence. The contribution of orgasm to sexual experience and its perceived value is obvious. It is the prime consummatory event for most individuals, with immediate suppressive effects on the AM.

Thus, the AM leads to initiation of sexual activity and is also involved in maintaining and continuing copulation, while in the process stimulation is received which contributes to the CM. Ejaculation is facilitated by increasing arousal and can be voluntarily delayed by copulatory tactics which decrease the amount of penile stimulation received. Regulating the amount of physical stimuli received is not, however, the only way that the conscious AM process can, by voluntary motor acts, influence the unconscious, involuntary CM. This can also be accomplished by purely cognitive acts, as exemplified by the use of erotic fantasy in facilitating orgasm and of distracting thoughts (e.g. mathematical!) to delay it.

The reader may have noticed by now a fairly close correspondence between the terms "libido" and "potency" on the one hand and our use of AM and CM, respectively, on the other. It should also be noted that the present analysis places the AM on a firmer basis than it can be in the animal literature, where it has to be defined obliquely in terms of initiation latencies for behavior.

Role of Androgen in Men

And now, finally, what about hormonal regulation? It has been seen that in animals both AM and CM are influenced importantly by androgen. There is no need to deny that man shares in this biological destiny. That testosterone lowers the AM threshold in man as it does in animals is obvious from the loss of libido in castrates and in cases of hypogonadism (76), either congenital or produced by such conditions as varicocele (77). Sexual imagery, a conscious manifestation of sexual arousal, is reputedly dependent on testosterone level (76), and drug-induced reduction of androgen levels suppresses it (78). The increased incidence and intensity of sexual interest at puberty and its relative decline in or after the sixth decade of life occur at times when circulating androgen is rising and falling, respectively. Unfortunately, individual correlations of these statistically parallel changes have not yet been adequately studied (77).

As to the CM, carefully collected scientific data are indeed hard to come by. Certainly, however, there are enough clinical data to establish that erection in men is androgen-dependent to an appreciable extent. Thus, erectile capacity is reduced or absent in castrate or hypogonadal men (76, 79) or after gonadal suppression with progestin (78) and can be restored with T treatment (80, 81). The largest amount of available data comes from the recent work in Europe on the antiandrogen cyproterone acetate. The effects of this steroidal drug on sexual arousal and behavior will be discussed later in relation to problems of interpretation. For the moment, it will suffice to point out that several studies have reported a dramatic decline in spontaneous waking and nocturnal erections (82).

In discussing animal sexual behavior, it was difficult to conceive of the relationship between androgenic effects on the AM and those on the CM. In the human context, however, the connection seems clearer, at least conceptually. In our view of the interaction between erection and arousal, these two effects of androgen could stem from a single action on a central arousal system involved in a positive feedback relationship with the erection mechanism. A neural basis for this connection can be suggested: the MPO—a structure sensitive to androgen and necessary for sexual arousal in rodent, carnivore, and primate species—is a "nodal point" for the cerebral system involved in controlling erection, at least in the squirrel monkey (48). (But the system is widely represented in the brain, so that destruction of the MPO does not abolish sexual reflexes (35). Although this viewpoint need not exclude a peripheral role of androgen, it places emphasis on central mechanisms, at least in subjects with well developed penile structure, which is relatively little affected by post-pubertal castration.)

The situation is far more complex, however, when we consider the neural mechanism underlying the final consummation of copulation which is a major part of the CM. The increase in IF or EL produced by T in rats (70) and mice (71) is without any known parallel in man, although EL increases with advancing age in humans (74). However, the real problem of understanding the CM in the context of human sexuality is that the culmination of copulation is seen to consist of at least four components (74): 1) the expulsion of semen, itself a complex process; 2) a set of rhythmic muscular contractions, including but not limited to those involved in emission, e.g., the anal sphincter contracts in synchrony with the muscles of emission; 3) a variety of other events, including detumescence and extragenital vascular and myotonic effects; and 4) the conscious experience of orgasm. The relationships between the first three, on the one hand, and the fourth, on the other, are unclear. Although these various components are usually simultaneous, or nearly so, they can be separated, although to an extent which is not yet clarified. It does appear that seminal expulsion without the usual manifestations of orgasm may occur in some cases of nocturnal emission and in premature ejaculation. It can also result from certain drug or electroshock treatments and in neurologic disorders (83). The converse is also found with certain drug treatments or with training in coitus reservatus, as practiced in the Eastern Tantric traditions and by the Oneida sect in this country. It is not clear, however, to what extent these examples involve retrograde emission into the bladder, or what the quality of these orgasms are.

Now the only element in this complex series of events which is known to be controlled by androgen is the production of semen, absent in the castrate. Yet the maintenance of copulatory behavior in men and animals long after the semen-producing glands are atrophied suggests that this element is quite unimportant in the context of behavior and experience, although there is a dearth of nonanecdotal data on subjective experience of sex after castration. Thus, as concerns the nature and mechanism of the effects of androgen on the CM, much remains to be learned in man, as is the case in animals.

Research and Clinical Applications in Man

Despite these many indications that androgen is a limiting factor in the sexuality of men, with the exception of severe hypogonadism, no reliable association has yet been demonstrated between circulating androgen level and any aspect of human sexuality. This is true of normal mature subjects; several studies have recently reported no correlation (84) or even a negative one (85) between orgasm frequency and circulating T level. This matter seems of sufficient conceptual and clinical interest to merit detailed consideration.

First, one has to point to the inadequacy of research in the area of human sex-behavioral endocrinology, both in regard to numbers of studies which have been conducted and to their degree of sophistication. Generally, the behavioral data are anecdotal or based on superficial questioning of subjects without sufficient attention to the complexity of the psychological factors involved. To establish correlations in either normal mature subjects during early development, senescence, or in cases of impotence, so many factors other than hormone level are involved that a large population must be surveyed if underlying hormone-behavior correlations are to be found. Such studies have not been performed. When correlations are sought in supposedly pathological or deviant situations, irrelevant aspects of behavior are often studied, so that it is not surprising when no correlations result.

A case in point is homosexuality. After Kolodny et al. (86) published their study providing evidence of low T levels in homosexual men, a flurry of reports appeared from other laboratories on this subject. Results varied (87), but all had in common the failure to confirm the conclusions of Kolodny et al. (86). There is, in fact, no biological reason to assume that this particular manifestation of human sexuality is androgen-related. The only animal data supporting hormonal control of the direction of sexual preferences (88) are not very convincing, although they have been used to promote biological "treatments" for human homosexuality. As to other sexual problems, Saba et al. (89), for example, failed to find abnormal T levels in a small series of patients with erotic aggressiveness, exhibitionism, paedophilia, etc., which again is hardly surprising. It is a reasonable expectation that androgen level may be related to quantitative aspects, but not necessarily to qualitative manifestations of sexuality (e.g., its direction), which are more likely cognitive in origin, according to current concepts.

A second reason for the failure to find T-behavior correlations may well be that, just as in the rat, the maximal effect of androgen on human sexual behavior is reached at a blood level considerably below that found in the normal man. Thus, one might expect that correlations with blood T would be found only in populations with a reasonably high incidence of hypogonadism (e.g., in senescence). Appropriate dose-response studies in subjects with low blood androgen and behavioral problems will be necessary to clarify this issue.

Finally, it is not unlikely that in men, total androgen level is a misleading parameter because of the presence of high affinity sex hormone-binding globulin (not found in male rats). Because presumably only the free (unbound) moiety of the hormone is effective, if binding were to vary significantly in a given study, the relevant correlations could be obscured. In aging males, binding protein levels may be increased by the rise in circulating estrogen, because estrogen treatment can have this effect in men (90). Preliminary data have been collected (10) suggesting this may be the case in the aging rhesus monkey (which has the binding protein).

Research in human behavioral endocrinology can isolate and assess the identifiable biological determinants in human sexuality but not without very careful analysis of its central dimension—the element of conscious experience.

But misleading as the application of animal models to human behavior can be, clinical investigators can also go astray by lack of attention to the details and intricacies of knowledge derived from animal research. Both errors can be seen in research on the use of cyproterone acetate for the reduction of human libido, mostly in cases of sexual behavior defined as criminal. The considerable body of research in this area clearly shows the effectiveness of the treatment, but this literature often shows insufficient attention to cognitive factors. It is not clear to what extent the effect of the antiandrogen relates to coercive pressures on individuals who may have to choose between drug therapy and prison, a scientific consideration quite apart from questions of the ethical justifiability of such treatments. Secondly, it is not at all clear that the view of cyproterone action which seems to dominate the thinking of leading investigators—an anti-androgenic effect on hormone-sensitive brain mechanisms controlling behavior (82, 91)—is correct. The acetylated ester, which is clinically used, is a strongly progestational steroid which has no antiandrogenic effects on the sexual behavior of adult, sexually experienced rats (47). Men treated with this agent have depressed androgen levels. Although the available data do not always show maintained decreases in gonadotropins (82, 91), the most reasonable interpretation of the effects of this drug is still that it produces a functional castration due, at least in part, to suppression of the pituitary-testicular axis by the negative feedback (progestational) action.

But apart from clinical considerations, it would be of basic interest to our understanding of sexuality with its immense influence on human life to establish as precisely as possible what the limits of androgenic effects on human behavior are and the nature and dynamics of the effects. This research can be done without contravening ethical standards, using hypogonadal and castrated subjects and normal volunteers. Detailed reports on sexual activity and subjective experience with and without androgen treatment should be collected and interpreted by trained behavioral scientists. In addition, it is important to record vascular and other sexual response variables including those measured by Masters and Johnson (74), but with the addition of more sophisticated measurements of blood flow, EEG recording, etc. Much information can be obtained by using masturbation to avoid problems arising from intercourse in the laboratory setting, although one would think that the latter type of study could be feasible a decade after the publication of Masters and Johnson's work. The apparent hesitation of investigators to continue this line of work and the resulting severe lack of reputable data in the area is a major, but hopefully a surmountable, barrier to our understanding of the role of hormones and other biologic factors in human sexuality.

REFERENCES

1. Dewsbury, D. A. (1975). Diversity and adaptation in rodent copulatory behavior. Science 190:947.

2. Beach, F. A., and Fowler, H. (1959). Individual differences in the response of male rats to androgen. J. Comp. Physiol. Psychol. 52:50.
3. Barfield, R. J., and Geyer, L. A. (1975). The ultrasonic postejaculatory vocalization and the postejaculatory period of the male rat. J. Comp. Physiol. Psychol. 88:723.
4. Hart, B. L. (1974). Gonadal androgen and sociosexual behavior of male mammals: a comparative analysis. Psychol. Bull. 81:383.
5. Bloch, G. J., and Davidson, J. M. (1968). Effects of adrenalectomy and prior experience on postcastrational sex behavior in the male rat. Physiol. Behav. 3:461.
6. Warren, R. P., and Aronson, L. R. (1956). Sexual behavior in castrated-adrenalectomized hamsters maintained on DCA. Endocrinology 58:293.
7. Beach, F. A. (1967). Cerebral and hormonal control of reflexive mechanisms involved in copulatory behavior. Physiol. Rev. 47:289.
8. Emery, D. E., and Sachs, B. D. (1975). Ejaculatory pattern in female rats without androgen treatment. Science 190:484.
9. Young, W. C. (1961). The hormones and mating behavior. In W. C. Young (ed.), Sex and Internal Secretions, pp. 1173–1239. Williams & Wilkins, Baltimore.
10. Robinson, J. A., Scheffler, G., Eisele, S. G., and Goy, R. W. (1975). Effects of age and season on sexual behavior and plasma testosterone and dihydrotestosterone concentrations of laboratory-housed male rhesus monkeys (Macaca mulatta). Biol. Reprod. 13:203.
11. Gordon, T. P., Rose, R. M., and Bernstein, I. S. (1976). Seasonal rhythm in plasma testosterone levels in the rhesus monkey (Macaca mulatta): a three year study. Horm. Behav. 7:229.
12. Damassa, D. A., Smith, E. R., and Davidson, J. M. The relationship between circulating testosterone levels and sexual behavior. Horm. Behav. In press.
13. Davidson, J. M. (1972). Hormones and reproductive behavior. In H. Balin and S. R. Glasser (eds.), Reproductive Biology, pp. 877–918. Excerpta Medica, Amsterdam.
14. Davidson, J. M., Damassa, D. A., Smith, E. R., and Cheung, C. (1976). Feedback control of gonadotropin secretion in the male rat. In C. Spilman (ed.), Regulatory Mechanisms of Male Reproductive Physiology, pp. 151–168. Excerpta Medica, Amsterdam.
15. Södersten, P., Damassa, D. A., and Smith, E. R. Sexual behavior in developing male rats. Horm. Behav. In press.
16. Berndtson, W. E., and Desjardins, C. (1974). Circulating LH and FSH levels and testicular function in hamsters during light deprivation and subsequent photoperiodic stimulation. Endocrinology 95:195.
17. Morin, L. P., Fitzgerald, K. M., Rusak, B., and Zucker, I. Circadian organization and neural mediation of hamster reproductive rhythms. Psychoneuroendocrinology. In press.
18. Sandler, M., and Gessa, G. L. (eds.) (1975). Sexual Behavior: Pharmacology and Biochemistry. Raven Press, New York.
19. Zigmond, R. E. (1975). Binding, metabolism and action of steroid hormones in the central nervous system. In L. L. Iverson, S. D. Iverson, and S. Snyder (eds.), Handbook of Psychopharmocology. Plenum Press, New York. Sec. 1, Vol. 5, Chapter 5, pp. 239–328.
20. Naftolin, F., Ryan, K. J., and Petro, Z. (1972). Aromatization of androstenedione by the anterior hypothalamus of adult male and female rats. Endocrinology 90:295.
21. McDonald, P., Beyer, C., Newton, F., Brien, B., Baker, R., Tan, J. S., Sampsom, C., Kitching, P., Greenhill, R., and Pritchard, D. (1970). Failure of 5α-dihydrotestosterone to initiate sexual behavior in the castrated male rat. Nature 227:964.
22. Larsson, K., and Södersten, P. (1973). Sexual behavior in male rats treated with estrogen in combination with dihydrotestosterone. Horm. Behav. 4:289.
23. Christensen, L. W., and Clemens, L. G. (1975). Blockade of testosterone-induced mounting behavior in the male rat with intracranial application of the aromatization inhibitor, androst-1,4,6-triene - 3,17-dione. Endocrinology 97:1545.
24. Yahr, P. (1975). Personal communication.
25. Quadagno, D. M., and Ho, G. K. W. (1975). The reversible inhibition of steroid-induced sexual behavior by intracranial cycloheximide. Horm. Behav. 6:19.

26. Eibergen, R. D., and Caggiula, A. R. (1973). Ventral midbrain involvement in copulatory behavior of the male rat. Physiol. Behav. 10:435.
27. Valenstein, E. (1973). Brain Control. John Wiley and Sons, New York.
28. Caggiula, A. R. (1969). Stability of behavior produced by electrical stimulation of the rat hypothalamus. Brain Behav. Evol. 2:343.
29. Heimer, L., and Larsson, K. (1964). Drastic changes in the mating behavior of male rats following lesions in the junction of diencephalon and mesencephalon. Experientia 20:460.
30. Barfield, R. J., Wilson, C., and McDonald, P. G. (1975). Sexual behavior: extreme reduction of postejaculatory refractory period by midbrain lesions in male rats. Science 189:147.
31. Klüver, H., and Bucy, P. C. (1939). Preliminary analysis of functions of the temporal lobes in monkeys. Arch. Neurol. Psychiatr. 42:979.
32. Malsbury, C. W., and Pfaff, D. W. (1974). Neural and hormonal determinants of mating behavior in adult male rats. In L. V. DiCara (ed.), Limbic and Autonomic Nervous Systems Research, pp. 86–136. Plenum Press, New York.
33. Powers, J. B., and Winans, S. S. (1974). Vomeronasal organ: critical role in mediating sexual behavior of the male hamster. Science 187:961.
34. Heimer, L., and Larsson, K. (1966). Impairment of mating behavior in male rats following lesions in the preoptic-anterior hypothalamic continuum. Brain Res. 3:248.
35. Hart, B. L. (1974). The medial preoptic-anterior hypothalamic area and sociosexual behavior of male dogs: a comparative neuropsychological analysis. J. Comp. Physiol. Psychol. 68:328.
36. Hitt, J. C., Byron, D. M., and Modianos, D. T. (1973). Effects of rostral medial forebrain bundle and olfactory tubercle lesions upon sexual behavior of male rats. J. Comp. Physiol. Psychol. 82:30.
37. Paxinos, G., and Bindra, D. (1973). Hypothalamic and midbrain neural pathways involved in eating, drinking, irritability, aggression and copulation in rats. J. Comp. Physiol. Psychol. 82:1.
38. Caggiula, A. R., Antelman, S. M., and Zigmond, M. J. (1973). Disruption of copulation in male rats after hypothalamic lesions: a behavioral anatomical and neurochemical analysis. Brain Res. 59:273.
39. Giantonio, G. W., Lund, N. L., and Gerall, A. A. (1970). Effect of diencephalic and rhinencephalic lesions on the male rat's sexual behavior. J. Comp. Physiol. Psychol. 73:38.
40. Hart, B. L. (1973). Reflexive behavior. In G. Bermant, (ed.), Perspectives on Animal Behavior, pp. 171–193. Scott, Foresman, Glenview, Illinois.
41. Adler, N., and Bermant, G. (1966). Sexual behavior of male rats: effects of reduced sensory feedback. J. Comp. Physiol. Psychol. 61:240.
42. Aronson, L. R., and Cooper, M. L. (1968). Desensitization of the glans penis and sexual behavior in cats. In M. Diamond (ed.), Perspectives in Reproduction and Sexual Behavior, pp. 51–82. University Press, Bloomington, Indiana.
43. Herbert, J. (1973). The role of the dorsal nerves of the penis in the sexual behavior of the male rhesus monkey. Physiol. Behav. 10:293.
44. Bermant, G., and Davidson, J. M. (1974). Biological Bases of Sexual Behavior. Harper and Row, New York.
45. Beach, F. A., and Holz, A. M. (1946). Mating behavior in male rats castrated at various ages and injected with androgen. J. Exp. Zool. 101:91.
46. Beach, F. A., Noble, R. G., and Orndoff, R. K. (1969). Effects of perinatal androgen treatment on responses of male rats to gonadal hormones in adulthood. J. Comp. Physiol. Psychol. 68:490.
47. Bloch, G. J., and Davidson, J. M. (1971). Behavioral and somatic responses to the antiandrogen cyproterone (1,2α-methylene-6-chloro-Δ-17α-hydroxy-progesterone). Horm. Behav. 2:11.
48. MacLean, P. D., and Ploog, D. W. (1962). Cerebral representation of penile erection. J. Neurophysiol. 25:29.
49. Rodgers, C., and Alheid, G. (1972). Relationship of sexual behavior and castration to tumescence in the male rat. Physiol. Behav. 9:581.

50. Baum, M. J., Södersten, P., and Vreeburg, J. T. M. (1974). Mounting and receptive behavior in the ovariectomized female rat: influence of estradiol, dihydrotestosterone, and genital anesthetization. Horm. Behav. 5:175.

51. Södersten, P., Gray, G., Damassa, D. A., Smith, E. R., and Davidson, J. M. (1975). Effects of a nonsteroidal antiandrogen on sexual behavior and pituitary-gonadal function in the male rat. Endocrinology 97:1468.

52. Gray, G. D., and Davidson, J. M. (1976). Effects of the nonsteroidal antiandrogen flutamide on sexual behavior in male rats. Manuscript in preparation.

53. Hart, B. L. (1968). Alteration of quantitative aspects of sexual reflexes in spinal male dogs by testosterone. J. Comp. Physiol. Psychol. 66:726.

54. Hart, B. L. (1973). Effects of testosterone propionate and dihydrotestosterone on penile morphology and sexual reflexes of spinal male rats. Horm. Behav. 4:239.

55. Hart, B. L., and Haugen, C. M. (1968). Activation of sexual reflexes in male rats by spinal implantation of testosterone. Physiol. Behav. 3:735.

56. Johnston, P., and Davidson, J. M. (1972). Intracerebral androgens and sexual behavior in the male rat. Horm. Behav. 3:345.

57. Davidson, J. M. (1966). Activation of the male rat's sexual behavior by intracerebral implantation of androgen. Endocrinology 79:783.

58. Smith, E. R., Damassa, D. A., and Davidson, J. M. Plasma testosterone and sexual behavior following intracerebral implantation of testosterone propionate in the castrated male rat. Horm. Behav. In press.

59. Christensen, L. W., and Clemens, L. G. (1974). Intrahypothalamic implants of testosterone or estradiol and resumption of masculine sexual behavior in long-term castrated male rats. Endocrinology 95:984.

60. Davidson, J. M., and Trupin, S. (1975). Neural mediation of steroid-induced sexual behavior in rats. In M. Sandler and G. L. Gessa (eds.), Sexual Behavior: Pharmacology and Biochemistry, pp. 13–20. Raven Press, New York.

61. Andrew, R. J. (1974). Arousal and the causation of behavior. Behavior 51:135.

62. Henderson, L. (1972). On mental energy. Br. J. Psychol. 63:1.

63. Beach, F. A. (1956). Characteristics of masculine sex drive. In M. R. Jones (ed.), The Nebraska Symposium of Motivation, pp. 1–32. University of Nebraska Press, Lincoln, Nebraska.

64. Crowley, W. R., Popolow, H. B., and Ward, O. B., Jr. (1973). From dud to stud: couplatory behavior elicited through conditioned arousal in sexually inactive male rats. Physiol. Behav. 10:391.

65. Sachs, B. D., and Barfield, R. J. Functional organization of male copulatory behavior in the rat. In J. S. Rosenblatt, R. Hinde, E. Shaw and C. G. Beer (eds.), Advances in the Study of Behavior. Academic Press, New York. In press.

66. Chen, J. J., and Bliss, D. K. (1974). Effects of sequential preoptic and mammillary lesions on male rat sexual behavior. J. Comp. Physiol. Psychol. 87:841.

67. Lisk, R. D. (1966). Increased sexual behavior in the male rat following lesions in the mammillary region. J. Exp. Zool. 161:129.

68. Harris, V. S., and Sachs, B. D. (1975). Copulatory behavior in male rats following amygdaloid lesions. Brain Res. 86:514.

69. Singer, J. J. (1972). Anogenital explorations and testosterone propionate-induced sexual arousal in rats. Behav. Biol. 7:743.

70. Davidson, J. M. (1969). Hormonal control of sexual behavior in adult rats. In G. Raspe (ed.), Advances in the Biosciences, Vol. 1, pp. 119–139.

71. McGill, T. E., Albelda, S. M., Bible, H. H., and Williams, C. L. Inhibition of the ejaculatory reflex in B6D2F$_1$ mice by testosterone propionate. Behav. Biol. In press.

72. Cooper, K. K., and Aronson, L. R. (1974). Effects of castration on neural afferent responses from the penis of the domestic cat. Physiol. Behav. 12:93.

73. Kinsey, A. C., Pomeroy, W. B., Martin, C. E., and Gebhard, P. H. (1953). Sexual Behavior in the Human Female. W. B. Saunders, Philadelphia.

74. Masters, W. H., and Johnson, V. E. (1966). Human Sexual Response. Little Brown, Boston.

75. Cherney, E. F., and Bermant, G. (1970). Role of stimulus female novelty in the

rearousal of copulation in male laboratory rats (*Rattus norwegicus*). Anim. Behav. 18:567.

76. Money, J. (1961). Sex hormones and other variables in human eroticism. *In* W. C. Young (ed.), Sex, and Internal Secretions, Vol. II, Ed. 3, pp. 1383–1400. Williams and Wilkins, Baltimore.

77. Comhaire, F., and Vermeulen, A. (1975). Plasma testosterone in patients with varicocele and sexual inadequacy. J. Clin. Endocrinol. Metab. 40:824.

78. Money, J., Wiedeking, C., Walker, P., Migeon, C., Meyer, W., and Borgaonkar, D. (1975). 47, XYY 46,XY males with antisocial and/or sex-offending behavior: antiandrogen therapy plus counseling. Psychoneuroendocrinology 1:165.

79. Bremer, J. (1959). Asexualization. MacMillan, New York.

80. Beumont, P. J. V., Bancroft, J. H. J., Beardwood, C. J., and Russell, G. F. M. (1972). Behavioural changes after treatment with testosterone: case report. Psychol Med. 2:70.

81. Miller, N., Hubert, G., and Hamilton, J. B. (1938). Mental and behavioral changes following male hormone treatment of adult castration, hypogonadism and psychic impotence. Proc. Soc. Exp. Biol. Med. 38:538.

82. Horn, H. J. (1974). Administration of antiandrogens in hypersexuality and sexual deviations. *In* O. Eichler, A. Farah, H. Herken, and A. D. Welch (eds.), Androgens II and Antiandrogens, pp. 543–562. Springer-Verlag, New York.

83. Beach, F. A., Westbrook, W. H., and Clemens, L. G. (1966). Comparisons of the ejaculatory responses—men and animals. Psychosom. Med. 28:749.

84. Raboch, J., and Starka, L. (1973). Reported coital activity of men and levels of plasma testosterone. Arch. Sex. Behav. 2:309.

85. Kraemer, H. C., Becker, H. G., Brodie, H. K. H., Doering, C. H., Moos, R. H., and Hamburg, D. A. (1976). Orgasmic frequency and plasma testosterone levels in normal human males. Arch. Sex. Behav. 5:125.

86. Kolodny, R. C., Masters, W. H., Hendryx, J., and Toro, G. (1971). Plasma testosterone and semen analysis in male homosexuals. N. Engl. J. Med. 285:1170.

87. Brodie, H. K. H., Gartrell, N., Doering, C., and Rhue, T. (1974). Plasma testosterone levels in heterosexual and homosexual men. Am. J. Psychiatr. 131:82.

88. Dörner, G., and Hinz, G. (1968). Induction and prevention of male homosexuality by androgen. J. Endocrinol. 40:387.

89. Saba, P., Salvadorini, F., Galeone, F., Pellicano, C., and Rainer, E. (1975). Antiandrogen treatment in sexually abnormal subjects with neuropsychiatric disorders. *In* M. Sandler and G. L. Gessa (eds.), Sexual Behavior: Pharmacology and Biochemistry, pp. 197–204. Raven Press, New York.

90. Murray, M. A. F., Bancroft, J. H. J., Anderson, D. C., Tennent, T. G., and Carr, P. J. (1975). Endocrine changes in male sexual deviants after treatment with antiandrogens, oestrogens or tranquilizers. J. Endocrinol. 67:179.

91. Graf, K. J., Brotherton, J., and Neumann, F. (1974). Clinical uses of antiandrogens. *In* E. Eichler, A. Farah, H. Herken, and A. D. Welch (eds.), Androgens II and Antiandrogens, pp. 485–497. Springer-Verlag, New York.

International Review of Physiology
Reproductive Physiology II, Volume 13
Edited by Roy O. Greep
Copyright 1977 University Park Press Baltimore

8
Growth Hormone and the Regulation of Somatic Growth

J. L. KOSTYO AND O. ISAKSSON

Department of Physiology
Emory University
Atlanta, Georgia

The final expression of the reproductive process is the growth and maturation of the offspring into an organism, which is itself capable of reproducing. The growth and maturation process of vertebrates, particularly in the postnatal period, is carefully regulated by a variety of hormones. Those concerned most intimately with the growth of the soma are pituitary growth hormone, thyroxin, and insulin. Other hormones, such as the tropic pituitary hormones and the sex steroids, are more particularly involved in the control of growth processes in specific target tissues. The following discussion will focus exclusively on pituitary growth hormone and current thinking regarding its role in the regulation of somatic growth.

Although pituitary growth hormone has widespread effects on many aspects of body metabolism, its stimulatory actions on the formation of proteins and nucleic acids are obviously the means by which its growth-promoting property is expressed. Some insight into the cellular basis for these effects of growth hormone on protein and nucleic acid synthesis have come from experiments conducted primarily on hypophysectomized laboratory animals or with the isolated tissues or organs of such animals. It has been shown that the acute administration of growth hormone to hypophysectomized rats results in the stimulation of protein synthesis in tissues such as skeletal muscles and the liver (1), an effect that appears to be the result of an increase in the activity of ribosomes engaged in the translation process (2–4). The effect of the hormone on ribosomal activity is not immediate, however. There is a lag period of approximately 30 min after the administration of growth hormone before protein synthesis is stimulated (1). This lag period does not signify an indirect effect of growth hormone on muscle or liver, however. Indeed, growth hormone added in vitro stimulates protein synthesis in the isolated rat diaphragm muscle (5), in the isolated perfused rat liver (6), and in the isolated perfused rat heart (7, 8), indicating that it influences protein synthesis in these tissues directly. Moreover, a lag period similar to that observed in vivo occurs when growth hormone exerts its in vitro effect on protein synthesis in these isolated tissues (5–8). Thus, the existence of the lag period suggests that the activation of ribosomes in a growth hormone-stimulated tissue is the culmination of a series of molecular processes triggered by the hormone. The lag period presumably reflects the time required for these processes to occur. The change produced in the ribosomes resulting in enhanced activity appears to be rather stable. A single injection of growth hormone into a hypophysectomized rat produces a stimulation of protein synthesis that persists for nearly 24 hr (2, 9). Moreover, numerous studies have shown that ribosomes isolated from the tissues of growth hormone-treated animals retain an enhanced capacity to carry out peptide bond formation in vitro.

Growth hormone also stimulates the synthesis of the several varieties of RNA in cells. However, enhanced RNA synthesis occurs only after the hormone has stimulated ribosomal activity. Thus, the acute effect of growth hormone on protein synthesis does not appear to involve the prior synthesis of new RNA. On the other hand, some hours after exposure of tissues to growth hormone, there is an increase in number of protein-synthetic units in the cells, in part due to the acceleration of ribosomal RNA synthesis. Consequently, full expression of the protein anabolic effect of growth hormone does depend upon its stimulatory effect on RNA synthesis.

Knowledge is meager concerning the events that occur during the period between the interaction of growth hormone with its receptors and the stimulation of ribosomal activity. It has been demonstrated that the in vitro stimulatory effect of growth hormone on protein synthesis in the isolated rat diaphragm can be prevented by inhibitors of cyclic nucleotide phosphodiesterase (10, 11). This observation has suggested that one of the events involved in mediating the effect of growth hormone on ribosomal activation is a reduction in cyclic nucleotide level. Efforts to demonstrate a reduction in the cyclic adenosine $3':5'$-monophosphate (cyclic AMP) or cyclic guanosine $3':5'$-monophosphate (cyclic GMP) level in the rat diaphragm following its exposure to growth hormone have failed (12, 13). However, the in vitro release of cyclic AMP from the isolated rat diaphragm is retarded by exposure of the tissue to growth hormone for 30 min, suggesting that the hormone does reduce the amount of nucleotide available in the cells for release (14). Moreover, recent studies with the isolated perfused rat heart have demonstrated that the addition of growth hormone to the perfusate results in a depression in the levels of both cyclic AMP and cyclic GMP in the tissue, effects that occur within 10 min of exposure of the tissue to the hormone (8). Whether changes in cyclic nucleotide levels are actually involved in mediating the effect of growth hormone on protein synthesis remains to be clarified.

In contrast to its direct effects on many tissues of the body (muscle, liver, heart, fat, and lymphoid organs), the action of growth hormone on cartilage has been considered to be indirect. This conclusion has been based primarily upon the finding that growth hormone has only small or inconsistent effects on the protein metabolism of isolated rat cartilage when added in vitro, whereas serum from growth hormone-treated animals is quite effective in stimulating the isolated tissue (15). The growth hormone-dependent factor present in serum that stimulates cartilage metabolism was first called "sulfation factor," but was subsequently renamed "somatomedin" to direct attention to its putative role as a mediator of the actions of growth hormone.

Two other actions of growth hormone deserve mention here because of their apparent relationship to its protein anabolic effect. These are its stimulatory effects on the membrane transport of amino acids and sugars into various target tissues. It is well documented that either the administration of growth hormone to hypophysectomized rats or the in vitro addition of growth hormone to isolated tissues of such animals results in an increase in the transport of certain amino acids and sugars into the cells. These membrane transport effects of the hormone, like the acute effect on protein synthesis, occur only after a lag period of 20–30 min. The stimulatory effect of the hormone on amino acid transport, but not the effect on sugar transport, can be prevented by prior exposure of the cells to inhibitors of protein synthesis (16, 17). From these experiments, it was proposed that growth hormone stimulates the synthesis of protein involved in activating the amino acid transport mechanism (5), although such a mediating protein has never been identified. When hypophysectomized rats or the isolated tissues of such animals are used as the experimental preparation, the stimulatory effects of growth hormone on amino acid and sugar transport are transitory, having a time course of approximately 2–3 hr. The rates of amino acid and sugar transport then revert to the prestimulated level and cannot be stimulated again with growth hormone for a period of many hours (18, 19). The stimulatory phase of growth hormone's action on these transport processes can be prolonged by exposure of the tissues to inhibitors of RNA and protein synthesis. Thus, it has been postulated that growth hormone may stimulate the synthesis of protein involved in terminating its stimulatory effects on membrane transport (18, 19). Again, such proteins have not been identified.

Although efforts continue to elucidate the details of the protein anabolic action of growth hormone, the present extent of our understanding of the mechanism is that which has been briefly described. The reader is referred to several recent reviews for additional details (16, 17, 20–22). Much of the current work and thinking regarding growth hormone and its growth-promoting action is now focused on other issues. Of considerable interest is the relationship or coupling between the plasma level of growth hormone and the expression of its metabolic activities. Involved in this issue is the nature of the growth hormone present in the circulation and whether current methods of assay do indeed measure the physiologically active principle. Furthermore, there is now great interest in the interaction of growth hormone with receptors on its target cells and the initial events precipitated by the hormone-receptor interaction. Lastly, there is considerable current effort being devoted to clarification of the role of somatomedin in mediating the effect of growth hormone on its target tissues. The discussion that follows will deal with each of these issues.

CONCENTRATION OF CIRCULATING
GROWTH HORMONE AND PROTEIN ANABOLIC ACTIVITY

Many hormones influence metabolic processes in an episodic fashion. Some stimulus accelerates the secretion of the hormone, thereby elevating its level in the circulation. As a consequence of this rise in circulating hormone level, hormone-receptor interactions occur in sufficient number in or on target cells to produce a change in their metabolic activity. This change in metabolism persists for some time, but may subside before the next burst of hormone secretion. In contrast, some hormones exert a so-called "permissive" action on their target tissues. In this instance, the mere presence of the hormone in the blood, even at its basal level, is sufficient to maintain target cells in a particular metabolic state, usually one that will permit them to respond metabolically to other factors or hormones. At the moment it is not clear whether the action of growth hormone on protein anabolism, and hence growth, is episodic or permissive in nature.

When a sensitive radioimmunoassay for human growth hormone (hGH) first became available more than a decade ago, it soon became clear that the concentration of the hormone in the blood of humans was rather variable and readily affected by certain metabolic and emotional factors. In 1965, Glick and co-workers (22) summarized their own work and that of others dealing with measurements of circulating growth hormone in the human. Samples of blood obtained from resting, fasted adults were usually found to contain only a few nanograms of hGH per ml of plasma, and in some instances, the level was too low to be detected by the radioimmunoassay. However, some individuals maintained on bed-rest occasionally had marked elevations in the concentration of circulating hGH. Factors that were found to stimulate the secretion of hGH in the adult included exercise, stresses of various kinds, hypoglycemia induced by insulin or other agents, and the administration of certain amino acids such as arginine. Similar observations were subsequently made on monkeys (23, 24). In comparison to the human and the monkey, the rat was found to have a much more variable level of circulating growth hormone, when populations of animals were sampled. A number of investigators reported circulating rat growth hormone (rGH) levels for adult animals ranging from zero to several hundred nanograms per ml of plasma (the values given in the literature are often not directly comparable, because a variety of standards varying in potency have been used in the radioimmunoassays). Moreover, growth hormone secretion by the rat was found to be affected rather differently by factors that stimulate growth hormone secretion in the primate. For example, stress and insulin-induced hypoglycemia were shown to depress rather than raise the level of rGH in the blood (25, 26).

Some attempt has been made to correlate the level of circulating growth hormone with its action on growth processes by comparing the concentration of growth hormone in the blood at various stages of life with the rate of growth. One complication in making this correlation is that vertebrates, including man, are not continually sensitive to the growth-promoting action of growth hormone. It is generally accepted (27–30) that growth hormone has little or no influence on the growth of fetal tissues. The tissues gradually develop sensitivity to growth hormone during postnatal life, and eventually, somatic growth becomes virtually dependent upon growth hormone. In the case of the human, it is generally accepted that the infant does not require growth hormone to grow (28) but that sensitivity to the hormone begins to develop sometime during the first year or two of life. Thereafter, growth will not occur at the normal rate in the absence of the hormone. Similarly, fetal and newborn rats are not sensitive to the growth-promoting action of growth hormone. Tissues of the rat begin to respond to the hormone after the first few days of life (29, 30), and the hormone becomes absolutely essential for somatic growth after the first few weeks of life. Interestingly, the concentration of circulating growth hormone appears to be the greatest during the period of life when the tissues are not responsive to the hormone. In the case of the human, hGH first appears in fetal blood at the time when acidophils appear in the developing pituitary gland. The concentration of hGH then rises to values above 100 ng/ml by mid-pregnancy (31) and then gradually declines. Measurements of circulating growth hormone in premature children and in infants less than a year old have revealed levels generally higher than those found in adults (22), whereas older children have hGH levels in the same range as adults (22, 32, 33). Studies on the rat have indicated essentially the same pattern. Growth hormone appears initially in the blood on the 19th day of gestation and reaches a peak value on the 21st day. Its level then declines during the first few days of life (34–36).

The impression given by the foregoing observations is that circulating growth hormone is minimal during the phase of life when somatic growth is most dependent upon the hormone. Recent work attempting to define the daily pattern of growth hormone secretion in man and the rat indicates that this impression is probably not correct. When serial blood samples have been taken from human subjects over a 24-hr period under carefully controlled conditions, it has been found that episodes of growth hormone secretion occur, resulting in peaks of growth hormone in the blood that far exceed the so-called basal level (less than 10 ng/ml). Although occasional peaks of circulating growth hormone occur in some adults during the awake period of the day, a burst of growth hormone secretion occurs within 90 min of the onset of slow wave sleep (37–41). The concentration of hGH in the plasma may rise to 30–40 ng/ml or greater as a result of this burst of secretory activity. Other bursts of growth hormone secretion may follow the initial one, but the resultant peaks of growth hormone produced are usually smaller and again associated with periods of slow wave sleep (39). There is little doubt that this daily burst of secretory activity is associated with the onset of sleep, because it nearly always coincides with the period of deep sleep, even when sleep is delayed for some hours or the awake-sleep cycle of an individual is reversed (37, 39). Children, like adults, also have episodes of growth hormone secretion. However, peaks of circulating hGH occur more frequently than in the adult and are often seen during both the awake and sleep periods (42). There is also some suggestion that more frequent episodes of enhanced growth hormone secretion occur in the adolescent than in the prepubertal child (43). Recent findings indicate that growth hormone secretion is also episodic in the rat (44–46). Tannenbaum and Martin (45) sampled blood of large (400 g) "nonstressed" male rats through a venous cannula over a 24-hr period and found that episodes of enhanced growth hormone secretion occurred in these animals at approximately 3.5-hr intervals. The peaks of rGH often reached 200 ng/ml. In the intervals between the bursts of rGH secretion, the plasma level of the hormone often fell below 1 ng/ml. Repeated testing of individual animals indicated that their particular patterns of growth hormone secretion were consistent from one day to the next, although there was some variation in the timing of the secretory bursts among various animals of a group. The episodic pattern of growth hormone secretion was not significantly influenced by the light-dark cycle, feeding, or activity (45). An essentially similar pattern of growth hormone secretion was also found in much younger (100 g) female rats (46). In this study, groups of "non-stressed" animals were decapitated at hourly intervals over a 24-hr period. Seven peaks of growth hormone were found, the peak and nadir values being similar to those reported by Tannenbaum and Martin (45). Thus, it would appear that the rat, which continues to grow throughout its life, experiences frequent episodes of enhanced growth hormone secretion much like the growing child.

These episodes of growth hormone secretion, as some investigators (38, 40) have implied, may be the critical events that regulate the growth process. The very low basal or resting level of growth hormone in the blood may have no physiological significance, even though it is often referred to by many investigarors as the "physiological concentration" of growth hormone. On the other hand, when the growth hormone level rises in the blood during a burst of hormone secretion, a hormone concentration may be reached that is sufficient to produce the number of hormone-receptor interactions required to stimulate the protein anabolic process. During the interval when the growth hormone concentration has returned to the basal level, the stimulatory effect of the hormone on protein metabolism may begin to decay or subside. Thus, it is conceivable that the individual may experience cycles of protein anabolic activity driven by the episodes of enhanced growth hormone secretion. In the rapidly growing organism, these cycles of anabolic activity may be more frequent, as are the bursts of growth hormone secretion, resulting in a more sustained state of protein anabolism. Unfortunately, there is little compelling evidence to support or refute this view at the present time. There are no systematic studies correlating the level of growth hormone in the blood with specific metabolic responses. In particular, it is not known whether transient changes in circulating growth hormone within the range of the peaks and nadirs recorded in intact organisms can produce concomitant changes in anabolic activity. The only related evidence comes from studies of in vitro effects of growth hormone on isolated tissues of the rat (16, 17, 20). In many instances, in vitro effects of growth hormone have been obtained with concentrations of the hormone that are similar to the levels achieved in the intact rat during bursts of hormone secretion. On the other hand, effects have usually not been seen with concentrations of the hormone that fall within the range of the basal or resting level (less than 20 ng/ml). Of course, the relevance of this evidence may be questioned on the grounds that the tissues used in in vitro experiments are not in a "physiological state." In any event, crucial experiments demonstrating whether or not there is coupling between the bursts of growth hormone secretion and protein anabolic activity remain to be done. Should it happen that the episodes of enhanced growth hormone secretion prove to be the physiologically significant events in the regulation of the growth process, then much of the work in the literature, in which limited sampling techniques were used to document levels of circulating growth hormone in various clinical and experimental situations, will have rather restricted value.

NATURE OF CIRCULATING GROWTH HORMONE

The foregoing discussion concerning the relationship between the level of circulating growth hormone and the expression of its protein anabolic activity is predicated upon the assumption that the growth hormone measured in blood by radioimmunoassay and, more recently, by radioreceptor assay, is indeed the physiologically active entity. The radioimmunoassay and the radioreceptor assay involve the displacement of purified *pituitary* growth hormone from binding sites on antibodies or cell membranes by the unknown material undergoing assay. Hence, the substance(s) detected in blood by these assays must bear some chemical resemblance to the native growth hormone molecule present in the pituitary gland. However, even a peptide fragment comprising the NH_2-terminal 134 amino acid residues of the hGH molecule will cross-react to a substantial degree in the hGH radioimmunoassay (40–50% that of native 191-residue hGH) (47). Moreover, large fragments of the growth hormone molecule can be virtually devoid of growth-promoting activity or have activity several times that of the native pituitary hormone, depending upon the portion of the amino acid sequence that is present in the fragment (48, 49). Thus, it is possible that the growth hormone measured in blood by the radioimmunoassay and the radio-receptor assay might actually be modified or fragmented forms of the pituitary hormone with negligible growth-promoting activity. The physiologically active species, i.e., the molecule that interacts with tissue receptors and *evokes the growth response,* might not be detected by these assays. Unfortunately, the radioimmunoassay and the radioreceptor assay reveal nothing about the biological activity of the moieties being detected, i.e., whether the substances measured are capable of triggering the chemical reactions that lead to enhanced protein synthesis and growth. The critical issue then is what is the physiologically important form of circulating growth hormone. Is it the substance(s) measured by the radioimmunoassay and the radioreceptor assay? If so, is this circulating hormone identical with the native hormone found in the pituitary gland or is it a modified or fragmented form of the native molecule? Conversely, is the circulating form of growth hormone a fragment that cannot be recognized by the radioimmunoassay and the radioreceptor assay? At the moment, the situation is far from clear.

There have been several recent attempts to determine the molecular size of the growth hormone present in blood. These studies have involved the gel filtration of plasma on Sephadex G-75 or G-100, followed by radioimmunoassay or radioreceptor assay of the effluent fractions to detect the presence of cross-reacting substances. Using human plasma, most investigators have found primarily two such substances (50, 51). The predominant one (in terms of amount present) appears to be about the size of pituitary hGH (22,000 daltons) and has been called "little growth hormone." The other is approximately twice as large and has been called "big growth hormone." Storage or repeated freezing and thawing of big growth hormone results in the generation of some little growth hormone, suggesting that big growth hormone is a noncovalent aggregate. Its exact nature, i.e., whether it is a dimer of little growth hormone or little growth hormone, bound to some other peptide, is not clear. Occasionally, small amounts of cross-reacting material larger than big growth hormone have been detected. With the exception of one report (52), most investigators using the gel filtration method have not found immunoreactive substances in the blood that are smaller than little growth hormone. Thus, there does not appear to be an immunoreactive species of growth hormone in the blood that is substantially smaller than native pituitary hGH. However, the exact chemical nature of little growth hormone is unknown. Considering the limited resolving power of the gel filtration method used in these studies, it could be a large fragment or a modified form of the native hGH molecule, rather than the native hormone itself. In any event, it is not known whether the little growth hormone and the big growth hormone detected in blood have any biological activity.

Some effort has been made recently to study the correlation between the amount of growth hormone that is detected in blood by radioimmunoassay and the amount that can be detected by various bioassays for the hormone. The basic premise of these studies is that if the predominant biologically active species of growth hormone in the blood is identical with or quite similar to the native growth hormone isolated from the pituitary gland, then one would expect the circulating species to possess essentially the same ratio of biological to immunological potency as native pituitary growth hormone. Prior to the advent of radioimmunoassay, several attempts were made to measure the amount of growth hormone activity present in the blood of rats and man (53). Because of the small amount of growth hormone present in the blood and the relative insensitivity of the bioassays for growth hormone available at the time (e.g., the tibia assay in hypophysectomized rats), most studies on human blood were conducted with the blood of individuals suffering from gigantism or acromegaly. The general finding in these early studies was that the concentration of growth hormone in the blood of both rats and man was in the *microgram* range. As noted earlier, latter-day estimates by radioimmunoassay of the concentration of growth hormone in the blood rarely exceed $0.1-0.2$ $\mu g/ml$ of plasma, even during peaks of growth hormone secretion. Thus, a comparison of these early estimates of the amount of growth hormone in blood by bioassay with current estimates by radioimmunoassay suggests that there is a considerable discrepancy between biological activity and immunological activity.

Recently, Ellis and Grindeland (53) have reinvestigated this question by using careful methods to prepare concentrated fractions of both rat and human blood for bioassay by the tibia test and for radioimmunoassay. These investigators found that the concentrated plasma of normal rats had 50 times more growth hormone-like activity in the tibia test than expected from the amount of rGH detected in the plasma by radioimmunoassay. When concentrated plasma of normal humans was used, a substantial discrepancy was also found between biological and immunological activity (200:1). Ellis and Grindeland (53) also administered concentrated rat plasma to hypophysectomized rats along with an antiserum against native pituitary rGH and found that the plasma was still capable of stimulating growth of the tibial cartilage. In control studies, the rGH antiserum was fully effective in blocking the stimulatory effect of native pituitary rGH on the growth of the tibial cartilage. From their work, these investigators concluded that there is a form of growth hormone present in small amounts in the blood that cross-reacts with antibodies to native pituitary growth hormone and another form that is present in high amounts that possesses growth-promoting activity but is not cross-reactive with antibodies against the native hormone. This latter substance is believed to be produced from native growth hormone by some transformation process occurring in the pituitary or the peripheral circulation. Attempts (49, 53) to isolate this "biologically active" form of growth hormone from plasma concentrates have yielded some fractions with activity in the tibia test but no activity in the growth hormone radioimmunoassay. Interestingly, these biologically active fractions have apparent molecular weights greatly in excess of that of native pituitary growth hormone. This suggests that the biologically active species is either aggregated or complexed to some circulating protein. In any case, it is not clear if the active species is actually quite large or if aggregation or complex formation has resulted as a consequence of concentration and further chemical manipulation of the plasma. In this connection, the possibility could be raised that the biologically active species of growth hormone studied by Ellis and Grindeland is indeed similar to native pituitary growth hormone. Through aggregation or complex formation the hormone's in vivo half-life might be greatly extended, whereas its cross-reactivity in the radioimmunoassay might be attenuated. This would yield a substance having a much higher ratio of biological activity to immunological activity, relative to the parent native hormone.

One alternative approach to the study of the correlation between the biologically active and immunoreactive growth hormone in blood has given rather different results than those reported by Ellis and Grindeland. Kostyo and Wilhelmi (49), Stewart et al. (54), and Reagan et al. (55) compared the biological and immunological activities of the growth hormone present in the blood of rats bearing the transplantable somatotropic-mammotropic tumor MtTW15. As measured by radioimmunoassay, the blood of these rats contains substantial quantities of rGH (1–200 µg/ml of plasma) due to the great secretory activity of the tumor. Using a sensitive in vitro bioassay for growth hormone, these investigators were able to measure the growth hormone-like activity of the plasma of the tumor-bearing rats at dilutions of 30- to 200-fold. The bioassay used depends upon the ability of growth hormone to stimulate 3-O-methyl-glucose transport into the isolated diaphragm of the hypophysectomized rat. With this assay, it was found that the ratio of biologically active to immunoreactive growth hormone was essentially 1:1. In related experiments, Kostyo and Wilhelm (49) and Stewart et al. (54) administered small amounts of unconcentrated plasma of the tumor-bearing animals to hypophysectomized rats and then measured the ability of their costal cartilage to incorporate thymidine into DNA in vitro, a process known to be stimulated by pituitary growth hormone. The plasma of the tumor-bearing animals stimulated thymidine incorporation to the extent expected on the basis of the amount of rGH estimated to be present in the plasma by radioimmunoassay. Moreover, the effects of the plasma were blocked by the concomitant administration of antibodies against pituitary rGH. Thus, these experiments confirmed the good correlation between the biological and immunological activities of the rGH present in the blood of the tumor-bearing animals and suggested that in these animals, the biologically active form of growth hormone in the blood, is quite similar to, if not identical with, native pituitary growth hormone. Whether the discrepancy between the findings of Ellis and Grindeland and those of Stewart et al. is due to methodological differences (testing of concentrated versus diluted plasma; different bioassays) remains to be established. On the other hand, if native pituitary growth hormone is indeed modified during the process of secretion or while it circulates in the blood, the tumor-bearing animal may be incapable of transforming significant amounts of the rGH that it produces. In that event, the growth hormone present in its blood should be biologically and immunologically similar to native rGH.

Clearly, much remains to be learned about the nature of circulating growth hormone.

GROWTH HORMONE-RECEPTOR INTERACTION

The initial event in the action of growth hormone is its interaction with tissue receptors, the process that generates the chain of reactions leading to accelerated protein synthesis and growth. Knowledge of the mechanism and immediate consequences of the growth hormone-receptor interaction is meager, because this subject has received little attention from investigators until quite recently. The current availability of highly purified preparations of growth hormone and the development of techniques for the radioactive labeling of polypeptide hormones under conditions that preserve biological activity (56, 57) have now made studies of this process feasible.

It is now widely accepted, although not unequivocally established, that the receptors for polypeptide hormones are located in the plasma membranes of their target cells. With the use of growth hormone labeled with radioactive iodine as a tracer, it has been possible to detect specific binding sites for the hormone on crude membrane fractions prepared from the livers of rabbits and rats (58, 59). Specific binding sites for growth hormone have also been observed on human lymphocytes cultured from a patient with chronic myelogenous leukemia (60, 61) and on circulating human lymphocytes from normal adults (62). The binding of growth hormone to these sites has been found to be both time- and temperature-dependent. With cultured lymphocytes, binding equilibrium was reached considerably more slowly when the temperature was decreased from 37°C to 15°C. On the other hand, maximal binding capacity increased at lower incubation temperatures (61). It has also been demonstrated (61) that treatment of cultured lymphocytes with trypsin to an extent that did not affect cell viability caused a marked loss of growth hormone-binding capacity. This observation suggests that the binding sites for growth hormone on these cells is dependent upon the integrity of some protein elements of the plasma membrane.

Unfortunately, there is only limited evidence favoring the view that these so-called specific binding sites on plasma membranes are functionally important, i.e., that they are indeed physiological receptors for growth hormone. It has been shown with cultured lymphocyte preparations that significant interaction of growth hormone with these binding sites occurs at in vitro concentrations of the hormone that are known to evoke physiological responses in vivo (61). Moreover, binding activity appears to be closely related to the growth-promoting activity of a particular growth hormone preparation, e.g., growth hormone preparations with low biological activity show weak binding activity

(61). Also, Lesniak and co-workers (63) found that the incubation of cultured human lymphocytes with hGH for 24 hr resulted in an 85% decrease in the binding capacity of these cells for hGH. Continued incubation of these cells in the absence of hGH restored their capacity to bind the hormone. Restoration of binding capacity was inhibited by cycloheximide, suggesting that protein synthesis is required to regenerate the binding sites. Thus, it would appear that hGH added in vitro can influence the number of binding sites for itself on the cultured lymphocyte. Unfortunately, it is not known whether this hormone-induced alteration in binding capacity is associated with any change in tissue responsiveness to growth hormone from a metabolic point of view. In contrast to these results with the cultured lymphocyte, Herington et al. (59) reported recently that high circulating levels of growth hormone and prolactin in the rat are associated with an *increase* in the capacity of liver membranes to bind hGH. Again, the physiological significance of this observation is unclear, because the metabolic responsiveness of the liver to hGH, under these circumstances, was not assessed. Clearly, further work is needed to establish a quantitative relationship between the binding of growth hormone to tissue receptors and the cellular effects produced in the tissue as a consequence of the binding or interactive process.

One interesting aspect of the relationship between the binding of growth hormone to its receptors and the triggering of metabolic events is the length of time that the hormone must occupy receptors in order to generate a metabolic response. Must growth hormone be present on the receptors continually in order for metabolic activity to be stimulated, or is only a brief interaction with the receptors sufficient to produce a complete metabolic response in the target cell? In the case of insulin, it appears that certain of its metabolic effects are only expressed while the hormone occupies its receptors. For example, Glieman et al. (64), using isolated fat cells of the rat have found that the presence of insulin on cellular receptors and the stimulation of lipid synthesis are closely coupled; dissociation of the hormone from the cells is accompanied by an immediate decline in the stimulation of lipid synthesis. In contrast, recent observations in our laboratory (65) suggest that only a brief encounter between growth hormone and its receptors is required to produce sustained metabolic alterations in the cells of target tissues. In the experiments involved, rGH molecules bound to receptors on the isolated diaphragm of the hypophysectomized rat were displaced with antibodies against rGH at various time intervals following interaction of the hormone with its receptors. The tissue was then washed at $0°C$ to remove hormone-antibody complexes and finally tested for the ability to transport 3-O-methylglucose. Stimulation of 3-O-methylglucose transport was taken as an indicator of successful hormone-receptor interaction. When rGH was bound to tissue receptors at $0°C$ and then displaced with antibodies without raising the temperature to permit active metabolism, there was no stimulation of 3-O-methylglucose transport. Similarly treated controls, in which hormone bound to receptors was not displaced with antibodies, gave a full stimulation of sugar transport. When metabolism was permitted to occur at 37°C for even a brief period (1–10 min) before removal of the hormone from the receptors, a full stimulation of membrane transport was subsequently produced. These experiments suggest that the growth hormone-receptor interaction triggers some temperature-dependent reaction(s) that leads to the generation of the metabolic response. Once this reaction has taken place, growth hormone-receptor interaction is no longer required for the response to be propagated. These studies also strongly imply that the functional receptors for growth hormone are located on the outer surface of the plasma membrane, because, under the appropriate conditions, the effect of tissue-bound hormone can be neutralized by antibodies added in vitro.

SOMATOMEDIN

In 1957, Salmon and Daughaday (66) demonstrated that the in vitro incorporation of radioactive sulfate into the chondroitin sulfate of isolated cartilage of hypophysectomized rats was depressed and could be restored to normal by the prior treatment of the animals with growth hormone. They further showed that the addition of serum from normal or growth hormone-treated rats to the incubation medium stimulated sulfate uptake by the cartilage of hypophysectomized rats. On the other hand, the addition of growth hormone to the incubation medium produced only small and inconsistent stimulatory effects on the sulfation process. It was concluded from these studies that serum contains a growth hormone-dependent factor that regulates chondroitin sulfate formation in cartilage. The factor was given the operational designation "sulfation factor" (SF). Subsequent studies exploring the influence of SF on cartilage revealed that it has a generalized effect on anabolic processes in this tissue (e.g., stimulatory effects on amino acid uptake, protein synthesis, RNA synthesis, DNA synthesis). Sulfation factor was also found in the serum of man, and there appeared to be a good correlation between its level and the growth hormone status of the individual. For example, the level of SF was found to be decreased in the blood of hypopituitary dwarfs and elevated in the blood of acromegalics (67–69). Moreover, treatment of hypopituitary dwarfs with growth hormone returned the level of SF toward normal.

Tissues other than cartilage were also found to respond to sulfation factor. For example, Salmon and DuVall (70) demonstrated that a partially purified preparation of SF made from rat serum stimulated the incorporation of radioactive amino acids into the protein of the isolated diaphragm of the hypophysectomized rat when added in vitro. A partially purified preparation of human SF was also found to exert in vitro insulin-like effects on isolated rat adipose tissue (71–73) and to stimulate amino acid and sugar transport and protein synthesis in the isolated diaphragm of the hypophysectomized rat (74). In the latter case, the in vitro effects of the human SF preparations on the rat diaphragm were qualitatively similar to the effects of insulin on this tissue and not like the in vitro effects of growth hormone itself. The finding that SF preparations had widespread effects on a variety of tissues lead Daughaday and associates (75) to propose that sulfation factor is the mediator for the actions of growth hormone on its target tissues. Accordingly, they proposed that the term "sulfation factor" be changed to "somatomedin," a term which implies that the factor mediates the actions of growth hormone on the soma.

During the last few years, extensive efforts have been made to isolate and chemically characterize somatomedin (76, 77). Progress has been hampered because of the low yields of material that have been obtained. One finding to emerge from these efforts is that somatomedin does not appear to be a single substance, but rather a family of small peptides having molecular weights in the range of 6,000–8,000. At least three apparently different somatomedin peptides have been identified, based upon net charge and behavior in various bioassays. These are 1) somatomedin A, a neutral peptide that stimulates the incorporation of sulfate into isolated chick cartilage, 2) somatomedin B, an acidic peptide that stimulates the incorporation of thymidine into the DNA of cultured human glial cells, and 3) somatomedin C, a basic peptide that stimulates both sulfate and thymidine incorporation into isolated rat cartilage (78). Recent studies (77) now suggest that somatomedins A and C may each be active in stimulating metabolic processes in both chick and rat cartilage. In any event, it can be concluded with some certainty that no single substance can account for the "sulfation activity" of whole serum.

There is little question that the level of somatomedin in the serum of both animals and man is partially growth hormone-dependent and that somatomedin can exert a number of metabolic effects in various in vitro bioassay systems. The critical question is whether somatomedin is a (or *the*) mediator of the growth-promoting effect of growth hormone on target tissues. The main experimental evidence supporting the hypothesis that the effects of growth hormone are mediated by somatomedin is the observation that growth hormone added in vitro produces only small or inconsistent effects on isolated cartilage metabolism, whereas serum from normal or growth hormone-treated animals or man produces quantitatively larger effects. The fact that a number of investigators (66, 69, 70, 79) found small but significant in vitro effects of growth hormone on sulfate incorporation into cartilage is of considerable interest in this connection but has received little discussion in the literature. Because the results of the initial experiments (66) were interpreted to mean that the action of growth hormone on cartilage is indirect, little effort was made in subsequent studies to determine whether the incubation conditions used were optimal for detecting direct effects of growth hormone on cartilage. In this connection, it is quite interesting that Almqvist (69) was able to demonstrate a significant in vitro effect of growth hormone on radioactive sulfate incorporation into rat costal cartilage when serum from normal or hypophysectomized humans was present in the incubation medium.

If the hypothesis that somatomedin mediates the action of growth hormone on cartilage or other tissues is correct, then one would expect that the administration of growth hormone to an experimental subject would produce an increase in the serum level of somatomedin prior to or coincident with the appearance of metabolic effects of the hormone. This important requirement has not been met, however. Recently, Phillips et al. (80) found that the incorporation of radioactive sulfate into cartilage isolated from hypophysectomized rats that had received low doses of growth hormone was markedly stimulated 30 hr after the start of treatment. However, there was no elevation in the somatomedin level in the serum of these animals. In man, a considerable discrepancy has also been found between the time of onset of metabolic effects after the injection of growth hormone and the time when an increase occurs in the somatomedin level in the serum. For example, a prompt hypoglycemia is produced when normal or hypopituitary individuals are given an intravenous injection of hGH (81), whereas an increase in somatomedin in the serum cannot be detected earlier than 3 hr following administration of the hormone (82). Somatomedin levels comparable to those found in normal humans are not produced in hypopituitary individuals until 24 hr following the administration of growth hormone. It might be argued that the metabolic effects of growth hormone result from the generation of somatomedin in the target tissues themselves and that some time must elapse before significant amounts of somatomedin leak from the cells and accumulate in the serum. However, if this is indeed the case, then one might expect the metabolic effects of growth hormone and somatomedin to be qualitatively similar. They are not. As noted earlier, the in vitro effects of partially purified human somatomedin on the isolated rat diaphragm are qualitatively similar to the effects of insulin and not growth hormone (74). Moreover, in a recent study (83), a partially purified preparation of rat somatomedin was found, unlike growth hormone, to produce a sustained in vitro insulin-like effect on glucose oxidation by isolated rat adipose tissue.

There is also a discrepancy between the dose of growth hormone required to stimulate cartilage metabolism and that needed to stimulate the generation of somatomedin. For example, Phillips et al. (80) recently demonstrated that the in vitro incorporation of sulfate by costal cartilage was significantly enhanced when the cartilage was obtained from hypophysectomized rats that had received 2 doses of 10 μg of growth hormone over a 30-hr period. Higher doses of growth hormone (0.1–1 mg) produced correspondingly greater effects on sulfate incorporation. However, only those rats treated with a dose of 1 mg of growth hormone had elevated levels of somatomedin in the serum, when compared to saline-treated controls. It has also been a consistent finding that the doses of growth hormone required to restore somatomedin levels to normal in hypopituitary dwarfs are much greater than those required for "catch-up" growth in these individuals (84).

If somatomedin does mediate the growth-promoting effect of growth hormone, one might expect it to stimulate growth in the hypophysectomized test animal. Until recently, the availability of only minute amounts of partially purified somatomedin precluded the exploration of its in vivo effects. In 1975, Uthne (85) reported that the daily administration of a partially purified preparation of human somatomedin to hypophysectomized rats failed to stimulate growth. Certainly, further experiments are essential to clarify this issue.

There can be no doubt that somatomedin is related in some way to the growth-promoting activity of growth hormone. On the basis of the evidence available, it is tempting to conclude that somatomedin is not a mediator of the growth-promoting effect of growth hormone. Rather, it may be a family of insulin-like peptides that are produced as a consequence of the protein anabolic effect of growth hormone on various tissues. An elevated level of somatomedin in plasma may merely reflect the protein anabolic state. In any event, the physiological role of somatomedin is far from clear.

REFERENCES

1. Kostyo, J. L., and Nutting, D. F. (1973). Acute in vivo actions of growth hormone on protein synthesis in various tissues of hypophysectomized rats and their relationship to the levels of thymidine factor and insulin in the plasma. Horm. Metab. Res. 5:167.
2. Kostyo, J. L., and Rillema, J. A. (1971). In vitro effect of growth hormone on the number and activity of ribosomes engaged in protein synthesis in the isolated rat diaphragm. Endocrinology 88:1054.
3. Garren, L. D., Richardson, A. P., Jr, and Crocco, R. M. (1962). Studies on the role of ribosomes in the regulation of protein synthesis in hypophysectomized and thyroidsectomized rats. J. Biol. Chem. 242:650.
4. Korner, A. (1968). Anabolic action of growth hormone. Ann. N. Y. Acad. Sci. 148:408.
5. Kostyo, J. L. (1968). Rapid effects of growth hormone on amino acid transport and protein synthesis. Ann. N. Y. Acad. Sci. 148:389.
6. Jefferson, L. S., and Korner, A. (1967). A direct effect of growth hormone on the incorporation of precursors into proteins and nucleic acids of perfused rat liver. Biochem. J. 104:826.
7. Ahrén, K., Hjalmarson, Å., and Isaksson, O. (1969). Effects of growth hormone on amino acid transport and protein synthesis in the isolated working rat heart. Acta Physiol. Scand. 76:23A.
8. Mowbray, J., Davies, J. A., Bates, D. J., and Jones, C. J. (1975). Growth hormone, cyclic nucleotides and the rapid control of translation in heart muscle. Biochem. J. 152:583.
9. Florini, J. F., and Brewer, C. B. (1966). Amino acid incorporation into protein by cell-free systems from rat skeletal muscle. V. Effects of pituitary growth hormone on activity of ribosomes and ribonucleic acid polymerase in hypophysectomized rats. Biochemistry 5:1870.
10. Payne, S. J., and Kostyo, J. L. (1970). Inhibition by theophylline of the stimulatory effects of growth hormone on amino acid transport and protein synthesis in muscle. Endocrinology 87:1186.
11. Rillema, J. A., Kostyo, J. L., and Gimpel, L. P. (1973). Inhibition of metabolic effects of growth hormone by various inhibitors of cyclic nucleotide phosphodiesterase. Biochim. Biophys. Acta 297:527.
12. Isaksson, O., Gimpel, L. P., Ahren, K., and Kostyo, J. L. (1974). Growth hormone and cyclic AMP in rat diaphragm muscle. Acta Endocrinol. (Kbh) (Suppl. 191) 77:73.
13. Ahrén, K., Albertsson-Wikland, K., Isaksson, O., and Kostyo, J. (1976). Cellular

mechanisms of the acute stimulatory effect of growth hormone. *In* A. Pecile and E. E. Müller (eds.), Growth Hormone and Related Peptides, pp. 94–103. Excerpta Medica Foundation, Amsterdam.

14. Kostyo, J. L., Gimpel, L. P., and Isaksson, O. (1975). In vitro effects of growth hormone on cyclic AMP metabolism in the isolated rat diaphragm. Adv. Metab. Disord. 8:249.

15. Daughaday, W. H. (1971). Sulfation factor regulation of skeletal growth: a stable mechanism dependent on intermittent growth hormone secretion. Am. J. Med. 50:277.

16. Kostyo, J. L. (1973). Amino acid transport and protein synthesis. *In* S. A. Berson and R. S. Yalow (eds.), Methods in Investigative and Diagnostic Endocrinology, pp. 279–291. North-Holland, Amsterdam.

17. Goodman, H. M. (1973). Carbohydrate metabolism. *In* S. A. Berson and R. S. Yalow (eds.), Methods in Investigative and Diagnostic Endocrinology, pp. 273–276. North-Holland, Amsterdam.

18. Ahrén, K., and Hjalmarson, A. (1968). Early and late effects of growth hormone on transport of amino acids and monosaccharides in the isolated rat diaphragm. *In* A. Pecile and E. E. Müller (eds.), Growth Hormone, pp. 143–152. Excerpta Medica, Amsterdam.

19. Goodman, H. M. (1968). Growth hormone and the metabolism of carbohydrate and lipid in adipose tissue. Ann. N. Y. Acad. Sci. 148:419.

20. Kostyo, J. L., and Nutting, D. F. (1974). Growth hormone and protein metabolism. *In* E. Knobil and W. H. Sawyer (eds.), Handbook of Physiology, Section 7, Volume IV, pp. 187–210. American Physiological Society, Washington, D.C.

21. Talwar, G. P., Pandian, M. R., Kumar, N., Hanjan, S. N. S., Saxena, R. K., Krishnaraj, R., and Gupta, S. L. (1975). Mechanism of action of pituitary growth hormone. Recent Prog. Horm. Res. 31:141.

22. Glick, S. M., Roth, J., Yalow, R. S., and Berson, S. A. (1965). The regulation of growth hormone secretion. Recent Prog. Horm. Res. 21:241.

23. Knobil, E., and Meyer, V. (1968). Observations on the secretion of growth hormone, and its blockade in the rhesus monkey. Ann. N. Y. Acad. Sci. 148:459.

24. Glick, S. M. (1968). Normal and abnormal secretion of growth hormone. Ann. N. Y. Acad. Sci. 148:471.

25. Takahashi, K., Daughaday, W. H., and Kipnis, D. M. (1971). Regulation of immunoreactive growth hormone secretion in male rats. Endocrinology 88:909.

26. Seggie, J. A., and Brown, G. M. (1975). Stress response pattern of plasma corticosterone, prolactin, and growth hormone in the rat following handling or exposure to novel environment. Can. J. Physiol. Pharmacol. 53:629.

27. Seckel, H. P. G. (1960). Concepts relating the pituitary growth hormone to somatic growth of the normal child. Am. J. Dis. Child. 99:349.

28. Ducharme, J. R., and Grumbach, M. M. (1961). Studies on the effects of human growth hormone in premature infants. J. Clin. Invest. 40:243.

29. Nutting, D. F. (1975). The development of sensitivity of rats to growth hormone. Presented at the International Symposium on Growth Hormone and Related Peptides, September 17–20, Milan.

30. Albertsson-Wikland, K., and Isaksson, O. (1976). Development of responsiveness of young normal rats to growth hormone. Metabolism 25:747.

31. Kalan, S. L., Grumbach, M. M., and Shepard, T. H. (1972). The ontogenesis of human fetal hormones. I. Growth hormone and insulin. J. Clin. Invest. 51:3080.

32. Greenwood, F. C., Hunter, W. M., and Marrian, V. J. (1964). Growth hormone levels in children and adolescents. Br. Med. J. I:25.

33. Utiger, R. D. (1964). Extraction and radioimmunoassay of growth hormone in human serum. J. Clin. Endocrinol. Metab. 24:60.

34. Rieutort, M. (1974). Pituitary content and plasma levels of growth hormone in foetal and weanling rats. J. Endocrinol. 60:261.

35. Blázquez, E., Simon, F. A., Blázquez, M., and Foà, P. P. (1974). Changes in serum growth hormone levels from foetal to adult age in the rat. Proc. Soc. Exp. Biol. Med. 147:780.

36. Strosser, M. T., and Mialhe, P. (1975). Growth hormone secretion in the rat as a function of age. Horm. Metab. Res. 7:275.
37. Takahashi, Y., Kipnis, D. M., and Daughaday, W. H. (1968). Growth hormone secretion during sleep. J. Clin. Invest. 47:2079.
38. Honda, Y., Takahashi, K., Takahashi, S., Azumi, K., Irie, M., Sakuma, M., Tsushima, T., and Shizume, K. (1969). Growth hormone during nocturnal sleep in normal subjects. J. Clin. Endocrinol. Metab. 29:20.
39. Sassin, J. F., Parker, D. C., Mace, J. W., Gotlin, R. W., Johnson, L. C., and Rossman, L. G. (1969). Human growth hormone release: relation to slow-wave sleep and sleep-waking cycles. Science 165:513.
40. Parker, D. C., Sassin, J. F., Mace, J. W., Gotlin, R. W., Rossman, L. G. (1969). Human growth hormone release during sleep: electroencephalographic correlation. J. Clin. Endocrinol. Metab. 29:871.
41. Weitzman, E. D., Boyar, R. M., Kapen, S., and Hellman, L. (1975). The relationship of sleep and sleep stages to neuroendocrine secretion and biological rhythms in man. Recent Prog. Horm. Res. 31:399.
42. Plotnick, L. P., Thompson, R. J., Kowarski, A., de Lacerda, L., Migeon, C. J., and Blizzard, R. M. (1975). Circadian variation of integrated concentration of growth hormone in children and adults. J. Clin. Endocrinol. Metab. 40:240.
43. Finkelstein, J. W., Roffwarg, H. P., Boyer, R. M., Kream, J., and Hellman, L. (1972). Age-related changes in the twenty-four hour spontaneous secretion of growth hormone. J. Clin. Endocrinol. Metab. 35:665.
44. Dunn, J. D., Schindler, W. J., Hutchins, M. D., Scheving, L. E., and Turpen, C. (1973). Daily variation in rat growth hormone concentration and the effect of stress on periodicity. Neuroendocrinology 13:69.
45. Tannenbaum, G. S., and Martin, J. B. (1976). Evidence for an endogenous ultradian rhythm governing growth hormone secretion in the rat. Endocrinology 98:562.
46. Isaksson, O., Nutting, D. F., Kostyo, J. L., and Reagan, C. R. The daily pattern of growth hormone secretion in the young rat. Manuscript in preparation.
47. Kostyo, J. L., Mills, J. B., Reagan, C. R., Rudman, D., and Wilhelmi, A. E. (1976). The nature of fragments of human growth hormone produced by plasmin digestion. In A. Pecile and E. E. Müller (eds.), Growth Hormones and Related Peptides, pp. 33–40. Excerpta Medica Foundation, Amsterdam.
48. Kostyo, J. L. (1974). The search for the active core of pituitary growth hormone. Metabolism 23:885.
49. Kostyo, J. L., and Wilhelmi, A. E. (1976). Conference on the structure-function relationships of pituitary growth hormone: a report. Metabolism 25:105.
50. Goodman, A. D., Tannenbaum, R., Wright, D. R., Trimble, K. D., and Rabinowitz, D. (1974). Existence of "big" and "little" forms of immunoreactive growth hormone in human plasma. In D. Rabinowitz and J. Roth (eds.), Heterogeneity of Polypeptide Hormones, pp. 48–56. Academic Press, New York.
51. Gordon, P., Lesniak, M. A., Hendricks, C. M., Roth, J., McGuffin, W., and Gavin, J. R. (1974). Human growth hormone: characterization of plasma and pituitary components and application of a new radioreceptor assay. In D. Rabinowitz and J. Roth (eds.), Heterogeneity of Polypeptide Hormones, pp. 57–64. Academic Press, New York.
52. Bala, R. M., Ferguson, K. A., and Beck, J. C. (1970). Plasma biological and immunoreactive human growth hormone-like activity. Endocrinology 87:506.
53. Ellis, S., and Grindeland, R. E. (1973). Dichotomy between bio and immunoassayable growth hormone. In S. Raiti (ed.), Advances in Human Growth Hormone Research, pp. 409–433. U. S. Government Printing Office, Washington, D. C.
54. Stewart, J., Reagan, C. R., and Kostyo, J. L. Correlation between the biologic and immunologic activities of growth hormone circulating in rats bearing MtTW15 tumors. Manuscript in preparation.
55. Reagan, C. R., Harper, J. S., and Kostyo, J. L. (1974). Detection of biological activity of growth hormone in plasma of rats bearing MtTW15 tumors. Fed. Proc. 33:417.
56. Hunter, W. M., and Greenwood, F. C. (1962). Preparation of iodine-131 labeled human growth hormone of high specific activity. Nature 194:495.

57. Thorell, J. I., and Johansson, B. G. (1971). Enzymatic iodination of polypeptides with ^{125}I to high specific activity. Biochim. Biophys. Acta. 251:363.
58. Tsushima, T., and Friesen, H. G. (1973). Radioreceptor assay for growth hormone. J. Clin. Endocrinol. Metab. 37:334.
59. Herington, A. C., Phillips, L. S., and Daughaday, W. H. (1976). Pituitary regulation of human growth hormone binding sites in rat liver membranes. Metabolism 25:341.
60. Lesniak, M. A., Roth, J., Gordon, P., and Gavin, J. R. (1973). Human growth hormone radioreceptor assay using cultured human lymphocytes. Nature (New Biol.) 241:20.
61. Lesniak, M. A., Gordon, P., Roth, J., and Gavin, J. R. (1974). Binding of ^{125}I human growth hormone to specific receptors in human cultured lymphocytes: characterization of the interaction and a sensitive radioreceptor assay. J. Biol. Chem. 249:1661.
62. Eshet, R., Manheimer, S., Chobsieng, P., and Laron, Z. (1975). Human growth hormone receptors in human circulating lymphocytes. Horm. Metab. Res. 7:352.
63. Lesniak, M. A., Bianco, A. R., Roth, J., and Gavin, J. R. (1974). Regulation by hormone of its receptors on cells: studies with insulin and growth hormone. Clin. Res. 22:343A.
64. Glieman, J., Gammeltoft, S., and Vinton, J. (1975). Insulin receptors in fat cells: relationship between binding and activation. Israel J. Med. Sci. 11:656.
65. Isaksson, O., Reagan, C., and Kostyo, J. (1976). Use of antibodies to modify the interaction of growth hormone (GH) with isolated muscle. Fed. Proc. 35:783.
66. Salmon, W. D., and Daughaday, W. H. (1957). A hormonally controlled serum factor which stimulates sulfate incorporation by cartilage in vitro. J. Lab. Clin. Med. 49:825.
67. Daughaday, W. H., Salmon, W. D., and Alexander, F. (1959). Sulfation factor activity of sera from patients with pituitary disorders. J. Clin. Endocrinol. Metab. 19:743.
68. Almqvist, S. (1960). Studies on sulfation factor (SF) activity of human serum: effect of human growth hormone on SF levels in pituitary dwarfism. Acta Endocrinol. (Kbh.) 35:381.
69. Almqvist, S. (1961). Studies on sulfation factor (SF) activity of human serum: evaluation of an improved method for the in vitro bioassay of SF and the effect of glutamine and human growth hormone in the system. Acta Endocrinol. (Kbh.) 36:31.
70. Salmon, W. D., and DuVall, M. R. (1970). In vitro stimulation of leucine incorporation into muscle and cartilage protein by a serum factor with sulfation activity: differentiation of effects from those of growth hormone and insulin. Endocrinology 87:1168.
71. Hall, K., and Uthne, K. (1968). Human growth hormone and sulfation factor. In A. Pecile and E. E. Müller (eds.), Proceedings of the Second International Symposium on Growth Hormone, pp. 192–198. Excerpta Medica, Amsterdam.
72. Underwood, L. E., Hintz, R. L., Voina, S. J., and Van Wyk, J. J. (1972). Human somatomedin, the growth hormone-dependent sulfation factor, is lipolytic. J. Clin. Endocrinol. Metab. 35:194.
73. Van Wyk, J. J., Underwood, L. E., Hintz, R. L., Voina, S. J., and Weaver, R. P. (1973). Chemical properties and some biological effects of human somatomedin. In S. Raiti (ed.), Advances in Human Growth Hormone Research, pp. 25–49. U. S. Government Printing Office, Washington, D.C.
74. Uthne, K., Reagan, C. R., Gimpel, L. P., and Kostyo, J. L. (1974). Effects of human somatomedin preparations on membrane transport and protein synthesis in the isolated rat diaphragm. J. Clin. Endocrinol. Metab. 39:548.
75. Daughaday, W. H., Hall, K., Raben, M. S., Salmon, W. D., Van den Brande, J. L., and Van Wyk, J. J. (1972). Somatomedin: proposed designation for sulfation factor. Nature 235:107.
76. Sievertsson, H., Fryklund, L., Uthne, K., Hall, K., and Westermark, B. (1975). Isolation and chemistry of human somatomedins A and B. Adv. Metab. Disord. 8:47.
77. Van Wyk, J. J., Underwood, L. E., Baseman, J. B., Hintz, R. L., Clemmons, D. R., and Marshall, R. N. (1975). Explorations of the insulin-like and growth-promoting properties of somatomedin by membrane radioreceptor assays. Adv. Metab. Disord. 8:127.
78. Hall, K., Takano, K., Fryklund, L., and Sievertsson, H. (1975). Somatomedins. Adv. Metab. Disord. 8:19.
79. Ito, Y., Tokumura, K., and Endo, H. (1959). The stimulative effect of pituitary growth

hormone on S^{35} -sulfate incorporation in the chick embryo femur in tissue culture. Endocrinol. Jpn. 6:68.

80. Phillips, L. S., Herington, A. C., and Daughaday, W. H. (1973). Hormone effects on somatomedin action and somatomedin generation. *In* S. Raiti (ed.), Advances in Human Growth Hormone Research, pp. 50–75. U. S. Government Printing Office, Washington, D. C.

81. Adamson, U., and Cerasi, E. (1976). Acute suppressive effect of human growth hormone on basal insulin secretion in man. Acta Endocrinol. (Kbh.) 79:474.

82. Hall, K. (1971). Effect of intravenous administration of human growth hormone on sulphation factor activity in serum of hypopituitary subjects. Acta Endocrinol. (Kbh.) 66:491.

83. Schwartz, J., and Goodman, H. M. (1976). Comparison of the delayed actions of growth hormone and somatomedin on adipose tissue metabolism. Endocrinology 98:730.

84. Van den Brande, J. L., and Du Caju, M. L. V. (1973). Plasma somatomedin activity in children with growth disturbances. *In* S. Raiti (ed.), Advances in Human Growth Hormone Research, pp. 98–126. U. S. Government Printing Office, Washington, D. C.

85. Uthne, K. (1975). Preliminary studies of somatomedin *in vitro* and *in vivo* in rats. Adv. Metab. Disord. 8:115.

Index